Condensed Matter Physics in the Prime of the 21st Century

Phenomena, Materials, Ideas, Methods

43rd Karpacz Winter School of Theoretical Physics

Condensed Matter Physics in the Prime of the 21st Century

Phenomena, Materials, Ideas, Methods

Lądek Zdrój, Poland 5 – 11 February 2007

editor

Janusz Jędrzejewski

University of Wrocław, Poland

World Scientific

NEW JERSEY • LONDON • SINGAPORE • BEIJING • SHANGHAI • HONG KONG • TAIPEI • CHENNAI

Published by

World Scientific Publishing Co. Pte. Ltd.

5 Toh Tuck Link, Singapore 596224

USA office: 27 Warren Street, Suite 401-402, Hackensack, NJ 07601

UK office: 57 Shelton Street, Covent Garden, London WC2H 9HE

British Library Cataloguing-in-Publication Data
A catalogue record for this book is available from the British Library.

CONDENSED MATTER PHYSICS IN THE PRIME OF THE 21ST CENTURY
Phenomena, Materials, Ideas, Methods

ISBN-13 978-981-270-944-8
ISBN-10 981-270-944-4

Printed in Singapore by World Scientific Printers

PREFACE

The Winter Schools of Theoretical Physics in Karpacz (or briefly, Karpacz Winter Schools) are now well known in the world community of physicists as scientific and cultural events that have been created by a few generations of theoretical physicists from the University of Wrocław, Poland. The Schools have been organized by the Institute of Theoretical Physics of the University of Wrocław (ITP UWr) annually for 44 years. Only once, in February 1982, did the School, being at its final stage of organization, have to be canceled because of the martial law imposed in Poland two months earlier.

Till the early nineties of the 20th century, the Schools served as a meeting ground for physicists from West and East and were held at University houses in Karpacz (Sudeten Mountains). For many researchers and physics students in Poland and other Eastern Block countries, they offered a unique opportunity to discuss physics with outstanding scientists — including Nobel Prize winners — from Western Europe, both Americas, Japan and the Soviet Union. For more information about the Karpacz Winter Schools, the reader is invited to consult the ITP UWr web page (www.ift.uni.wroc.pl).

Reflecting diverse research directions pursued by members of ITP UWr, the topics of the Schools have been varied, almost periodically, from the physics of elementary particles and high energy physics to mathematical physics and field theory to condensed matter physics and statistical physics. Also the character of the Schools has varied, oscillating between a conference with many specialized talks addressed to advanced researchers, and a school with lectures addressed to young adepts of physics.

The 43rd School, entitled "Condensed Matter Physics in the Prime of XXI Century: Phenomena, Materials, Ideas, Methods" was held at Lądek Zdrój, Poland (one of the recent substitutes for Karpacz), between 5th and 11th February 2007. It was designed and organized by representatives of the three largest institutions in Wrocław that educate and employ physicists: Janusz Jędrzejewski (University of Wrocław), Romuald Lemański (Institute of Low Temperature and Structure Research of the Polish Academy of Sciences), and Arkadiusz Wójs (Wrocław University of

Technology). This team was joined by Manuel Richter (Leibniz-Institut für Festkörper- und Werkstoffforschung Dresden e.V.) from neighbouring Dresden, Germany.

As planned, the 43rd School was a genuine school. The lectures delivered by 13 lecturers — renowned specialists — were organized into coherent series of a few talks per topic. A number of modern, promising, dynamically-developing research directions in condensed matter physics were presented to participants. The topics included: dynamical mean-field theory, density functional theory, quantum spin models, frustrated quantum magnets, quantum dots, quantum gates, spintronics, carbon nanostructures, thermal transport, and warm dense matter.

This volume constitutes a collection of articles written by 11 lecturers, specialists in a number of modern, promising, dynamically-developing research directions in condensed matter/solid state theory. The lectures are concerned with phenomena, materials and ideas, discussing theoretical and experimental features, as well as with methods of calculations. It should be emphasized that the lectures are addressed primarily to young researchers. Due to the style of presentation and extensive lists of references, the lectures should prove helpful in acquainting oneself with the subjects discussed in this volume. Concerning methods, the reader will find up-to-date presentations of methods of carrying out efficient calculations for electronic systems (most powerful, versatile and rapidly developing methods of dynamical mean-field theory, density-functional theory, finite frequency theory for thermoelectric transport) and quantum spin systems (Jordan-Wigner fermionization, localized-magnon and localized-electron states in frustrated lattices), together with applications in describing phenomena and designing new materials. Concerning phenomena and materials, systems of quantum dots, quantum gates, semiconductor materials for spintronics, and last but not least, unusual characteristics of warm dense matter, are neatly discussed.

One of the lecturers, professor Bengt Lundqvist (Göteborg, Sweden), who delivered lectures entitled "Density-Functional Theory of Dense and Sparse Matter" has kindly sent us pdf files of his lecture notes with a permission to display them at the School's web page.

I am grateful to my younger colleagues from ITP UWr, who undertook the burden of dealing with: financial matters, Dr. Artur Duda; the School's correspondence, Dr. Volodymyr Derzhko; WLAN at the School's site, Mr Szymon Owczarek; the School's web page, Mr Łukasz Andrzejewski.

It is my pleasure to thank professors David Blaschke and Zygmunt Petru from ITP UWr for their invaluable advice at various stages of the organization process.

Of course no school can be organized without financial means. The organizers are particularly grateful to our traditional and generous sponsors of Karpacz Winter Schools: the University of Wrocław and the Committee on Physics of the Polish Academy of Sciences. Financial support from the Institute of Low Temperature and Structure Research of the Polish Academy of Sciences (Wrocław) and of the Institute of Physics of the Wrocław University of Technology is gratefully acknowledged.

In the name of all the organizers, I take the opportunity to thank once more all the lecturers for their excellent lectures and for writing articles to this volume. I express my sincere thanks to all the participants from various research institutions in Europe for creating the ambiance of the School.

This volume, the School's poster and badges, as in numerous previous Schools, have been prepared in camera-ready form by Mrs Anna Jadczyk, to whom the organizers address special thanks.

Janusz Jędrzejewski
Head of the Organizing Committee
XLIII Karpacz Winter School of Theoretical Physics

ORGANIZING COMMITTEES

JANUSZ JĘDRZEJEWSKI	Institute of Theoretical Physics of the University of Wrocław
ROMUALD LEMAŃSKI	Institute of Low Temperature and Structure Research of the Polish Academy of Sciences
MANUEL RICHTER	Leibniz-Institut fur Festk orper- und Werkstoffforschung Dresden e.V.
ARKADIUSZ WÓJS	Institute of Physics of the Wrocław University of Technology

Secretary:

Volodymyr Derzhko	Institute of Theoretical Physics of the University of Wrocław

Treasurer:

Artur Duda	Institute of Theoretical Physics of the University of Wrocław

ORGANIZED BY

Institute of Theoretical Physics of the University of Wrocław
Institute of Low Temperature and Structure Research
Polish Academy of Sciences in Wrocław
Institute of Physics of the Wrocław University of Technology
Leibniz-Institute for Solid State and Materials Research Dresden

CONTENTS

Preface v

Organizing Committees ix

Dynamical Mean-Field Theory for Correlated Lattice Fermions 1
K. Byczuk

Jordan-Wigner Fermionization and the Theory of Low-Dimensional Quantum Spin Models. Dynamic Properties 35
O. Derzhko

Quantum Computing with Electrical Circuits: Hamiltonian Construction for Basic Qubit-Resonator Models 89
M.R. Geller

Coherent Control and Decoherence of Charge States in Quantum Dots 119
P. Machnikowski

Basics of Spintronics: From Metallic to All-Semiconductor Magnetic Tunnel Junctions 159
J.A. Majewski

Physics of Carbon Nanostructures 201
V.A. Osipov

Quantum Molecular Dynamics Simulations for Warm Dense Matter and Applications in Astrophysics 223
R. Redmer, N. Nettelmann, B. Holst, A. Kietzmann and M. French

Correlated Systems on Geometrically Frustrated Lattices: From Magnons to Electrons 237
J. Richter and O. Derzhko

Full-Potential Local-Orbital Approach to the Electronic Structure of Solids and Molecules 271
M. Richter, K. Koepernik and H. Eschrig

Theory of Dynamical Thermal Transport Coefficients in Correlated Condensed Matter 293
B. Sriram Shastry

Carrier Concentration Induced Ferromagnetism in Semiconductors 327
T. Story

Index 353

DYNAMICAL MEAN-FIELD THEORY FOR CORRELATED LATTICE FERMIONS

K. BYCZUK

(1) Theoretical Physics III, Center for Electronic Correlations and Magnetism, Institute for Physics, University of Augsburg, 86135 Augsburg, Germany

(2) Institute of Theoretical Physics, Warsaw University, ul. Hoża 69, 00-681 Warszawa, Poland

Dynamical mean-field theory (DMFT) is a successful method to investigate interacting lattice fermions. In these lecture notes we present an introduction into the DMFT for lattice fermions with interaction, disorder and external inhomogeneous potentials. This formulation is applicable to electrons in solids and to cold fermionic atoms in optical lattices. We review here our investigations of the Mott-Hubbard and Anderson metal-insulator transitions in correlated, disordered systems by presenting selected surprising results.

1. Introduction

The exceptional properties of strongly correlated electron systems have fascinated physicists for several decades already.[1–10] New correlated electron materials and unexpected correlation phenomena are discovered every year. Often the properties of those systems are influenced by disorder. Also inhomogeneous external potentials or layers and interfaces are present in particular experimental realizations, e.g see in [11]. Unfortunately, real materials and even model systems with strong electronic correlations and randomness or inhomogeneities are notoriously hard to investigate theoretically because standard approximations are invalid in the most interesting parameter regime – that of intermediate coupling. Here the recently developed dynamical mean-field theory (DMFT)[12–21] has proved to be an almost ideal mean-field approximation since it may be used at arbitrary coupling. For this reason the DMFT has been successfully employed in the investigation of electronic correlation effects in theoretical models and even real materials.[19–21]

In the present lecture notes we provide an introduction into the DMFT for correlated and disordered electron systems in the inhomogeneous external potential. We start with a brief description of the correlated and disordered electron systems and their models. The DMFT is introduced from the practical point of view. The emphasis is put on the physical interpretation rather then on a formal mathematical structure of the theory. Different approaches to deal with randomness in many-body systems are discussed on the physical ground. The second part of these notes present selected applications of the DMFT for correlated and disordered systems based on our research.

2. Correlation and correlated electron systems

2.1. *Correlations*

The word *correlation* comes from Latin and means "with relation". It implies that at least two objects are in a relation with each other. In mathematics and statistics or natural sciences the correlation has a rigorous meaning.[22] Namely, two random variables x and y are correlated if the average (expectation value) $\langle xy\rangle$ cannot be written as a product of the averages $\langle x\rangle\langle y\rangle$, i.e. explicitly:

$$\langle xy\rangle \neq \langle x\rangle\langle y\rangle\,. \tag{1}$$

In other words, the covariance $cov(x,y) \equiv \langle (x-\langle x\rangle)(y-\langle y\rangle)\rangle \neq 0$. In such a case the probability distribution function $P(x,y)$ cannot be expressed as a product of marginal distribution functions $P_x(x)$ and $P_y(y)$. Of course, the lack of correlations, i.e. $\langle xy\rangle = \langle x\rangle\langle y\rangle$, does not automatically imply that x and y are independent random variables.

To illustrate the concept of correlations imagine a crowd of pedestrians going on a sidewalk in different directions. For each person the walk is deterministic. However, for an observer the walk of a single person X is described by a random variable - the position $\mathbf{r}_X(t)$ at a given time t. Of course, to avoid collisions people make the trajectories mutually dependent. In a statistical analysis of such a system the observer finds that the correlation function of two positions is not factorized $\langle \mathbf{r}_X(t)\mathbf{r}_Y(t')\rangle \neq \langle \mathbf{r}_X(t)\rangle\langle \mathbf{r}_Y(t')\rangle$. It means that the pedestrian motion in a dense crowd is correlated. The same holds true for a motion of cars or airplanes, for example. Dealing with correlations in modelling of such complex systems is a difficult task and requires very sophisticated mathematical techniques.[23]

2.2. *Weakly correlated many-particle systems*

In condensed matter physics we want to model systems containing moles of particles, i.e. of the order 10^{23} in each cube centimeter. A typical course of many-body physics starts from discussing the noninteracting particles.[24] The goal is to find the probability distribution function $f(\mathbf{k})$ describing how particles are distributed over different momenta $\mathbf{k}$. The classical particles are uncorrelated in the absence of interaction and the distribution function is given by the Maxwell-Boltzmann formula. Quantum mechanics introduces new idea: the indistinguishability of identical particles. Then, the distribution function for fermions is given by the Fermi-Dirac formula and for bosons by the Bose-Einstein one. Even then in quantum mechanics, when the interaction is absent, the distribution functions at different momenta are still uncorrelated.

In reality electrons, atoms, and/or protons in condensed matter (liquids, solids, ultracold gases of atoms, nuclei, or nuclear matter in stars) interact with each other by long-range Coulomb and/or short-range strong (nuclear) forces. Often the long-range Coulomb interaction is effectively screened and only the short-range potentials remain. Even then models of such many-body systems are not exactly solvable. An approximation which neglects correlations between the relevant variables is often employed. For example, it is assumed for the density-density correlation function that

$$\langle \rho(\mathbf{r},t)\rho(\mathbf{r}',t')\rangle \approx \langle \rho(\mathbf{r},t)\rangle\langle \rho(\mathbf{r}',t')\rangle\,, \tag{2}$$

where $\rho(\mathbf{r},t)$ is a density of the particles and the expectation values $\langle ...\rangle$ are taken with respect to the quantum mechanical and thermal states. Such factorizing approximation is known as a *Weiss mean-field* type or as a *Hartree-Fock mean-field* type approximation.[25] In these types of theories particles are kept independent from each others and their mutual interaction is included by an average mean-field potential. Saying differently, the independent particle propagates in a mean-field potential created by all other particles.

Nowadays the density functional theory (DFT) is routinely used to explain or predict properties of various interacting many-body systems.[26] In principle, this is an exact, mathematical theory of many particles. However, any practical calculation within DFT uses a mean-field type approximation. The most common one is the local density approximation (LDA) which uses the average charge and spin density distributions to define electric and magnetic mean-field potentials.[27] Although approximate, the DFT with LDA gives a very accurate description, even on a quantitative level, of many bulk systems as well as molecules and atoms.[26]

2.3. *Strongly correlated many-particle systems*

There are various classes of systems, however, containing either transition metal elements (e.g., Ni, La_2CuO_4, V_2O_3, or NiS_2Se),[8] or Latinate and Actinide elements (e.g., CeCu2Si2, UBe13, or UPt3, commonly named as heavy-fermions),[28,29] which are not described by a theory which neglects electronic correlations. These systems are named *strongly correlated electron systems.* Many examples of correlated electron systems, including high - temperature superconductors with coper-oxygen planes, are reviewed and discussed in Ref. [8].

The canonical example of correlated electron system is V_2O_3.[30] Its phase diagram is presented in Fig. 1. The DFT within LDA predicts that this system is a metal with half-filled conducting band. In experiments however, changing the temperature between 200 and 400K at ambient pressure (see left panel in Fig. 1) one find a huge (8 orders of magnitude) drop of the resistivity at around 150K (see right panel in Fig. 1). The observed phenomenon is an example of the Mott-Hubbard metal insulator transition (MIT) driven by the electronic correlations. The low-temperature paramagnetic Mott insulator is not describe by the DFT with LDA. Also the high-temperature correlated metallic phase is different from that predicted by DFT within LDA, in particular the effective electron mass is strongly enhanced.[10]

The strongly correlated electrons are usually found in systems with partially filled d- or f-orbitals. To understand this we consider a hopping probability amplitude t_{ij} between two sites i and j on a crystal lattice. It is expressed by the overlap matrix element containing the one-particle part of the Hamiltonian $\hat{T}$, i.e. $t_{ij} = \langle i|\hat{T}|j\rangle$, where $|i\rangle$ are Wannier localized wave functions centered at sites i. These hopping amplitudes determine the total band-width W and the average kinetic energy of the electrons. The mean time τ spent by the electron on a given atomic orbital is inversely proportional to the band-width, i.e. $\tau \sim \hbar/W$.[32] For narrow band systems, such as those with partially filled d- or f-orbitals, this mean time τ is large. Hence, the effects due to the interaction with other electrons at the same orbital become very important. The dynamics of a single electron must be correlated with the other electrons. Any theory, which neglects those correlations, fails to explain the existence of Mott insulators and Mott-Hubbard MIT.

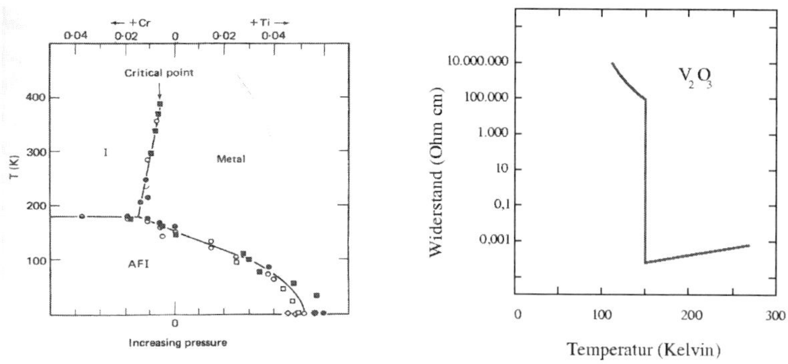

Fig. 1. Left: Phase diagram of V_2O_3 on pressure-temperature (p-T) plane.[30] The low temperature phase is a long-range antiferromagnetic insulator which disappears only at high pressures. At higher temperatures, when the system is paramagnetic one finds a spectacular meta-insulator transition when changing p at constant T or vice versa. Note that substitution of Cr or Ti in position of V acts like internal pressure as long as the system remains isoelectronic. Right: Drop of the resistivity by 8 orders of magnitudes when the metal-insulator transition occurs.[31]

Optical lattices (crystal type structures made of standing laser lights) filled with neutral fermionic atoms (e.g., ^{6}Li, ^{40}K, or ^{171}Yb) provide another experimental systems for studying correlated lattice fermions.[33] The relevant system parameters in these artificial crystals can be tuned as desired and various possible phases can be investigated. It is now very rapidly developing field of research, where ideas of condensed matter theory comes into the quantum optics and laser physics.[34–36]

2.4. *Correlated fermions and inhomogeneous potentials*

Inhomogeneous external potentials are present both in cold fermionic atoms loaded into optical lattices and in real materials with correlated electrons. Cold atoms are trapped inside the magneto-optical potentials, usually of ellipsoid-like shapes in space.[37] In case of solids, the experiments are often made with the presence of external gates, which produce space-dependent inhomogeneous electric potentials. One of the example is the quantum point contact, which is a narrow, smooth constriction inside a bulk system.[38] In addition, layers, interfaces, and surfaces play very important role and are extensively studied,[11] as for example in the heterostructres made of $LaTiO_3$ (Mott insulator) and $SrTiO_3$ (band insulator).[39]

3. Disorder and disordered electron systems

In correlated electron materials it is a rule rather than an exception that the electrons, apart from strong interactions, are also subject to disorder. The disorder may result from non-stoichiometric composition, as obtained, for example, by doping of manganites ($La_{1-x}Sr_xMnO_3$) and cuprates ($La_{1-x}Sr_xCuO_4$),[8] or in the disulfides $Co_{1-x}Fe_xS_2$ and $Ni_{1-x}Co_xS_2$.[40] In the first two examples, the Sr ions create different potentials in their vicinity which affect the correlated d electrons/holes. In the second set of examples, two different transition metal ions are located at random positions, creating two different atomic levels for the correlated d electrons. In both cases the random positions of different ions break the translational invariance of the lattice, and the number of d electrons/holes varies. As the composition changes so does the randomness, with $x = 0$ or $x = 1$ corresponding to the pure cases. With changing composition the system can undergo various phase transitions. For example, FeS_2 is a pure band insulator which becomes a disordered metal when alloyed with CoS_2, resulting in $Co_{1-x}Fe_xS_2$. This system has a ferromagnetic ground state for a wide range of x with a maximal Curie temperature T_c of 120 K. On the other hand, when CoS_2 (a metallic ferromagnet) is alloyed with NiS_2 to make $Ni_{1-x}Co_xS_2$, the Curie temperature is suppressed and the end compound NiS_2 is a Mott-Hubbard antiferromagnetic insulator with Néel temperature $T_N = 40$ K.

The transport properties of real materials are also strongly influenced by the electronic interaction and randomness.[41] In particular, Coulomb correlations and disorder are both driving forces behind metal–insulator transitions connected with the localization and delocalization of particles. While the Mott–Hubbard MIT is caused by the electronic repulsion,[4] the Anderson MIT is due to coherent backscattering of non-interacting particles from randomly distributed impurities.[42] Furthermore, disorder and interaction effects are known to compete in subtle ways.[41,43,44]

4. Models for correlated, disordered lattice fermions with inhomogeneous potentials

4.1. *Hubbard model*

As we discussed above, in narrow-band systems the interaction between two electrons occupying the same orbital can play a dominant role. Therefore in theoretical modeling, we take into account this local, on-site interaction. In addition, there is a hopping of the electrons between different lattice sites. The simplest lattice model to describe this situation is provided by the Hubbard Hamiltonian[45]

$$H = \sum_{ij\sigma} t_{ij} c^{\dagger}_{i\sigma} c_{j\sigma} + \sum_{i\sigma} V_i \, n_{i\sigma} + U \sum_{i} n_{i\uparrow} n_{i\downarrow} \,, \tag{3}$$

where $c^{\dagger}_{i\sigma}$ and $c_{i\sigma}$ are the fermionic creation and annihilation operators of the electron with spin $\sigma = \pm 1/2$ at the lattice site i, $n_{i\sigma} = c^{\dagger}_{i\sigma} c_{i\sigma}$ is the particle number operator with eigenvalues 0 or 1, and t_{ij} is the probability amplitude for an electron hopping between lattice sites i and j. The second term describes the additional external potential V_i, which breaks the ideal lattice symmetry. For homogeneous systems we set $V_i = 0$. Due to the third, two-body term, two electrons with opposite spins at the same site increase the system energy by $U > 0$. In (3) only a local part of the Coulomb interaction is included and other longer-range terms are neglected for simplicity.

The hopping and the interacting terms in the Hamiltonian (3) have different, competing effects in homogeneous systems. The first, kinetic part drives the particles to be delocalized, spreading along the whole crystal as Bloch waves. Then, the one-particle wave functions strongly overlap with each other. The second, interacting term keeps the particles staying apart from each other by reducing the number of double occupied sites. In particular, when the number of electrons N_e is the same as the number of lattice sites N_L the interacting term favors a ground state with all sites being single occupied. Then the overlap of one-particle wave functions is strongly reduced.

The occupation of a single site fluctuates in time when both kinetic and interacting terms are finite. The lattice site i can be either empty $|i, 0\rangle$, or single occupied $|i, \sigma\rangle$, or double occupied with two electrons with opposite spins $|i, 2\rangle$, as shown in Fig. 2. The time evolution depends on the ratio U/t and on the average number of electrons per site $n = \langle \sum_{i\sigma} n_{i\sigma} \rangle / N_L$. As we will explain latter, the DMFT keeps this local dynamics exactly at each site, which is a key point for describing Mott insulators and Mott-Hubbard MIT.

4.2. *Models for external inhomogeneous potential*

The additional inhomogeneous term V_i in the Hamiltonian (3) allows us to model different physical systems. Cold fermionic atoms in optical lattices are trapped by the magneto-optical potential,[37] which is very well described

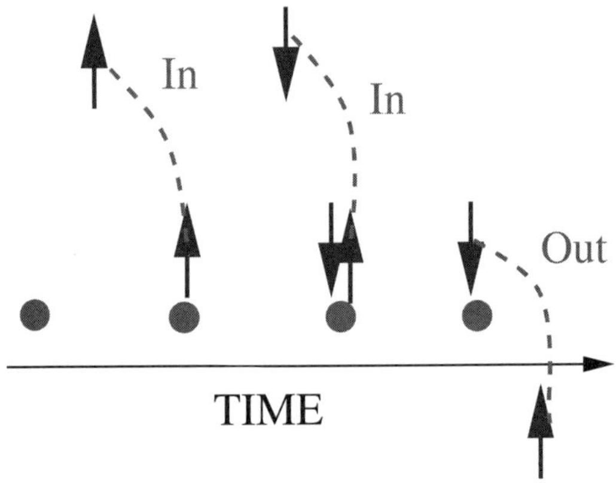

Fig. 2. Evolution of a quantum state at a single lattice site for the electrons described by the Hubbard model (3). The lattice site i can be either empty $|i,0\rangle$, or single occupied $|i,\sigma\rangle$, or double occupied with two electrons with opposite spins $|i,2\rangle$.

by a three dimensional ellipsoid

$$V_i = k_x(R_i^x - R_0^x)^2 + k_y(R_i^y - R_0^y)^2 + k_z(R_i^z - R_0^z)^2 \,, \tag{4}$$

centered at the site $\mathbf{R}_0$ and parameterized by three numbers $k_i > 0$. Another choice of the external inhomogeneous potential would describe, for example, a quantum point contact,[38] which is a narrow constriction along the z-direction. Here we model it by a three dimensional hyperboloid potential

$$V_i = -k_x(R_i^x - R_0^x)^2 - k_y(R_i^y - R_0^y)^2 + k_z(R_i^z - R_0^z)^2 \,. \tag{5}$$

Interlayers and thin films or surfaces are modelled by the potential V_i, which changes in a step-wise manner along one selected dimension.[11]

4.3. *Anderson model*

For modelling disordered electrons a similar strategy is usually employed as for the interacting problem. We assume that the particles can hop on a regular lattice but the atomic energy ϵ_i at each lattice site is a random variable. The minimal model describing such a case is given by the Anderson Hamiltonian[42]

$$H = \sum_{ij,\sigma} t_{ij} c_{i\sigma}^{\dagger} c_{j\sigma} + \sum_{i\sigma} V_i \, n_{i\sigma} + \sum_{i\sigma} \epsilon_i \, n_{i\sigma} \,, \tag{6}$$

where the meanings of the first two terms are the same as in the Eq. (3). In fact, this is a one-body Hamiltonian without any two-body interaction.

The effect of disorder on the system is taken into account through a local random term, for which we have to assume a probability distribution function (PDF).

Typically we consider the uncorrelated quenched disorder with

$$\mathcal{P}(\epsilon_1, ..., \epsilon_{N_L}) = \prod_{i=1}^{N_L} P(\epsilon_i)\,, \tag{7}$$

where $P(\epsilon_i)$ is a normalized PDF for the atomic energies ϵ_i. The quenched disorder means that $P(\epsilon_i)$ is time independent. The atomic energies are randomly distributed over the lattice but then are fixed and cannot be changed. This is different from the annealed disorder where the random atomic energies are supposed to change in time.

4.4. *Models for disorders*

If $P(\epsilon_i) = \delta(\epsilon_i)$, where $\delta(x)$ is a delta-Dirac function, the system is *pure*, i.e. without any disorder. For binary alloy disorder we assume the PDF has the form

$$P(\epsilon_i) = x\delta\left(\epsilon_i + \frac{\Delta}{2}\right) + (1-x)\delta\left(\epsilon_i - \frac{\Delta}{2}\right)\,, \tag{8}$$

where Δ is the energy difference between the two atomic energies, while x and $1-x$ are concentrations of the two alloy atoms. At $x = 0$ or 1 the system is non-disordered even when Δ is finite. Therefore the proper measure of the disorder strength would be a combination $\delta \equiv x(1-x)\Delta$.[46] This model of disorder is applicable to binary alloy A_xB_{1-x} systems, e.g. Ni_xFe_{1-x}. Another choice is a model with the continuous PDF describing continuous disorder. Here we use the box-type PDF

$$P(\epsilon_i) = \frac{1}{\Delta}\Theta(\frac{\Delta}{2} - |\epsilon_i|)\,, \tag{9}$$

with Θ as the step function. The parameter Δ is a measure of the disorder strength. Physics described by these two PDFs is qualitatively different as we shall see further. However, the use of a different continuous, normalized function for the PDF would bring about only quantitative changes.

4.5. *Anderson-Hubbard model*

To describe both correlations and disorder we simply merge together these two Hamiltonians (3) and (6) obtaining so called the Anderson-Hubbard

Hamiltonian

$$H = \sum_{ij,\sigma} t_{ij} c_{i\sigma}^{\dagger} c_{j\sigma} + \sum_{i\sigma} V_i \, n_{i\sigma} \sum_{i\sigma} \epsilon_i \, n_{i\sigma} + U \sum_i n_{i\uparrow} n_{i\downarrow} \,. \tag{10}$$

This model is a working horse for us to investigate the competition between correlations and disorder in lattice fermions.[47–50]

4.6. *Anderson-Falicov-Kimball model*

We also investigate here the spinless Anderson-Falicov-Kimball Hamiltonian

$$H = \sum_{ij} t_{ij} c_i^{\dagger} c_j + \sum_i V_i \, c_i^{\dagger} c_i + \sum_i \epsilon_i \, c_i^{\dagger} c_i + U \sum_i f_i^{\dagger} f_i c_i^{\dagger} c_i \,, \tag{11}$$

where $c_i^{\dagger}$ ($f_i^{\dagger}$) and c_i (f_i) are fermionic creation and annihilation operators for *mobile* (*immobile*) fermions at a lattice site i. Furthermore, t_{ij} is the hopping amplitude for mobile particles between sites i and j, and U is the local interaction energy between mobile and immobile particles occupying the same lattice site. The atomic energy ϵ_i is again a random, independent variable which describes the local, *quenched disorder* affecting the motion of mobile particles. Note that imobile fermions are thermodynamically coupled to mobile particles and therefore they act as an annealed disorder.

The Falicov-Kimball Hamilonian without disorder was introduced to model the f- and s-electrons in rare-earth solids, see for review [51], and later generalized for systems with quenched disorder.[52] It can also be realized experimentally using optical lattices filled with light (Li) and heavy (Rb) fermionic atoms, where the relevant parameters are fine tuned by changing the external potentials and magnetic field around the Feschbach resonance.[53] Pure Falicov-Kimball model was studied within the DMFT[51] and also within exact approaches, e.g. [54], or Monte Carlo simulations.[55] This model can also be generalized to include many orbitals[51,56] and exchange interactions.[57]

5. Average over disorder

5.1. *Average and most probable value*

Imagine now a system, described by one of the above Hamiltonians, with a given and fixed distribution of atomic energies over the lattice sites: $\{\epsilon_1, ..., \epsilon_{N_L}\}$. Even if by some means we solve the Hamiltonian, results will depend on this particular realization of the disorder. Different distributions of atoms will give different results. Usually, when the system is infinitely large ($N_L \to \infty$) we take an arithmetic average of the physical quantity (observable) $O(\epsilon_1, ..., \epsilon_{N_L})$ over infinitely may realizations of the disorder, i.e.

$$\langle O(\epsilon)\rangle = \int \prod_{i=1}^{N_L} d\epsilon_i P(\epsilon_i) O(\epsilon_1, ..., \epsilon_{N_L}), \tag{12}$$

in accord with the Central Limit Theorem[58] for infinitely many, independent random variables $O(\epsilon_1, ..., \epsilon_{N_L})$. Such methodology holds only if the system is self-averaging. It means that the sample-to-sample fluctuations

$$D_{N_L}(O) = \frac{\langle O^2\rangle - \langle O\rangle^2}{\langle O\rangle^2} \tag{13}$$

vanish when $N_L \to \infty$.

An example of the non-self-averaging system is an Anderson insulator, where the one-particle wave functions are exponentially localized in a finite subsystem.[42] In other words, during the dynamical evolution a quantum state cannot penetrate the full phase space, probing all possible random distributions. Here we are faced with a problem how to describe such systems. In principle, we should investigate the PDF for a given physical observable $O(\epsilon_1, ..., \epsilon_{N_L})$[59] and find the *most probable value* of the observable $O(\epsilon_1, ..., \epsilon_{N_L})$, i.e. such value where the PDF is maximal. This value will represent *typical behavior* of the system. It requires to have a very good statistics, which is based on many (perhaps infinitely many) samples. Such requirement is hardly to achieve, in particular in correlated electron systems, because the relevant Hilbert space is too large to be effectively dealt with. Therefore in practice, we look at a generalized mean which gives the best approximation to this most probable value, e.g. see in [50].

5.2. *Generalized mean*

Generalized f-mean for a single random variable x with a PDF given by $p(x)$ is defined as follows

$$\langle x\rangle_f = f^{-1}(\langle f(x)\rangle), \tag{14}$$

where f and f^{-1} is a function and its inverse.[60,61] The average inside the function f^{-1} is the arithmetic mean with respect to $p(x)$. The geometric mean is obtained when $f(x) = \ln x$.

If $f(x) = x^p$ is a power function of x, the corresponding generalized average is known as a Hölder mean[61] or p-mean

$$\langle x \rangle_p = \left[\int dx p(x) x^p \right]^{\frac{1}{p}} , \tag{15}$$

which is parameterized by a single number p. Like the arithmetical mean, the Hölder mean:

- is a homogeneous function of x, and
- have a block property, i.e. $\langle xy \rangle_p = \langle x \rangle_p \langle y \rangle_p$ if x and y are independent random variables.

In particular, it can be seen that $p = -\infty$ gives the possible minimum of x, $p = -1$ gives the harmonic mean, $p = 0$ gives the geometric mean, $p = 1$ is the arithmetic mean, $p = 2$ gives the quadratic mean, and $p = \infty$ gives the maximum of possible x.

Which mean gives the best approximation, in particular when disorder is very strong and drives the system from the self-averaging limit, is a matter of experience and tries. For example, many biological and social processes are modelled by the log-normal PDF, for which the geometrical mean gives exactly the most probable value of the random variable.[62,63] The same is also true for a system of conductors in series with random resistances.

6. Static mean-field theory

6.1. *Exchange Hamiltonian*

Before reviewing the DMFT for correlated fermions we discuss the static (Weiss) mean-field approximation for magnetic systems. This helps to appreciate similarities and differences between the static and the dynamical mean-field theories.

At large U the Hubbard model (3) can be reduced, via the canonical transformation projecting out double occupied sites, to the t-J Hamiltonians with spin-exchange interactions between the electrons.[64] At half-filling the t-J model is exactly equivalent to the Heisenberg Hamiltonian for localized magnetic moments

$$H_{\text{exch}} = -\frac{1}{2} \sum_{ij} J_{ij} \mathbf{S}_i \cdot \mathbf{S}_j , \tag{16}$$

where $\mathbf{S}_i$ is a quantum mechanical spin operator and $J_{ij} = 4t_{ij}^2/U \ll t_{ij}$ is a kinetic exchange coupling.

6.2. *Static mean-field approximation*

The idea of any mean-field theory is to replace an unsolvable many-body Hamiltonian by a solvable one-body Hamiltonian $H_{\rm MF}$ containing an external fictitious field. In the case of the Heisenberg model (16), we rewrite the partition function as

$$Z = \mathrm{Tr}_{\mathbf{S}_i} e^{-\beta H_{\rm exch}} = \mathrm{Tr}_{\mathbf{S}_i} e^{-\beta H_{\rm MF}} , \tag{17}$$

where $\beta = 1/kT$ is the inverse of the temperature kT in energy units and Tr denotes the trace over the spin degrees of freedom. The one-body mean-field Hamiltonian $H_{\rm MF}$ is assumed to be

$$H_{\rm MF} = \sum_i \mathbf{B}_i^{\rm MF} \cdot \mathbf{S}_i + E_{shift} . \tag{18}$$

The interpretation of (18) is straightforward: localized moments interact only with the external magnetic (Weiss, molecular) mean-field $\mathbf{B}_i^{\rm MF}$. The transformation from the Hamiltonian (16) into the mean-field Hamiltonian (18) is exact from the formal point of view. However, the static Weiss mean-field $\mathbf{B}_i^{\rm MF}$ is not known yet and has to be found within an approximation scheme.

We determine the molecular field within the mean-field (decoupling) approximation $\mathbf{S}_i \cdot \mathbf{S}_j \approx \langle \mathbf{S}_i \rangle \cdot \mathbf{S}_j$. Hence, $\mathbf{B}_i^{\rm MF} = \sum_{j(i)} J_{ij} \langle \mathbf{S}_j \rangle_{H_{\rm MF}}$. Now the average spin $\langle \mathbf{S}_i \rangle$ is found by the self-consistent equation

$$\langle S^z \rangle_{H_{MF}} = \tanh\left(\beta J \langle S^z \rangle_{H_{MF}}\right) , \tag{19}$$

where the unknown quantity appears on both sides of it. In the last step we assumed also that the interaction is only between nearest neighbor moments and that the systems is homogeneous. We can solve Eq. (19) iteratively, in the l-th step we plug $\langle S^z \rangle_{H_{MF}}^l$ into the right hand side of (19) and determine new $\langle S^z \rangle_{H_{MF}}^{l+1}$ from the left hand side of (19). We shall repeat this until l-th and l+1-th results are numerically the same with pre-assumed accuracy.

The only approximation within this static mean-field theory is that we have neglected spatial spin-spin correlations, i.e. we have assumed explicitly that

$$\langle [\mathbf{S}_i - \langle \mathbf{S}_i \rangle] \cdot [\mathbf{S}_j - \langle \mathbf{S}_j \rangle] \rangle = 0 \Longrightarrow \langle \mathbf{S}_i \cdot \mathbf{S}_j \rangle = \langle \mathbf{S}_i \rangle \cdot \langle \mathbf{S}_j \rangle . \tag{20}$$

It means that a single spin interacts now with an average (mean) external magnetic field, produced by all other spins as is plotted schematically in Fig. 3.

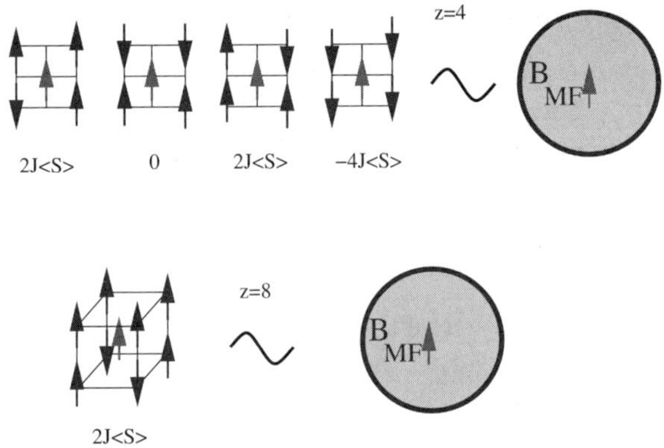

Fig. 3. Different spin configurations are replaced by a single central spin interacting with a mean magnetic field B^{MF}. The more spins interact with the central one the more accurate is such approximation.

6.3. *Large dimensional limit*

This approximation becomes exact when the single spin interacts with infinitely many other spins. For this either the exchange coupling $J_{ij} = t_{ij}^2/U$ has to be infinitely long-range or there has to be infinitely many nearest neighbors to a given site. The former would lead to hopping amplitudes with pathological properties. The latter is admissible in view of a physical realization. Indeed, note that for the two-dimensional simple cubic (sc) lattice the coordination number $z = 4$, for the three-dimensional simple cubic lattice $z = 6$, for the body centered cubic (bcc) lattice $z = 8$, whereas for the the three-dimensional face centered cubic lattice (fcc) $z = 12$. Large z limit corresponds to the existence of a small, dimensionless parameter $1/z$ in the theory, which vanishes when $z \to \infty$.

The best strategy is to construct a mean-field type theory which is exact in the $z \to \infty$ limit. In practice, a useful non-trivial theory is only obtained when we rescale the exchange coupling, $J = J^*/z$ where $J^* = \mathrm{const}$. In this case the static mean-magnetic field

$$\mathbf{B}^{\mathrm{MF}} = \sum_{j=1}^{z} J \langle \mathbf{S} \rangle_{H_{MF}} = \frac{J^*}{z} \sum_{j=1}^{z} \langle \mathbf{S} \rangle_{H_{\mathrm{MF}}} = J^* \langle \mathbf{S} \rangle_{H_{\mathrm{MF}}} \tag{21}$$

is finite (bounded) in $z \to \infty$ limit. Then the spatial correlations exactly vanish

$$\lim_{z\to\infty} \langle [\mathbf{S}_i - \langle \mathbf{S}_i \rangle] \cdot [\mathbf{S}_j - \langle \mathbf{S}_j \rangle] \rangle = 0 \Longrightarrow \lim_{z\to\infty} \langle \mathbf{S}_i \cdot \mathbf{S}_j \rangle = \langle \mathbf{S}_i \rangle \cdot \langle \mathbf{S}_j \rangle \,. \tag{22}$$

In principle, we can also find corrections to the $z \to \infty$ limit by applying a perturbation theory with respect to $1/z$. Hence, this static mean-field theory is a controlled approximation with well defined small parameter.

Unfortunately this is not the theory which we expect to have for the Hubbard model of correlated fermions. The static mean-field theory was constructed only for the t-J model, where $t \ll U$. On the other hand, as we discussed in the beginning sections, the interesting physics with Mott-Hubbard MIT occurs when $t \sim U$, i.e. in the intermediate coupling regime. Then the local dynamics seems to be important and we need a dynamical theory, which incorporates those local quantum fluctuations.

7. The Holy Grail for lattice fermions or bosons

In the lack of exact solutions to the interesting models (3,6,10,11) in two or three dimensions, we search for an approximate but comprehensive mean-field theory that:

- is valid for all values of parameters, e.g. U/t, the electron density, disorder strength Δ and concentration x, or temperature;
- is thermodynamically consistent, i.e. irrespectively in which way thermodynamical quantities are calculated, final results are identical;
- is conserving, i.e. the approximation must preserve the microscopic conservation laws of the Hamiltonian;
- has a small controlled (expansion) parameter z and the approximate theory is exact when $z \to \infty$;
- is flexible in application to various classes of correlated fermion models, with and without disorder, and useful to realistic, material specific calculations.

Among many different approximate theories for the Hubbard-like models of lattice fermions, the DMFT is the only one which satisfies all of those requirements.[11,17–21,51] Recently, the bosonic dynamical mean-field theory (B-DMFT) for correlated lattice bosons in normal and Bose-Einstein condensate phases has also been formulated.[65]

8. DMFT - practical and quick formulation

In this Section we discuss the comprehensive mean-field theory for the correlated lattice fermions with disorder and inhomogeneous external potentials. We present a quick and practical formulation of the DMFT equations, which emphasizes the mean-field character of the theory.

8.1. *Exact partition function, Green function, and self-energy*

We begin with the exact partition function for a given lattice problem, e.g. as that described by one of the Hamiltonians (3,6,10,11),

$$Z = \mathrm{Tr} e^{-\beta(H-\mu N)} = \prod_{\sigma\omega_n} \mathrm{Det}[-\hat{G}_\sigma(\omega_n)^{-1}] = \exp\left(\sum_{\sigma\omega_n} \mathrm{Tr}\ln[-\hat{G}_\sigma(\omega_n)^{-1}]\right). \tag{23}$$

This partition function is derived in the Appendix, cf. Eq. (50). Odd Matsubara frequencies $\omega_n = (2n+1)\pi/\beta$, corresponding to the imaginary time τ by the Fourier transform, keep track of the Fermi-Dirac statistics. The hat symbol in $\hat{G}$ reminds us that the Green function is an infinite matrix when expressed in the basis of one-particle wave functions.

Our goal is to formulate the DMFT for arbitrary discrete systems, where the lattice is not necessary of Bravais type.[66–68] Therefore we work with a general quantum numbers α, corresponding to the one-particle basis $|\alpha\rangle$ which diagonalizes the equations of motion, see the Eq. (48) in the Appendix. In particular, the present general formulation of the DMFT is applicable: to regular crystals, where the basis functions are Bloch waves with quasi-momenta **k** as proper quantum numbers, as well as to thin films, interlayers, and surfaces,[11,69–71] or to irregular constrictions for electrons and magneto-optical traps for cold atoms, or to infinite graphs, e.g. the Bethe tree with different hoppings.[72,73] We only demand that the lattice is infinitely large since we work directly in the thermodynamic limit.

According to the Dyson equation,[74] the one-particle Green function

$$G_{ij\sigma}(\tau) = -\langle T_\tau c_{i\sigma}(\tau) c^\dagger_{j\sigma}(0)\rangle, \tag{24}$$

where T_τ is a chronological operator, can be expressed exactly by the non-interacting Green function $G^0_{ij\sigma}(\omega_n)$ and the self-energy $\Sigma_{ij\sigma}(\omega_n)$, i.e.

$$G_{ij\sigma}(\omega_n)^{-1} = G^0_{ij\sigma}(\omega_n)^{-1} - \Sigma_{ij\sigma}(\omega_n). \tag{25}$$

Here the non-interacting Green function can be written in the $|\alpha\rangle$ basis $G^0_{\alpha\sigma}(\omega_n) = 1/(i\omega_n + \mu - \epsilon_\alpha)$, where ϵ_α are exact eigenvalues of the non-interacting and non-disorder part H_0 of the corresponding Hamiltonian, i.e. $H_0|\alpha\rangle = \epsilon_\alpha|\alpha\rangle$. Then, the partition function (23) is expressed by the exact self-energy

$$Z = \exp\left(\sum_{\sigma\omega_n} \mathrm{Tr}\ln[\hat{G}^0_\sigma(i\omega_n)^{-1} - \hat{\Sigma}_\sigma(\omega_n)]\right). \tag{26}$$

The problem would be solved exactly if we had only known the self-energy $\hat{\Sigma}_\sigma(\omega_n)$.

8.2. *DMFT approximation*

Unfortunately, the self-energy is not known so we have to determine it approximately. Within the DMFT for homogeneous systems we make the main assumption, namely, that the self-energy is *α-independent*, i.e.

$$\Sigma_{\alpha\alpha'\sigma}(\omega_n) = \Sigma_\sigma(\omega_n)\delta_{\alpha\alpha'} \,. \tag{27}$$

For the electrons on Bravais lattices it means that the self-energy is *momentum independent*. Correspondingly in the lattice space, the self-energy is *local*, i.e. only diagonal elements are non-vanishing

$$\Sigma_{ij\sigma}(\omega_n) = \sum_{\alpha\alpha'} \langle i|\alpha\rangle \Sigma_{\alpha\alpha'\sigma} \langle \alpha'|j\rangle = \Sigma_\sigma(\omega_n)\delta_{ij} \,, \tag{28}$$

where δ_{ij} is the lattice Kronecker delta, and we used the completeness of the α-basis, i.e. $1 = \sum_\alpha |\alpha\rangle\langle\alpha|$, together with the orthonormality property of the Wannier states.

For lattice systems with additional, non-uniform external potential V_i we have to make different, independent assumption, namely the *self-energy is local and depends explicitly on site indices*, i.e.

$$\Sigma_{ij\sigma}(\omega_n) = \Sigma_{i\sigma}(\omega_n)\delta_{ij} \,. \tag{29}$$

In general such an assumption does not imply that the self-energy in α-basis is diagonal or independent on α.

Within the DMFT we neglect space correlations by assuming that the self-energy is local, diagonal in the lattice space. The local, dynamical correlations, however, are included exactly by keeping the frequency dependence of $\Sigma_{i\sigma}(\omega_n)$, which translates into explicit time-dependence of the self-energy $\Sigma_{i\sigma}(\tau - \tau')$. Local dynamics is preserved by this approximation and therefore we use the name *dynamical mean-field theory*. The DMFT is the equilibrium theory expecting to describe systems in thermal equilibrium (or close to it, within linear response regime). The extension of the DMFT to non-equilibrium situations is a separate problem, currently being developed.[75]

8.3. *Local Green function*

The local Green function $G_{i\sigma}(i\omega_n) \equiv G_{ii\sigma}(i\omega_n)$ is given by the diagonal elements obtained from the matrix Dyson equation

$$G_{ij\sigma}(i\omega_n) = \left[G^0_{ij\sigma}(i\omega_n)^{-1} - \Sigma_{i\sigma}(i\omega_n)\delta_{ij}\right]^{-1} , \tag{30}$$

provided that we know all local self-energies on all lattice sites. This exact expression is applicable to study finite size systems. When the system is infinitely large we have to make other approximation to make the DMFT equation numerically tractable.

8.4. *Local approximation to Dyson equation*

The local (diagonal in a lattice space) Green function can be expressed solely by the local, non-interacting, pure density of states (LDOS) and the self-energy, namely

$$G_{i\sigma}(\omega_n) = \sum_{\alpha} \frac{|\langle\alpha|i\rangle|^2}{i\omega_n + \mu - \epsilon_\alpha - \Sigma_{i\sigma}(\omega_n)} = \int d\epsilon \frac{N_i^0(\epsilon)}{i\omega_n + \mu - \epsilon - \Sigma_{i\sigma}(\omega_n)}, \tag{31}$$

where $N_i^0(\epsilon) = \sum_\alpha |\langle\alpha|i\rangle|^2 \delta(\epsilon - \epsilon_\alpha)$ is the non-interacting LDOS. This expression is obtained by assuming that on all sites the self-energy $\Sigma_{i\sigma}(i\omega_n)$ is the same when calculating $G_{ii\sigma}(i\omega_n)$. For another site R_j we assume the same but now the self-energy should be $\Sigma_{j\sigma}(i\omega_n)$. We call this as a local Dyson equation approximation (LDEA). For a given problem with the external potential V_i the noninteracting, pure LDOS is determined once at the beginning for all lattice sites and then stored in the computer memory. In practice a system of arbitrary size can be studied within this local approximation. This LDOS is the same and site independent for homogeneous lattices.

This assumption is certainly valid as long as the external potential V_i is a slowly-varying function compared with other characteristic length scales, as a lattice constant a and the Fermi wave-length. This is a quasi-classical (Wigner) description where both the α-quantum number in ϵ_α and the R_i-position in $\Sigma_{i\sigma}(\omega_n)$ are used at the same time.

8.5. *Dynamical mean-field function*

The locality of the self-energy has far reaching consequences as we shall see now. We write the local Green function in a different form, i.e.

$$G_{i\sigma}(\omega_n) = \frac{1}{i\omega_n + \mu - \eta_{i\sigma}(\omega_n) - \Sigma_{i\sigma}(\omega_n)}, \tag{32}$$

where we introduced frequency dependent function $\eta_{i\sigma}(\omega_n)$. We can interpret $\eta_{i\sigma}(\omega_n)$ as *dynamical mean-field function.* It describes resonant broadenings of single-site levels due to coupling of a given site to the rest of the system. In different words, a coupling of the selected single site to all other sites is described in average by the local dynamical mean-field function $\eta_{i\sigma}(\omega_n)$.

8.6. *Self-consistency conditions*

The partition function is expressed now as a product of the partition functions determined on each lattice sites

$$Z = \prod_{i=1}^{N_L} Z_i = \prod_{i=1}^{N_L} \exp\left(\sum_{\sigma\omega_n} \ln[i\omega_n + \mu - \eta_{i\sigma}(\omega_n) - \Sigma_{i\sigma}(\omega_n)] \right) . \tag{33}$$

The mean-field function $\eta_{i\sigma}(\omega_n)$ looks formally like an site- and time- dependent potential. In the interaction representation, the unitary time evolution due to this potential is described by the local, time-dependent evolution operator[11,51]

$$U[\eta_{i\sigma}] = T_\tau e^{-\int_0^\beta d\tau \int_0^\beta d\tau' c_{i\sigma}^\dagger(\tau)\eta_{i\sigma}(\tau-\tau')c_{i\sigma}(\tau')} . \tag{34}$$

In the final step we write the partition function (33) as a trace of operators

$$Z = Z[\eta_{i\sigma}] = \prod_{i=1}^{N_L} \mathrm{Tr}\left[e^{-\beta(H_i^{\mathrm{loc}} - \mu N_i^{\mathrm{loc}})} U[\eta_{i\sigma}] \right] , \tag{35}$$

where H_i^{loc} is the local part of the lattice Hamiltonian operator and describes the interaction and/or disorder. Here N_i^{loc} is the local particle number operator. The Eq. (35) is our main result allowing us to determine the interesting, local Green function for a given dynamical mean-field $\eta_{i\sigma}(\omega_n)$. Indeed, the local Green function is obtained by taking a functional derivative of the logarithm from the partition function (35) with respect to $\eta_{i\sigma}(\omega_n)$,[11,51] i.e.

$$G_{i\sigma}(\omega_n) = -\frac{\partial \ln Z[\eta_{i\sigma}]}{\partial \eta_{i\sigma}(\omega_n)} . \tag{36}$$

In systems with disorder, the local part of the Hamiltonian H_i^{loc} has the random variable ϵ_i and therefore the local Green function (36) is also a random quantity. In such a case we determine the average local Green function by taking one of the mean, introduced above. For this we calculate the spectral function, the interacting LDOS, from (36), i.e.

$$A_{i\sigma}(\omega) = -\frac{1}{\pi}\mathrm{Im}G_{i\sigma}(\omega_n \to \omega + i0^+) , \tag{37}$$

and determine we the p-mean with respect to a given PDF

$$\langle A_{i\sigma}(\omega)\rangle_p = \left[\int d\epsilon P(\epsilon) A_{i\sigma}(\omega)^p \right]^{\frac{1}{p}} . \tag{38}$$

According to the spectral representation theorem[74] the averaged local Green function is

$$\langle G_{i\sigma}(\omega_n)\rangle_p = \int d\omega \frac{\langle A_{i\sigma}(\omega)\rangle_p}{i\omega_n - \omega}, \tag{39}$$

and should be used instead of (36) is systems with disorder.

Knowing $G_{i\sigma}(\omega_n)$ from the Eq. (36) or its average form from the Eq. (39) we find directly the self-energy from the Eq. (32), i.e.

$$\Sigma_{i\sigma}(\omega_n) = i\omega_n + \mu - \eta_{i\sigma}(\omega_n) - \frac{1}{G_{i\sigma}(\omega_n)}, \tag{40}$$

or

$$\Sigma_{i\sigma}(\omega_n) = i\omega_n + \mu - \eta_{i\sigma}(\omega_n) - \frac{1}{\langle G_{i\sigma}(\omega_n)\rangle_p}, \tag{41}$$

respectively.

The local Green function is determined for a given, fixed mean-field potential $\eta_{i\sigma}(\omega_n)$. To obtain a self-consistent solution we proceed iteratively by employing the Eq. (31) to obtain a new local Green function, which afterwards is used to determine a new mean-field function from Eq. (32). This should be done on all lattice sites in parallel if the system contains the external, inhomogeneous potential V_i. For Bravais lattices the self-energy is the same on each site so does the mean-field function $\eta_\sigma(\omega_n)$. Then the site index i is irrelevant and can be omitted. The iteration steps should be repeated until the l-th and l+1-th results are numerically the same with a pre-assumed accuracy.

9. Limit of large coordination number

The success of the DMFT can be ascribed to the fact that this theory provides an exact solution for non-trivial lattice Hamiltonians in the limit of large coordination number, i.e. $z \to \infty$. Similarly to the static mean-field theory for the exchange Hamiltonian, in the $z \to \infty$ limit the space correlation functions vanish. The remaining correlations are local but time dependent and completely taken into account by the DMFT.

To obtain a non-trivial theory in the $z \to \infty$ limit the hopping amplitudes in the lattice Hamiltonians have to be rescaled,[12] i.e. $t_{ij} = t^*_{ij}/\sqrt{d^{R_{ij}}}$, where R_{ij} is a distance between sites i and j obtained by counting the minimal number of links between them. Then the non-local hopping, the local interacting, and the local random parts of the Hamiltonians (3,6,10,11) are treated on equal footing.[12–14,19,76] It can be exactly shown that the self-energy is then local. Hence our quick, practical derivation becomes also an exact solution of the corresponding lattice problem. In addition, it can be shown that within this rescaling scheme, the lattice disordered systems are self-averaging and the use of arithmetic averaging is justified.[76] For uncorrelated disordered systems the DMFT is equivalent to the coherent potential approximation (CPA).[77]

Since the theory is exact in the large z limit it must be conserving and thermodynamically consistent. Also it must give a comprehensive description of full phase diagrams in all possible regimes of the model and external parameters.

We also mention here that DMFT is formulated in such a way as to describe phases with long-range orders.[19] Also the DMFT can be merged with realistic *ab initio* calculations, which is known as the LDA+DMFT approach.[20,21] These examples show flexibility of the DMFT and ability to describe real physical systems.

The DMFT set of equations have to be solved. The most difficult part is determination of the local Green function from the Eq. (35). Apart of the Falicov-Kimball model and similar ones, where there exist additional local conservation laws, the calculation of the partition function (35) and the following Green functions requires advanced numerical approaches. Different techniques are routinely used now and discussed in details in literature.[19] The results presented here were obtained by using Quantum Monte Carlo (QMC) simulations at finite temperatures[19] and Numerical Renormalization Group (NRG) at zero temperature.[78,79]

10. Surprising results from DMFT

The investigations of electronic correlations and their interplay with disorder by means of the DMFT has led us to the discovery of several unexpected properties and phenomena. Examples are: (i) a novel type of Mott-Hubbard metal insulator transition away from integer filling in the presence of binary alloy disorder;[48] (ii) an enhancement of the Curie temperature in correlated electron systems with binary alloy disorder;[47,49] and (iii) unusual effects of correlations and disorder on the Mott-Hubbard and Anderson MITs, respectively.[50,52] Below we describe and explain these often surprising results as an illustration of the theory introduced above.

10.1. *Metal-insulator transition at fractional filling*

The Mott-Hubbard MIT occurs upon increasing the interaction strength U in the models (10) and (11) if the number of electrons N_e is commensurate with the number of lattice sites N_L or, more precisely, if the ratio N_e/N_L is an odd integer. At zero temperature it is a continuous transition whereas at finite temperatures the transition is of first-order.[19,80] Surprisingly, in the presence of binary alloy disorder the MIT occurs at fractional filling.[48]

We describe this situation by using the Anderson-Hubbard model (10) with the distribution (8) which corresponds to a binary-alloy system composed of two different atoms A and B. The atoms are distributed randomly on the lattice and have ionic energies $\epsilon_{A,B}$, with $\epsilon_B - \epsilon_A = \Delta$. The concentration of A (B) atoms is given by $x = N_A/N_L$ $(1 - x = N_B/N_L)$, where N_A (N_B) is the number of the corresponding atoms.

From the localization theorem (the Hadamard–Gerschgorin theorem in matrix algebra) it is known that if the Hamiltonian (10), with a binary alloy distribution for ϵ_i, is bounded, then there is a gap in the single–particle spectrum for sufficiently large $\Delta \gg \max(|t|,\ U)$. Hence at $\Delta = \Delta_c$ the DOS splits into two parts corresponding to the lower and the upper alloy subbands with centers of mass at the ionic energies ϵ_A and ϵ_B, respectively. The width of the alloy gap is of the order of Δ. The lower and upper alloy subband contains $2xN_L$ and $2(1-x)N_L$ states, respectively.

New possibilities appear in systems with correlated electrons and binary alloy disorder.[48] The Mott–Hubbard metal insulator transition can occur at any filling $n = x$ or $1 + x$, corresponding to a half–filled lower or to a half–filled upper alloy subband, respectively, as shown schematically for $n = x$ in Fig. 4. The Mott insulator can then be approached either by increasing U when $\Delta \geq \Delta_c$ (alloy band splitting limit), or by increasing Δ when $U \geq U_c$ (Hubbard band splitting limit). The nature of the Mott insulator in the binary alloy system can be understood physically as follows. Due to the high energy cost of the order of U the randomly distributed ions with lower (higher) local energies ϵ_i are singly occupied at $n = x$ $(n = 1 + x)$, i.e., the double occupancy is suppressed. In the Mott insulator with $n = x$ the ions with higher local energies are empty and do not contribute to the low–energy processes in the system. Likewise, in the Mott insulator with $n = 1 + x$ the ions with lower local energies are double occupied implying that the lower alloy subband is blocked and does not play any role.

For $U > U_c(\Delta)$ in the Mott insulating state with binary alloy disorder one may use the lowest excitation energies to distinguish two different types of insulators. Namely, for $U < \Delta$ an excitation must overcome the energy gap between the lower and the upper Hubbard subbands, as indicated in Fig. 4. We call this insulating state an *alloy Mott insulator*. On the other hand, for $\Delta < U$ an excitation must overcome the energy gap between the

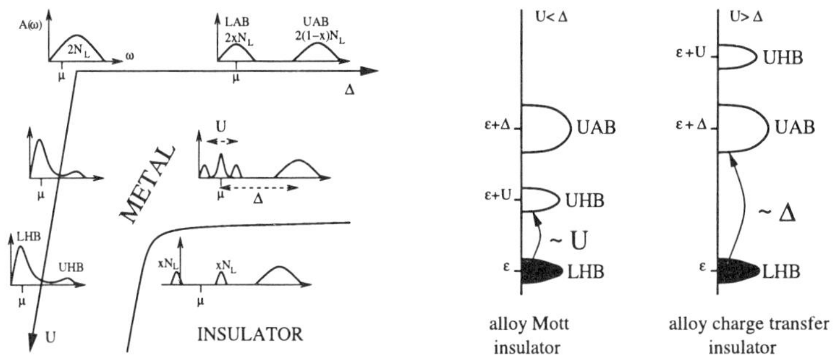

Fig. 4. Left: Schematic plot representing the Mott–Hubbard metal–insulator transition in a correlated electron system with the binary alloy disorder. The shapes of spectral functions $A(\omega)$ are shown for different interactions U and disorder strengths Δ. Increasing Δ at $U = 0$ leads to splitting of the spectral function into the lower (LAB) and the upper (UAB) alloy subbands, which contain $2xN_L$ and $2(1-x)N_L$ states respectively. Increasing U at $\Delta = 0$ leads to the occurrence of lower (LHB) and upper (UHB) Hubbard subbands. The Fermi energy for filling $n = x$ is indicated by μ. At $n = x$ (or $n = 1 + x$, not shown in the plot) the LAB (UAB) is half-filled. In this case an increase of U and Δ leads to the opening of a correlation gap at the Fermi level and the system becomes a Mott insulator. Right: Two possible insulating states in the correlated electron system with binary–alloy disorder. When $U < \Delta$ the insulating state is an alloy Mott insulator with an excitation gap in the spectrum of the order of U. When $U > \Delta$ the insulating state is an alloy charge transfer insulator with an excitation gap of the order of Δ; after Ref. [48]

lower Hubbard subband and the upper alloy–subband, as shown in Fig. 4. We call this insulating state an *alloy charge transfer insulator.*

In Fig. 5 we present a particular phase diagram for the Anderson-Hubbard model at filling $n = 0.5$ showing a Mott-Hubbard type of MIT with typical hysteresis.

10.2. *Disorder-induced enhancement of the Curie temperature*

Itinerant ferromagnetism in the pure Hubbard model occurs only away from half-filling and if the DOS is asymmetric and peaked at the lower edge.[81,82] While the Curie temperature increases with the strength of the electron interaction one would expect it to be lowered by disorder. However, our investigations show that in some cases the Curie temperature can actually be increased by binary alloy disorder.[47,49]

Indeed, the Curie temperature as a function of alloy concentration exhibits very rich and interesting behavior as is shown in Fig. 6. At some concentrations and certain values of U, Δ and n, the Curie temperature is

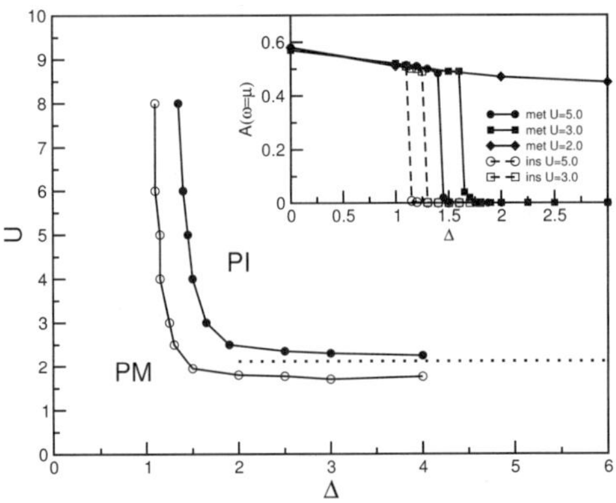

Fig. 5. Ground state phase diagram of the Hubbard model with binary–alloy disorder at filling $n = x = 0.5$. The filled (open) dots represent the boundary between paramagnetic metallic (PM) and paramagnetic insulating (PI) phases as determined by DMFT with the initial input given by the metallic (insulating) hybridization function. The horizontal dotted line represents U_c obtained analytically from an asymptotic theory in the limit $\Delta \to \infty$. Inset: hysteresis in the spectral functions at the Fermi level obtained from DMFT with an initial metallic (insulating) host represented by filled (open) symbols and solid (dashed) lines; after.[48]

enhanced above the corresponding value for the non-disordered case ($x = 0$ or 1). This is shown in the upper panel of Fig. 6 for $0 < x < 0.2$. The relative increase of T_c can be as large as 25%, as is found for $x \approx 0.1$ at $n = 0.7$, $U = 2$ and $\Delta = 4$ (upper panel of Fig. 6).

This unusual enhancement of T_c is caused by three distinct features of interacting electrons in the presence of binary alloy disorder:

i) The Curie temperature in the non-disordered case $T_c^{\text{p}} \equiv T_c(\Delta = 0)$, depends non-monotonically on band filling n.[81] Namely, $T_c^{\text{p}}(n)$ has a maximum at some filling $n = n^*(U)$, which increases as U is increased; see also our schematic plots in Fig. 7.

ii) As was described above, in the alloy-disordered system the band is split when $\Delta \gg W$. As a consequence, for $n < 2x$ and $T \ll \Delta$ electrons occupy only the lower alloy subband and for $n > 2x$ both the lower and upper alloy subbands are filled. In the former case the upper subband is empty while in the later case the lower subband is completely full. Effectively, one can therefore describe this system by a Hubbard model mapped onto the either lower or the upper alloy subband, respectively. The second subband plays a passive role. Hence, the situation corresponds to a *single* band with the *effective* filling $n_{\text{eff}} = n/x$ for $n < 2x$ and $n_{\text{eff}} = (n - 2x)/(1 - x)$ for $n > 2x$. It is then possible to determine T_c from the phase diagram of the Hubbard model without disorder.

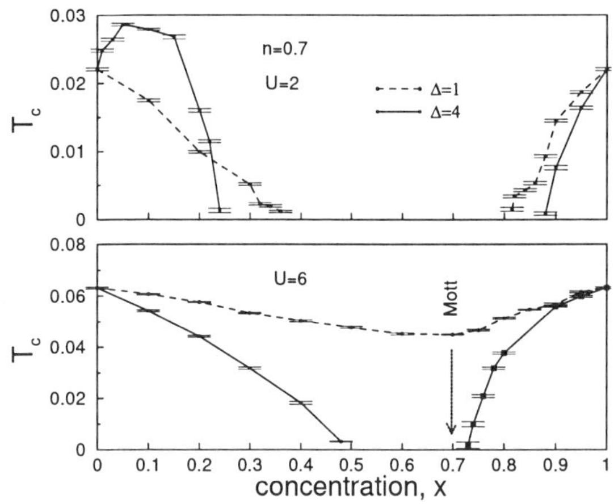

Fig. 6. Curie temperature as a function of alloy concentration x at $U = 2$ (upper panel) and 6 (lower panel) for $n = 0.7$ and disorder $\Delta = 1$ (dashed lines) and 4 (solid lines); after Refs. [47,49].

iii) The disorder leads to a reduction of $T_c^{\mathrm{p}}(n_{\mathrm{eff}})$ by a factor $\alpha = x$ if the Fermi level is in the lower alloy subband or $\alpha = 1 - x$ if it is in the upper alloy subband, i. e. we find

$$T_c(n) \approx \alpha T_c^{\mathrm{p}}(n_{\mathrm{eff}})\,, \tag{42}$$

when $\Delta \gg W$. Hence, as illustrated in Fig. 7, T_c can be determined by $T_c^{\mathrm{p}}(n_{\mathrm{eff}})$. Surprisingly, then, it follows that for suitable U and n the Curie temperature of a disordered system can be higher than that of the corresponding non-disordered system [cf. Fig. 7].

10.3. *Continuously connected insulating phases in strongly correlated systems with disorder*

The Mott-Hubbard MIT is caused by Coulomb correlations in the pure system. By contrast, the Anderson MIT, also referred to as Anderson localization, is due to coherent backscattering from randomly distributed impurities in a system without interaction.[42] It is therefore a challenge to investigate the effect of the *simultaneously* presence of interactions and disorder on electronic systems.[50,52] In particular, the question arises whether it will suppress or enlarge a metallic phase. And what about the Mott and Anderson insulating phases: will they be separated by a metallic phase? Possible scenarios are schematically plotted in Fig. 8.

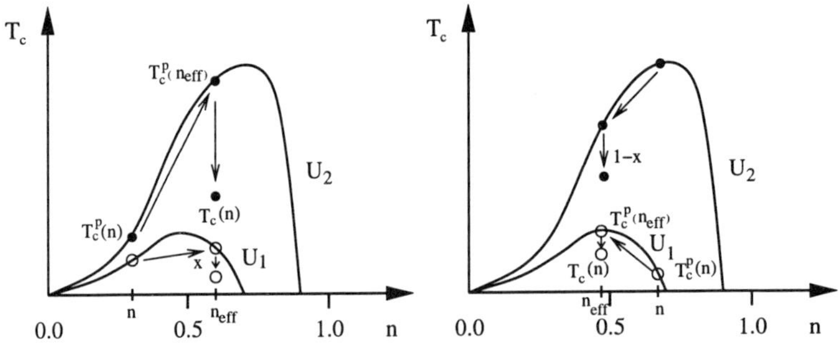

Fig. 7. Schematic plots explaining the filling dependence of T_c for interacting electrons with strong binary alloy disorder. Curves represent T_c^{P}, the Curie temperature for the pure system, as a function of filling n at two different interactions $U_1 \ll U_2$. Left: For $n < x$, T_c of the disordered system can be obtained by transforming the open (for U_1) and the filled (for U_2) point from n to $n_{\mathrm{eff}} = n/x$, and then multiplying $T_c^{\mathrm{P}}(n/x)$ by x as indicated by arrows. One finds $T_c(n) < T_c^{\mathrm{P}}(n)$ for U_1, but $T_c(n) > T_c^{\mathrm{P}}(n)$ for U_2. Right: For $n > x$, T_c of the disordered system can be obtained by transforming $T_c^{\mathrm{P}}(n)$ from n to $n_{\mathrm{eff}} = (n - 2x)/(1 - x)$, and then multiplying $T_c^{\mathrm{P}}[(n - 2x)/(1 - x)x]$ by $1 - x$ as indicated by arrows. One finds $T_c(n) > T_c^{\mathrm{P}}(n)$ for U_1, but $T_c(n) < T_c^{\mathrm{P}}(n)$ for U_2; after Ref. [47,49].

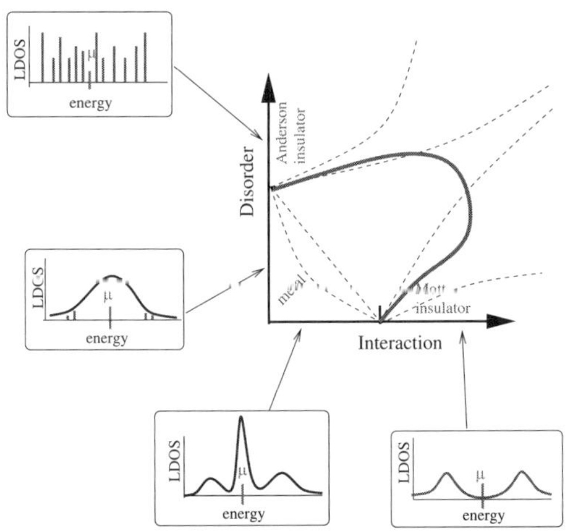

Fig. 8. Possible phases and phase transitions triggered by interaction and disorder in the same system. According to DMFT investigations the simultaneous presence of correlations and disorder enhances the metallic regime (thick line); the two insulating phases are connected continuously. Insets show different local density of states when disorder or interaction is switched off.

The Mott-Hubbard MIT is characterized by the opening of a gap in the density of states at the Fermi level. At the Anderson localization transition the character of the spectrum at the Fermi level changes from a continuous spectrum to a dense, pure point spectrum. It is plausible to assume that both MITs can be characterized by a single quantity, namely, the local density of states (LDOS). Although the LDOS is not an order parameter associated with a symmetry breaking phase transition, it discriminates between a metal and an insulator, which is driven by correlations and disorder, cf. insets to Fig. 8.

In a disordered system the LDOS depends on a particular realization of the disorder in the system. To obtain a full understanding of the effects of disorder it would therefore in principle be necessary to determine the entire probability distribution function of the LDOS, which is almost never possible. Instead one might try to calculate moments of the LDOS. This, however, is insufficient because the arithmetically averaged LDOS (first moment) stays finite at the Anderson MIT.[83] It was already pointed out by Anderson[42] that the "typical" values of random quantities, which are mathematically given by the most probable values of the probability distribution functions, should be used to describe localization. The *geometric* mean is defined by

$$A_{\text{geom}} = \exp\left[\langle \ln A(\epsilon_i) \rangle_{\text{dis}}\right] , \tag{43}$$

and differs from the arithmetical mean given by

$$A_{\text{arith}} = \langle A(\epsilon_i) \rangle_{\text{dis}} , \tag{44}$$

where $\langle F(\epsilon_i) \rangle_{\text{dis}} = \int d\epsilon_i \mathcal{P}(\epsilon_i) F(\epsilon_i)$ is an arithmetic mean of function $F(\epsilon_i)$. The geometrical mean gives an approximation of the most probable ("typical") value of the LDOS and vanishes at a critical strength of the disorder, hence providing an explicit criterion for Anderson localization.[42,84–86]

A non-perturbative framework for investigations of the Mott-Hubbard MIT in lattice electrons with a local interaction and disorder is provided by the dynamical mean-field theory (DMFT).[17,19] If in this approach the effect of local disorder is taken into account through the arithmetic mean of the LDOS[87] one obtains, in the absence of interactions, the well known coherent potential approximation (CPA),[76] which does not describe the physics of Anderson localization. To overcome this deficiency Dobrosavljević *et al.*[85] incorporated the geometrically averaged LDOS into the self-consistency cycle and thereby derived a mean-field theory of Anderson localization which reproduces many of the expected features of the disorder-driven MIT for non-interacting electrons. This scheme uses only one-particle quantities and is therefore easily incorporated into the DMFT for disordered electrons in the presence of phonons,[88] or Coulomb correlations.[50,52] In particular,

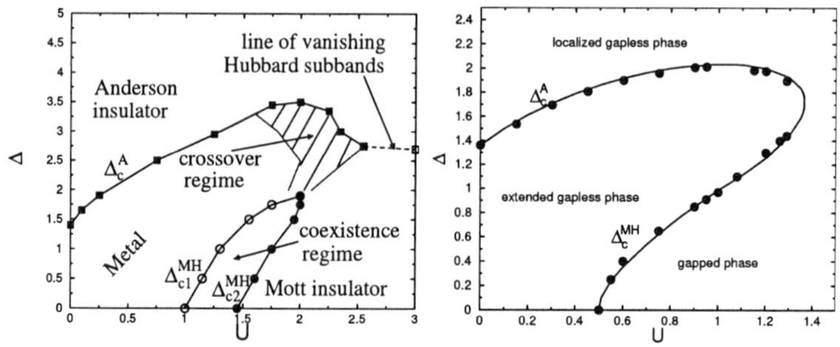

Fig. 9. Non-magnetic ground state phase diagram of the Anderson-Hubbard (left) and Anderson-Falicov-Kimball (right) models at half-filling as calculated by DMFT with the typical local density of states; after Refs. [50,52]

the DMFT with geometrical averaging allows to compute phase diagrams for the Anderson-Hubbard model (10) and the Anderson-Falicov-Kimball model (11) with the continuous probability distribution function (9) at half-filling.[50,52] In this way we found that, although in both models the metallic phase is enhanced for small and intermediate values of the interaction and disorder, metallicity is finally destroyed. Surprisingly, the Mott and Anderson insulators are found to be continuously connected. Phase diagrams for the non-magnetic ground state are shown in Fig. 9. The method based on geometric averaging described here was further investigated in details in Refs. [89–92].

11. Conclusions

The physics of correlated electron systems is known to be extremely rich. Therefore their investigation continues to unravel novel and often surprising phenomena, e.g. formation of kinks in the electronic dispersion relations.[93] The presence of disorder further enhances this complexity. Here we discussed several remarkable features induced by correlations with and without disorder, which came as a surprise when they were first discovered, but which after all have physically intuitive explanations. Behind these discoveries is the dynamical mean-field theory, which was described here for lattice quantum systems with interaction, disorder and external potentials.

Acknowledgments

It is a pleasure to thank R. Bulla, M. Eckstein, W. Hofstetter, A. Kauch, M. Kollar, and in particular D. Vollhardt for many discussions and collaborations involving different DMFT projects. This work was supported by the Sonderforschungsbereich 484 of the Deutsche Forschungsgemeinschaft.

Appendix

A very convenient method to calculate the fermionic partition function is the path-integral technique[74] invented originally by Feynman. The partition function is represented as a functional integral over anticommuting, time dependent functions

$$Z = Tre^{-\beta(H-\mu N)} = \int D[c_{i\sigma}^*, c_{i\sigma}]e^{-S[c_{i\sigma}^*,c_{i\sigma}]}\,, \tag{45}$$

where the action S is defined explicitly for a given Hamiltonian

$$S[c_{i\sigma}^*, c_{i\sigma}] = \int_0^\beta d\tau \sum_{i\sigma} c_{i\sigma}^*(\tau)(\partial_\tau - \mu)c_{i\sigma}(\tau) + \int_0^\beta d\tau H[c_{i\sigma}^*(\tau), c_{i\sigma}(\tau)]\,, \tag{46}$$

with the antisymmetric boundary condition $c_{i\sigma}(\tau+\beta) = -c_{i\sigma}(\tau)$ to keep the Fermi-Dirac statistics.

In the DMFT we are mainly interested in the one-particle Green function (propagator), which is simply expressed by the functional integral

$$\begin{aligned} G_{ij\sigma}(\tau-\tau') &= -\langle T_\tau c_{i\sigma}(\tau)c_{i\sigma}^\dagger(\tau')\rangle \\ &= -\frac{1}{Z}\int D[c_{i\sigma}^*, c_{i\sigma}]c_{i\sigma}(\tau)c_{i\sigma}^*(\tau')e^{-S[c_{i\sigma}^*,c_{i\sigma}]}\,. \end{aligned} \tag{47}$$

The Green function obeys the equation of motion

$$\left[(\partial_\tau - \mu)\hat{1} + \hat{H}\right]\hat{G}(\tau-\tau') = -\delta(\tau-\tau')\hat{1}\,, \tag{48}$$

where we used the matrix notation for $\hat{G}$ and the Hamiltonian operator $\hat{H}$, whereas $\hat{1}$ is a unit matrix.

With the help of Eq. (48), the exact partition function (45) can be expressed exactly by the Green function

$$Z = \int D[\bar{c}^*, \bar{c}]e^{\int_0^\beta d\tau \int_0^\beta d\tau' \bar{c}^*(\tau)\hat{G}^{-1}(\tau-\tau')\bar{c}(\tau')}\,, \tag{49}$$

where we used compact vector and matrix notations. The Gaussian functional integral is evaluated and we obtain

$$Z = \mathrm{Det}[-\hat{G}^{-1}] = e^{\mathrm{Tr}\ln[-\hat{G}^{-1}]}\,, \tag{50}$$

where the determinant and the trace are taken over relevant quantum numbers and over the imaginary time τ or Matsubara frequencies ω_n, depending which representation we use. This is equivalent because of the Fourier transform relation

$$\hat{G}(\tau) = \frac{1}{\beta} \sum_{\omega_n} e^{i\omega_n \tau} \hat{G}(\omega_n) , \tag{51}$$

where $\omega_n = (2n + 1)\pi/\beta$ are odd Matsubara frequencies. The formal expression (50) is our starting point in derivation the DMFT equations.

References

1. J.H. de Boer and E.J.W. Verwey, Proc. Phys. Soc. **49**, No. 4S, 59 (1937).
2. N.F. Mott, Proc. Phys. Soc. **49**, No. 4S, 57 (1937).
3. D. Pines, *The Many-Body Problem*, (W. A. Benjamin, Reading, 1962).
4. N.F. Mott, Proc. Phys. Soc. **A62**, 416 (1949); *Metal–Insulator Transitions*, 2nd edn. (Taylor and Francis, London 1990).
5. D. Vollhardt, *Proceedings of the International School of Physics Enrico Fermi, Course CXXI*, Eds. R. A. Broglia and J. R. Schrieffer, p. 31, (North-Holland, Amsterdam, 1994).
6. P. Fulde, *Electron Correlations in Molecules and Solids*, (Springer, Heidelberg, 1995).
7. F. Gebhard, *The Mott Metal-Insulator Transition*, (Springer, Heidelberg, 1997).
8. M. Imada, A. Fujimori and Y. Tokura, Rev. Mod. Phys. **70**, 1039 (1998).
9. P. Fazekas, *Lecture Notes on Electron Correlation and Magnetism*, (World Scientific, Singapore, 1999).
10. J. Spalek, Eur. J. Phys. **21**, 511 (2000).
11. J.K. Freericks, *Transport in multilayered nanostructures The dynamical mean-field theory approach*, (Imperial College Press, 2006).
12. W. Metzner and D. Vollhardt, Phys. Rev. Lett. **62**, 324 (1989).
13. E. Müller-Hartmann, Z. Phys. **B76**, 211 (1989).
14. V. Janiš, Z. Phys. **B83**, 227 (1991);
    V. Janiš and D. Vollhardt, Int. J. Mod. Phys. **6**, 731 (1992).
15. A. Georges and G. Kotliar, Phys. Rev. **B45**, 6479 (1992).
16. M. Jarrell, Phys. Rev. Lett. **69**, 168 (1992).
17. D. Vollhardt, in *Correlated Electron Systems*, Ed. V.J. Emery, World Scientific, Singapore, 1993, p. 57.
18. Th. Pruschke, M. Jarrell and J. K. Freericks, Adv. in Phys. **44**, 187 (1995).
19. A. Georges, G. Kotliar, W. Krauth and M. J. Rozenberg, Rev. Mod. Phys. **68**, 13 (1996).
20. G. Kotliar and D. Vollhardt, Physics Today **57**, No. 3 (March), 53 (2004).
21. G. Kotliar, S.Y. Savrasov, K. Haule, V.S. Oudovenko, O. Parcollet and C.A. Marianetti, Rev. Mod. Phys. **78**, 865 (2006).
22. Wikipedia, the free encyclopedia:
    http://en.wikipedia.org/wiki/Correlation .
23. E.g.: A. Seyfried, B. Steffen and Th. Lippert, Physica **A368**, 232 (2006).
24. K. Huang, *Statistical mechanics*, (John Wiley and Sons, Inc., 1963, 1987).

25. Wikipedia, the free encyclopedia: http://en.wikipedia.org/wiki/Mean_field_theory .
26. W. Kohn, Rev. Mod. Phys. **71**, 1253 (1999); Rev. Mod. Phys. **71**, S59 (1999); and references therein.
27. W. Kohn and L.J. Sham, Phys. Rev. **140**, A1133 (1965).
28. G.R. Stewart, Rev. Mod. Phys. **56**, 755 (1984).
29. A.C. Hewson, *The Kondo Problem to Heavy Fermions*, (Cambridge University Press, 1997).
30. D.B. McWhan, A. Menth, J.P. Remeika, W.F. Brinkman and T.M. Rice, Phys. Rev. **B7**, 1920 (1973).
31. D. Vollhardt, UniPress, Zeitschrift der Universität Augsburg, Band 3 und 4, p. 29 (1996); see also: http://www.physik.uni-augsburg.de/theo3/Research/research_fest.vollha.de.shtml.
32. Indeed, the group velocity v is estimated as $v = a/\tau$, where a is a lattice constant, and using the Heisenberg principle $W\tau \sim \hbar$ we find that $a/\tau \sim aW/\hbar$.
33. I. Bloch, Nature Phys. **1**, 23 (2005);
M. Köhl and T. Esslinger, Europhys. News **37**, 18 (2005).
34. W. Hofstetter, Advances in Solid State Physics **45**, 109 (2005).
35. M. Lewenstein, A. Sanpera, V. Ahufinger, B. Damski, A. Sen De and U. Sen, Advances in Physics **56**, 243 (2007).
36. I. Bloch, J. Dalibard and W. Zwerger, arXiv:0704.3011.
37. W.D. Phillips, Rev. Mod. Phys. **70**, 721 (1998).
38. H. van Houten and C.W.J. Beenakker, Physics Today **49** July, 22 (1996);
C.W.J.Beenakker and H. van Houten, Solid State Physics **44** (1991).
39. A. Ohtomo, D.A. Muller, J.L. Grazul and H.Y. Hwang, Nature **419**, 378 (2002);
M. Takizawa, H. Wadati, K. Tanaka, M. Hashimoto, T. Yoshida, A. Fujimori, A. Chikamtsu, H. Kumigashira, M. Oshima, K. Shibuya, T. Mihara, T. Ohnishi, M. Lippmaa, M. Kawasaki, H. Koinuma, S. Okamoto and A.J. Millis, Phys. Rev. Lett. **97**, 057601 (2006).
40. H.S. Jarrett *et al.*, Phys. Rev. Lett. **21**, 617 (1968);
G.L. Zhao, J. Callaway and M. Hayashibara, Phys. Rev. **B48**, 15781 (1993);
S.K. Kwon, S.J. Youn and B.I. Min, Phys. Rev. **B62**, 357 (2000);
T. Shishidou *et al.*, Phys. Rev. **B64**, 180401 (2001).
41. P.A. Lee and T.V. Ramakrishnan, Rev. Mod. Phys. **57**, 287 (1985);
D. Belitz and T.R. Kirkpatrick, Rev. Mod. Phys. **66**, 261 (1994).
42. P.W. Anderson, Phys. Rev. **109**, 1492 (1958).
43. S.V. Kravchenko *et al.*, Phys. Rev. **B50**, 8039 (1994);
D. Popović, A.B. Fowler and S. Washburn, Phys. Rev. Lett. **79**, 1543 (1997);
S.V. Kravchenko and M.P. Sarachik, Rep. Prog. Phys. **67**, 1 (2004);
H. von Löhneysen, Adv. in Solid State Phys. **40**, 143 (2000).
44. A.M. Finkelshtein, Sov. Phys. JEPT **75**, 97 (1983);
C. Castellani *et al.*, Phys. Rev. **B30**, 527 (1984);
M.A. Tusch and D.E. Logan, Phys. Rev. **B48**, 14843 (1993); ibid. **51**,

11940 (1995);
D.L. Shepelyansky, Phys. Rev. Lett. **73**, 2607 (1994);
P.J.H. Denteneer, R.T. Scalettar and N. Trivedi, Phys. Rev. Lett. **87**, 146401 (2001).
45. J. Hubbard, Proc. Roy. Soc. London **A281** 238 (1963); ibid. **277**, 237;
M.C. Gutzwiller, Phys. Rev. Lett. **10**, 159 (1963);
J. Kanamori, Prog. Theor. Phys. **30**, 275 (1963).
46. U. Yu, K. Byczuk and D. Vollhardt, in preparation.
47. K. Byczuk, M. Ulmke and D. Vollhardt, Phys. Rev. Lett. **90**, 196403 (2003).
48. K. Byczuk, W. Hofstetter and D. Vollhardt, Phys. Rev. **B69**, 045112 (2004).
49. K. Byczuk and M. Ulmke, Eur. Phys. J. **B45**, 449-454 (2005).
50. K. Byczuk, W. Hofstetter and D. Vollhardt, Phys. Rev. Lett. **94**, 056404 (2005).
51. J.K. Freericks and V. Zlatić, Rev. Mod. Phys. **75**, 1333 (2003).
52. K. Byczuk, Phys. Rev. **B71**, 205105 (2005).
53. K. Ziegler, arXiv:cond-mat/0611010.
54. Z. Gajek, J. Jedrzejewski and R. Lemanski, arXiv:cond-mat/9507122.
55. M.M. Maska and K. Czajka, Phys. Rev. **B74**, 035109 (2006).
56. V. Zlatic, J.K. Freericks, R. Lemanski and G. Czycholl, Phil. Mag. **B81**, 1443 (2001).
57. R. Lemanski, arXiv:cond-mat/0507474 and arXiv:cond-mat/0702415.
58. Wikipedia, the free encyclopedia:
http://en.wikipedia.org/wiki/Central_limit_theorem .
59. P.W. Anderson, Rev. Mod. Phys. **50**, 191 (1978).
60. P.S. Bullen, *Handbook of means and their inequalities*, (Kluwer Academic Publishers, 2003).
61. Wikipedia, the free encyclopedia:
http://en.wikipedia.org/wiki/Generalized_mean .
62. *Log-normal distribution–theory and applications*, Ed. E.L. Crow and K. Shimizu (Marcel Dekker, inc. 1988).
63. E.W. Montroll and M.F. Schlesinger, J. Stat. Phys. **32**, 209 (1983);
M. Romeo, V. Da Costa and F. Bardou, Eur. Phys. J. **B32**, 513 (2003).
64. K.A. Chao, J. Spalek and A. M. Oles, J. Phys. **C10**, L271 (1977).
65. K. Byczuk and D. Vollhardt, arXiv:0706.0839.
66. M.-T. Tran, Phys. Rev. **B73**, 205110 (2006); *ibid.* **76**, 245122 (2007).
67. R. Helmes, T.A. Costi, and A. Rosh, arXiv:0709.1669.
68. C. Töke, M. Snoek, I. Titvinidze, K. Byczuk, W. Hofstetter, in preparation.
69. M. Potthoff and W. Nolting, Phys. Rev. **B59**, 2549 (1999); Eur. Phys. J. **B8**, 555 (1999).
70. R. Bulla and M. Potthoff, Eur. Phys. J. **B13**, 257 (2000).
71. S. Okamoto and A.J. Millis, Phys. Rev. **B70**, 241104 (2004).
72. M. Eckstein, M. Kollar, K. Byczuk and D. Vollhardt, Phys. Rev. **B71**, 235119 (2005).
73. M. Kollar, M. Eckstein, K. Byczuk, N. Blümer, P. van Dongen, M.H. Radke de Cuba, W. Metzner, D. Tanaskovic, V. Dobrosavljevic, G. Kotliar and D. Vollhardt, Ann. Phys. (Leipzig) **14**, 642 (2005).

74. J.W. Negele and H. Orland, *Quantum many-particle systems*, (Perseus Book Publishing, L.L.C., 1998, 1988).
75. V. Turkowski and J.K. Freericks, arXiv:cond-mat/0612466.
76. R. Vlaming and D. Vollhardt, Phys. Rev. **B45**, 4637 (1992).
77. R.J. Elliott, J.A. Krumhansl and P.L. Leath, Rev. Mod. Phys. **46**, 465 (1974).
78. R. Bulla, Th. Costi and Th. Pruschke, arXiv:cond-mat/0701105.
79. W. Hofstetter, Advances in Solid State Physics **41**, 27 (2001).
80. J. Spalek, A. Datta and J.M. Honig, Phys. Rev. Lett. **59**, 728 (1987).
81. M. Ulmke, Eur. Phys. J. **B1**, 301 (1998).
82. J. Wahle, N. Blumer, J. Schlipf, K. Held and D. Vollhardt, Phys. Rev. **B58**, 12749 (1998).
83. D. Lloyd, J. Phys. **C2**, 1717 (1969);
D. Thouless, Phys. Reports **13**, 93 (1974);
F. Wegner, Z. Phys. **B44**, 9 (1981).
84. V. Dobrosavljević and G. Kotliar, Phys. Rev. Lett. **78**, 3943 (1997).
85. V. Dobrosavljević, A.A. Pastor and B.K. Nikolić, Europhys. Lett. **62**, 76 (2003).
86. G. Schubert, A. Weiße and H. Fehske, *High Performance Computing in Science and Engineering Garching 2004*, p. 237, Eds. A. Bode and F. Durst, (Springer Verlag, 2005).
87. V. Janiš and D. Vollhardt, Phys. Rev. **B46**, 15712 (1992);
M. Ulmke, V. Janiš and D. Vollhardt, Phys. Rev. **B51**, 10411 (1995).
88. F.X. Bronold, A. Alvermann and H. Fehske, Phil. Mag. **84**, 637 (2004).
89. A. Alvermann and H. Fehske, Eur. Phys. J. **B48**, 295 (2005).
90. M. Potthoff and M. Balzer, Phys. Rev. **B75**, 125112 (2007).
91. Yun Song, W.A. Atkinson and R. Wortis, arXiv:cond-mat/0609590.
92. A.M.C. Souza, D. de O. Maionchi and H.J. Herrmann, arXiv:cond-mat/0702355.
93. K. Byczuk, M. Kollar, K. Held, Y.-F. Yang, I.A. Nekrasov, Th. Pruschke and D. Vollhardt, Nature Physics **3**, 168 (2007).

JORDAN-WIGNER FERMIONIZATION AND THE THEORY OF LOW-DIMENSIONAL QUANTUM SPIN MODELS. DYNAMIC PROPERTIES

O. DERZHKO

Institute for Condensed Matter Physics
of the National Academy of Sciences of Ukraine
1 Svientsitskii Street, L'viv-11, 79011, Ukraine
E-mail: derzhko@icmp.lviv.ua

The Jordan-Wigner transformation is known as a powerful tool in condensed matter theory, especially in the theory of low-dimensional quantum spin systems. The aim of this chapter is to review the application of the Jordan-Wigner fermionization technique for calculating dynamic quantities of low-dimensional quantum spin models. After a brief introduction of the Jordan-Wigner transformation for one-dimensional spin one-half systems and some of its extensions for higher dimensions and higher spin values we focus on the dynamic properties of several low-dimensional quantum spin models. We start from the famous $s = 1/2$ XX chain. As a first step we recall well-known results for dynamics of the z-spin-component fluctuation operator and then turn to the dynamics of the dimer and trimer fluctuation operators. The dynamics of the trimer fluctuations involves both the two-fermion (one particle and one hole) and the four-fermion (two particles and two holes) excitations. We discuss some properties of the two-fermion and four-fermion excitation continua. The four-fermion dynamic quantities are of intermediate complexity between simple two-fermion (like the zz dynamic structure factor) and enormously complex multi-fermion (like the xx or xy dynamic structure factors) dynamic quantities. Further we discuss the effects of dimerization, anisotropy of XY interaction, and additional Dzyaloshinskii-Moriya interaction on various dynamic quantities. Finally we consider the dynamic transverse spin structure factor $S_{zz}(\mathbf{k},\omega)$ for the $s = 1/2$ XX model on a spatially anisotropic square lattice which allows one to trace a one-to-two-dimensional crossover in dynamic quantities.

1. Introduction (Spin models, dynamic probes etc.)

The subject of quantum magnetism dates back to 1920s. E. Ising[1] suggested a simplest model of a magnet as a collection of N spins which may acquire two values $\sigma = \pm 1$ and interact with nearest neighbors on a lattice as $\sum J\sigma_i\sigma_j$ and with an external magnetic field as $-h\sum\sigma_i$. To explain the properties of the model we have to calculate the partition function $Z = \mathrm{Tr}\exp(-\beta H)$ which yields the Helmholtz free energy per site $f = \lim_{N\to\infty}(-T\ln Z/N)$ (in what follows we set $k_{\mathrm{B}} = 1$ to simplify the notations). In one dimension the problem was solved by E. Ising. Later L. Onsager solved the square-lattice Ising model[2] and we know the solution in two dimensions.[3] There is no solution of the Ising model in three dimensions until now.

Another version of interspin interaction was suggested by P. A. M. Dirac and W. Heisenberg. The Heisenberg exchange interaction reads $\sum J\vec{\sigma}_i\cdot\vec{\sigma}_j = \sum J\left(\sigma_i^x\sigma_j^x + \sigma_i^y\sigma_j^y + \sigma_i^z\sigma_j^z\right)$ where the Pauli matrices $\vec{\sigma} = (\sigma^x, \sigma^y, \sigma^z)$ are defined as

$$\sigma^x = \begin{pmatrix} 0 & 1 \\ 1 & 0 \end{pmatrix}, \quad \sigma^y = \begin{pmatrix} 0 & -\mathrm{i} \\ \mathrm{i} & 0 \end{pmatrix}, \quad \sigma^z = \begin{pmatrix} 1 & 0 \\ 0 & -1 \end{pmatrix}. \tag{1}$$

Denoting the halves of the Pauli matrices as $s^\alpha = \sigma^\alpha/2$ (in what follows we set $\hbar = 1$ to simplify the notations) we consider the following Hamiltonian

$$H = \sum_{\langle i,j\rangle}\left(J^x s_i^x s_j^x + J^y s_i^y s_j^y + J^y s_i^z s_j^z\right) - h\sum_i s_i^z. \tag{2}$$

We note that the Hamiltonian of the anisotropic XYZ Heisenberg model (2) covers in the limiting cases some specific models like the Ising model ($J^x = J^y = 0$), the isotropic XY (or XX or $XX0$) model ($J^x = J^y$, $J^z = 0$), the anisotropic XY model ($J^x \neq J^y$, $J^z = 0$), the isotropic (XXX) Heisenberg model ($J^x = J^y = J^z$), and the Heisenberg-Ising (XXZ) model ($J^x = J^y = J$, $J^z = \Delta J$).

Again we would like to calculate the partition function Z of the spin-1/2 model (2). Unfortunately, this task is very complicated even in one dimension. Due to H. Bethe we know how to find the eigenstates of the spin-1/2 linear chain Heisenberg model.[4] The famous Bethe ansatz for the wave function has the form

$$|\psi\rangle = \sum_{1\leq n_1<\ldots<n_r\leq N} a(n_1,\ldots,n_r)s_{n_1}^-\ldots s_{n_r}^-|\uparrow\uparrow\ldots\uparrow\rangle,$$

$$a(n_1,\ldots,n_r) = \sum_{\mathcal{P}\in S_r}\exp\left(\mathrm{i}\sum_{j=1}^{r}k_{\mathcal{P}_j}n_j + \frac{\mathrm{i}}{2}\sum_{i<j}\theta_{\mathcal{P}_i\mathcal{P}_j}\right),$$

$$2\cot\frac{\theta_{ij}}{2} = \cot\frac{k_i}{2} - \cot\frac{k_j}{2}, \quad Nk_i = 2\pi\lambda_i + \sum_{j\neq i}\theta_{ij}, \tag{3}$$

where the sum in the definition of coefficients $a(n_1,\ldots,n_r)$, $\mathcal{P}\in S_r$, runs over all $r!$ permutations of the labels $\{1,2,\ldots,r\}$, $\mathcal{P}_j$ is the image of j under the permutation $\mathcal{P}$. For further details see, e.g., Ref. [5].

Let us briefly recall the quantities of interest in the statistical mechanical studies of the spin models. As we have mentioned already the thermodynamic quantities like the entropy, the specific heat, the magnetization etc. follow from the partition function $Z=\sum_\lambda \exp\left(-E_\lambda/T\right)=\operatorname{Tr}\exp\left(-\beta H\right)$, the sum runs over all states λ of the system with energy E_λ. Usually we are also interested in the equal-time spin correlation functions, e.g. $\langle \vec{s}_i\cdot\vec{s}_j\rangle$, $\langle(\ldots)\rangle=\operatorname{Tr}\left(\exp\left(-\beta H\right)(\ldots)\right)/Z$; their nonzero limiting values, (e.g., $\lim_{|i-j|\to\infty}\langle\vec{s}_i\cdot\vec{s}_j\rangle$) may indicate the existence of long-range order in the system.

Within a linear response regime we add to the Hamiltonian H_0 a small perturbation $H_0\to H_0-b(t)B$, where the external field $b(t)$ couples to the dynamical variable B, and observe a response of a dynamical variable A, $\langle A(t)\rangle-\langle A\rangle_0=\int_{-\infty}^{\infty}\mathrm{d}t'\chi_{AB}(t-t')b(t')$ with $\chi_{AB}(t-t')=\mathrm{i}\theta(t-t')\langle[A(t),B(t')]\rangle_0$ (here $\theta(x)$ is the Heaviside step function). The Fourier-transform of the dynamic susceptibility $\chi_{AB}(t-t')$, $\Re\chi_{AB}(\omega)+\mathrm{i}\Im\chi_{AB}(\omega)$, is the quantity which can be measured experimentally. We note that the real and imaginary parts of the dynamic susceptibility are connected via the dispersion (or Kramers-Kronig) relation. On the other hand, the imaginary part of the dynamic susceptibility can be expressed with the help of the fluctuation-dissipation theorem through another dynamic quantity, the dynamic structure factor. Thus, $S_{AA}(\omega)=\int_{-\infty}^{\infty}\mathrm{d}t\exp\left(\mathrm{i}\omega t\right)\langle A(t)A\rangle=2\Im\chi_{AA}(\omega)/\left(1-\exp(-\beta\omega)\right)$.

Usually, the operator A is constructed from the local operator of the considered system A_n as follows: $A_k=(1/\sqrt{N})\sum_{n=1}^{N}\exp(\mathrm{i}kn)A_n$. We can also rewrite the dynamic structure factor in the following forms

$$\begin{aligned}S_{AA}(k,\omega)&=\sum_{l=1}^{N}\exp\left(-\mathrm{i}kl\right)\int_{-\infty}^{\infty}\mathrm{d}t\exp\left(\mathrm{i}\omega t\right)\langle A_n(t)A_{n+l}(0)\rangle\\&=2\pi\sum_{\lambda,\lambda'}\frac{\exp\left(-\beta E_{\lambda'}\right)}{Z}\left|\langle\lambda'|A_k|\lambda\rangle\right|^2\delta\left(\omega-E_\lambda+E_{\lambda'}\right).\end{aligned}\qquad(4)$$

Sometimes it is convenient to make the following change in the first line in Eq. (4): $A_n(t)\to A_n(t)-\langle A\rangle$, $A_{n+l}(0)\to A_{n+l}(0)-\langle A\rangle$. We also note that in the zero-temperature limit $T=0$ (or $\beta\to\infty$) the second line in Eq. (4) becomes simpler, $S_{AA}(k,\omega)=2\pi\sum_\lambda\left|\langle\mathrm{GS}|A_k|\lambda\rangle\right|^2\delta\left(\omega-\omega_\lambda\right)$,

$\omega_\lambda = E_\lambda - E_{GS}$.

In what follows we discuss mainly the dynamic properties of spin-1/2 XY chains; just for this class of spin models application of the Jordan-Wigner fermionization approach is most fruitful. We notice here that recently it has been found that Cs_2CoCl_4 is a good realization of the spin-1/2 XX chain[6] and calculations of the dynamic quantities for the corresponding spin models might be important for the interpretation of the data from dynamic experiments.[7] As an example of earlier studies we may mention dynamic experiments on the spin-1/2 XX chain compound $PrCl_3$.[8]

The rest of this chapter is organized as follows. At first we briefly introduce the Jordan-Wigner transformation (Sec. 2) and concisely discuss some of its generalizations (Sec. 3). Then we consider in detail the dynamic structure factors for the spin-1/2 isotropic XY chain in a transverse field distinguishing the quantities which probe two-fermion, four-fermion and many-fermion excitations (Sec. 4). Next we examine the dynamics for two slightly more complicated chains: the dimerized isotropic XY chain (Sec. 5) and the XY chains with the Dzyaloshinskii-Moriya interaction (Sec. 6). The results obtained for one-dimensional XY spin models do not involve any approximation. This is not true in the two-dimensional case for which the Jordan-Wigner approach provides only approximate expressions for dynamic quantities. We illustrate the Jordan-Wigner fermionization approach in two dimensions examining some dynamic quantities for the square-lattice spin-1/2 isotropic XY model (Sec. 7). We end up with a brief summary (Sec. 8).

2. The Jordan-Wigner transformation

To be specific, we consider the one-dimensional spin $s = 1/2$ XXZ Heisenberg chain with the Hamiltonian

$$H = \sum_{n=1}^{N} J\left(s_n^x s_{n+1}^x + s_n^y s_{n+1}^y + \Delta s_n^z s_{n+1}^z\right) - h\sum_{n=1}^{N} s_n^z; \tag{5}$$

we imply either periodic or open boundary conditions in Eq. (5). Here the spin operators s_i^α satisfy the commutation relations $\left[s_i^\alpha, s_j^\beta\right] = \mathrm{i}\delta_{ij}\epsilon_{\alpha\beta\gamma}s_i^\gamma$, $\epsilon_{\alpha\beta\gamma}$ is the totally antisymmetric Levi-Civita tensor with $\epsilon_{xyz} = 1$. In particular, $\left[s_i^x, s_j^y\right] = \mathrm{i}\delta_{ij}s_i^z$ etc. Obviously, s^α can be viewed as the halves of the Pauli matrices (1). After introducing the spin raising and lowering operators (or the ladder operators) $s_n^\pm = s_n^x \pm \mathrm{i}s_n^y$ $(s_n^x = (s_n^+ + s_n^-)/2,$

$s_n^y = (s_n^+ - s_n^-)/2\mathrm{i})$ the Hamiltonian (5) becomes

$$H = \sum_n J\left(\frac{1}{2}\left(s_n^+ s_{n+1}^- + s_n^- s_{n+1}^+\right)\right.$$
$$\left. + \Delta\left(s_n^+ s_n^- - \frac{1}{2}\right)\left(s_{n+1}^+ s_{n+1}^- - \frac{1}{2}\right)\right) - h\sum_n\left(s_n^+ s_n^- - \frac{1}{2}\right). \quad (6)$$

We note that the spin raising and lowering operators satisfy commutation relations of Fermi type at the same site, i.e.

$$\{s_n^-, s_n^+\} = 1, \quad \{s_n^-, s_n^-\} = \{s_n^+, s_n^+\} = 0 \quad (7)$$

and of Bose type at different sites

$$[s_n^-, s_m^+] = [s_n^-, s_m^-] = [s_n^+, s_m^+] = 0, \quad n \neq m. \quad (8)$$

We may use the Jordan-Wigner transformation[9] to introduce Fermi operators according to the following formulas

$$c_1 = s_1^-, \quad c_n = (-2s_1^z)(-2s_2^z)\dots(-2s_{n-1}^z)\, s_n^-, \ n = 2, \dots, N, \quad (9)$$

$$c_1^\dagger = s_1^+, \quad c_n^\dagger = (-2s_1^z)(-2s_2^z)\dots(-2s_{n-1}^z)\, s_n^+, \ n = 2, \dots, N. \quad (10)$$

(Sometimes one can find in Eqs. (9), (10) instead of $-2s^z$ the identical expressions $1 - 2s^+s^- = \exp(\pm \mathrm{i}\pi s^+ s^-)$.) Really, the operators introduced always satisfy the Fermi commutation relations

$$\{c_n, c_m^\dagger\} = \delta_{nm}, \quad \{c_n, c_m\} = \{c_n^\dagger, c_m^\dagger\} = 0. \quad (11)$$

(To check this one has to note that $(-2s^z)^2 = 1$ and that $s^z s^\pm = -s^\pm s^z$.) The inverse transformation to the one given by Eqs. (9), (10) reads

$$s_1^- = c_1, \quad s_n^- = \exp\left(\pm \mathrm{i}\pi \sum_{j=1}^{n-1} c_j^\dagger c_j\right) c_n, \ n = 2, \dots, N, \quad (12)$$

$$s_1^+ = c_1^\dagger, \quad s_n^+ = \exp\left(\pm \mathrm{i}\pi \sum_{j=1}^{n-1} c_j^\dagger c_j\right) c_n^\dagger, \ n = 2, \dots, N. \quad (13)$$

Moreover, the Hamiltonian (6) in terms of the Fermi operators (9), (10) has the following form

$$H = \sum_n J\left(\frac{1}{2}\left(c_n^\dagger c_{n+1} - c_n c_{n+1}^\dagger\right)\right.$$
$$\left. + \Delta\left(c_n^\dagger c_n - \frac{1}{2}\right)\left(c_{n+1}^\dagger c_{n+1} - \frac{1}{2}\right)\right) - h\sum_n\left(c_n^\dagger c_n - \frac{1}{2}\right) \quad (14)$$

(we use $c_j^\dagger c_{j+1}^\dagger = s_j^+ \left(-2s_j^z\right) s_{j+1}^+ = s_j^+ s_{j+1}^+$ etc.). In the case of periodic boundary conditions implied for the spin Hamiltonian (6) the transformed Hamiltonian (14) obeys either periodic or antiperiodic boundary conditions depending on the parity of the number of fermions. However, in what follows the calculated quantities in the thermodynamic limit $N \to \infty$ will be insensitive to the boundary conditions implied (for further details see Ref. [10]).

From Eq. (14) it becomes clear that the spin-1/2 isotropic XY chain in a transverse (z) magnetic field with the Hamiltonian

$$H = \sum_n J \left(s_n^x s_{n+1}^x + s_n^y s_{n+1}^y\right) + \Omega \sum_n s_n^z \tag{15}$$

in the Jordan-Wigner picture is represented by the Hamiltonian

$$H = \sum_n \frac{J}{2} \left(c_n^\dagger c_{n+1} - c_n c_{n+1}^\dagger\right) + \Omega \sum_n \left(c_n^\dagger c_n - \frac{1}{2}\right) \tag{16}$$

and therefore is an exactly solvable model.[11,12] Moreover, the XY exchange interaction may be anisotropic; then the intersite interaction has the form

$$\begin{aligned} &J^x s_n^x s_{n+1}^x + J^y s_n^y s_{n+1}^y \\ &\to \frac{J}{2} \left(c_n^\dagger c_{n+1} - c_n c_{n+1}^\dagger\right) + \frac{\gamma}{2} \left(c_n^\dagger c_{n+1}^\dagger - c_n c_{n+1}\right) \end{aligned} \tag{17}$$

with $J = (J^x + J^y)/2$, $\gamma = (J^x - J^y)/2$. We can also consider an additional intersite interaction, the so-called Dzyaloshinskii-Moriya interaction, which does not spoil a simple fermionic Hamiltonian[13]

$$D \left(s_n^x s_{n+1}^y - s_n^y s_{n+1}^x\right) \to \frac{\mathrm{i}D}{2} \left(c_n^\dagger c_{n+1} + c_n c_{n+1}^\dagger\right). \tag{18}$$

Moreover, within the Jordan-Wigner fermionization approach we can examine rigorously some types of multi-spin interactions,[14] for example,

$$\begin{aligned} s_n^x s_{n+1}^z s_{n+2}^x + s_n^y s_{n+1}^z s_{n+2}^y &\to -\frac{1}{4} \left(c_n^\dagger c_{n+2} - c_n c_{n+2}^\dagger\right); \\ s_n^x s_{n+1}^z s_{n+2}^y - s_n^y s_{n+1}^z s_{n+2}^x &\to -\frac{\mathrm{i}}{4} \left(c_n^\dagger c_{n+2} + c_n c_{n+2}^\dagger\right). \end{aligned} \tag{19}$$

Within the frames of the Jordan-Wigner approach we can also generalize simple spin-1/2 XY chains assuming regularly alternating Hamiltonian parameters[15] or some types of random Hamiltonian parameters[16] and still face exactly solvable models.

On the other hand, as can be easily seen from Eq. (14) the Ising interaction between z spin components leads to interacting spinless fermions and as a result the advantages of fermionization are less evident. (Obviously, we can split the interaction term in the spirit of the Hartree-Fock approximation,[17] however, the resulting theory will be only an approximate one. On the other hand, in the low-energy limit we can bosonize the fermionic Hamiltonian obtaining an exact low-energy effective theory.[18–20]) We cannot examine rigorously within the Jordan-Wigner fermionization approach the case of the next-nearest-neighbor interaction since

$$s_n^x s_{n+2}^x + s_n^y s_{n+2}^y \to c_n^\dagger \left(1 - 2c_{n+1}^\dagger c_{n+1}\right) c_{n+2} \qquad (20)$$
$$- c_n \left(1 - 2c_{n+1}^\dagger c_{n+1}\right) c_{n+2}^\dagger \, .$$

It is worthwhile to note here that recently the Jordan-Wigner fermionization approach has been applied to the spin-1/2 isotropic XY model on a diamond chain,[21] however, the authors of that paper apparently missed some interaction terms in the fermionic Hamiltonian and their statement about rigorous results for such a model is wrong. Finally we note that an external magnetic field directed along x or y axes has an enormously complicated form in the Jordan-Wigner picture.

3. Generalization of the Jordan-Wigner transformation

The Jordan-Wigner fermionization is a powerful tool for the study of quantum spin chains. Since the late 1980s there were several attempts to extend this approach to two (and three) dimensions[22–25] as well as to spin values $s > 1/2$.[26–28] For a review on the two-dimensional Jordan-Wigner fermionization approach see also Ref. [29].

Bearing in mind the Jordan-Wigner transformation in one dimension as a guideline we consider in the two-dimensional case the following relation between spin $s = 1/2$ and Fermi operators

$$d_{\vec{i}} = \exp\left(-\mathrm{i}\alpha_{\vec{i}}\right) s_{\vec{i}}^-, \quad d_{\vec{i}}^\dagger = \exp\left(\mathrm{i}\alpha_{\vec{i}}\right) s_{\vec{i}}^+,$$
$$s_{\vec{i}}^- = \exp\left(\mathrm{i}\alpha_{\vec{i}}\right) d_{\vec{i}}, \quad s_{\vec{i}}^+ = \exp\left(-\mathrm{i}\alpha_{\vec{i}}\right) d_{\vec{i}}^\dagger,$$

$$\alpha_{\vec{i}} = \sum_{\vec{j}(\neq\vec{i})} B_{\vec{i}\vec{j}} d_{\vec{j}}^\dagger d_{\vec{j}}. \qquad (21)$$

Here d, $d^\dagger$ are the Fermi operators, the operators $s^\pm$ defined according to (21) commute at different sites if the c-number matrix $B_{\vec{i}\vec{j}}$ satisfies the relation

$$\exp\left(\mathrm{i}B_{\vec{i}\vec{j}}\right) = -\exp\left(\mathrm{i}B_{\vec{j}\vec{i}}\right). \qquad (22)$$

There are many choices of the matrix $B_{\vec{i}\vec{j}}$ which realize the two-dimensional Jordan-Wigner transformation. Following Y. R. Wang[23] we use the Cartesian coordinates $\vec{i} = (i_x, i_y)$ to construct a complex number $\tau_{\vec{i}} = i_x + \mathrm{i} i_y = |\tau_{\vec{i}}| \exp\left(\mathrm{i} \arg(\tau_{\vec{i}})\right)$ and then choose

$$B_{\vec{i}\vec{j}} = \arg\left(\tau_{\vec{j}} - \tau_{\vec{i}}\right) = \Im \ln\left(\tau_{\vec{j}} - \tau_{\vec{i}}\right) = \Im \ln\left(j_x - i_x + \mathrm{i}(j_y - i_y)\right). \quad (23)$$

Indeed, for such a choice Eq. (22) is satisfied, $\exp\left(\mathrm{i}B_{\vec{j}\vec{i}}\right) = \exp\left(\mathrm{i}\arg(\tau_{\vec{i}} - \tau_{\vec{j}})\right) = \exp\left(\mathrm{i}\left(\arg(\tau_{\vec{j}} - \tau_{\vec{i}}) \pm \pi\right)\right) = -\exp\left(\mathrm{i}B_{\vec{i}\vec{j}}\right)$. Another choice of the matrix $B_{\vec{i}\vec{j}}$ has the following form[24]

$$B_{\vec{i}\vec{j}} = \pi\left(\theta\left(i_x - j_x\right)\left(1 - \delta_{i_x,j_x}\right) + \delta_{i_x,j_x}\theta\left(i_y - j_y\right)\left(1 - \delta_{i_y,j_y}\right)\right); \quad (24)$$

here $\theta(x)$ is the Heaviside step function (see also Ref. [29]).

After performing the Jordan-Wigner transformation (21) for the two-dimensional spin-1/2 XXZ Heisenberg Hamiltonian one gets

$$H = \sum_{\langle\vec{i},\vec{j}\rangle}\left(\frac{J_{\vec{i}\vec{j}}}{2}\left(d_{\vec{i}}^{\dagger}\exp\left(\mathrm{i}\left(\alpha_{\vec{j}} - \alpha_{\vec{i}}\right)\right)d_{\vec{j}} + d_{\vec{i}}\exp\left(\mathrm{i}\left(\alpha_{\vec{i}} - \alpha_{\vec{j}}\right)\right)d_{\vec{j}}^{\dagger}\right)\right.$$
$$\left. + J_{\vec{i}\vec{j}}\Delta\left(d_{\vec{i}}^{\dagger}d_{\vec{i}} - \frac{1}{2}\right)\left(d_{\vec{j}}^{\dagger}d_{\vec{j}} - \frac{1}{2}\right)\right) \quad (25)$$

with

$$\alpha_{\vec{j}} - \alpha_{\vec{i}} = \int_{\vec{i}}^{\vec{j}} \mathrm{d}\vec{r} \cdot \vec{A}(\vec{r}),$$
$$\vec{A}(\vec{r}) = \vec{\nabla}\alpha_{\vec{r}} = -\sum_{\vec{r}'(\neq\vec{r})} \frac{\vec{n}_z \times (\vec{r}' - \vec{r})}{(\vec{r}' - \vec{r})^2} d_{\vec{r}'}^{\dagger} d_{\vec{r}'} \quad (26)$$

(we have used Eq. (23) for $\alpha_{\vec{r}}$ (21)). We need further approximations to proceed with statistical mechanics calculations for the Hamiltonian (25). Within the mean-field description one assumes $d_{\vec{r}}^{\dagger}d_{\vec{r}} \to \langle d_{\vec{r}}^{\dagger}d_{\vec{r}}\rangle = \langle s_{\vec{r}}^{z}\rangle + 1/2 \to 1/2$. We expect such an approximation to be valid in the case of zero magnetic field. For the mean-field description in the case of nonzero magnetic field and an analysis of the magnetization processes in the spin system see Ref. [30]. We also note that a more sophisticated (self-consistent site-dependent) mean-field treatment has been suggested as well.[25]

After adopting the mean-field approach we face the problem of particles in a magnetic field with the flux per elementary plaquette $\Phi_0 = \pi$. We may change the gauge preserving the flux per elementary plaquette to make the

Hamiltonian more convenient for further calculations. For example, for a square lattice we have

$$H = \sum_{\langle \vec{i},\vec{j}\rangle} \left(\frac{J_{\vec{i}\vec{j}}}{2} \left(d^{\dagger}_{\vec{i}} d_{\vec{j}} - d_{\vec{i}} d^{\dagger}_{\vec{j}} \right) + J_{\vec{i}\vec{j}} \Delta \left(d^{\dagger}_{\vec{i}} d_{\vec{i}} - \frac{1}{2} \right) \left(d^{\dagger}_{\vec{j}} d_{\vec{j}} - \frac{1}{2} \right) \right)$$
$$J_{i_x,i_y;i_x+1,i_y} = -J\,,$$
$$J_{i_x,i_y;i_x,i_y+1} = J_{i_x+1,i_y;i_x+2,i_y} = J_{i_x+1,i_y;i_x+1,i_y+1} = J\,. \tag{27}$$

In the one-dimensional case when either vertical or horizontal bonds vanish the Hamiltonian (27) transforms into Eq. (14) (with $h = 0$).

Recently A. Kitaev has suggested a new exactly solvable two-dimensional quantum spin model.[31] This is a spin-1/2 model on a honeycomb lattice with interactions between different components of neighboring spins along differently directed bonds. An alternative representation of the honeycomb lattice is a brick-wall lattice (see Fig. 1). The Hamiltonian of

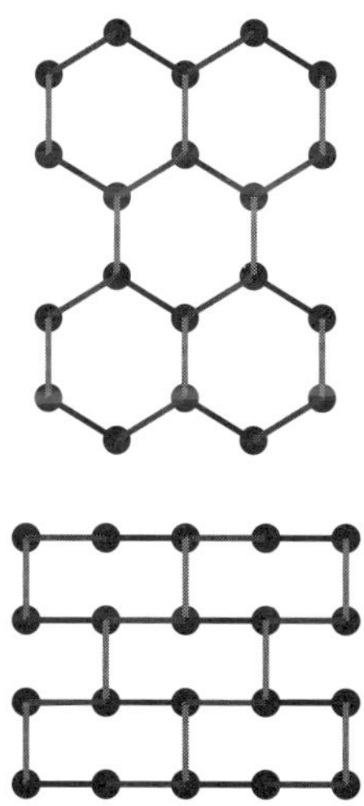

Fig. 1. A honeycomb lattice (up) with its equivalent brick-wall lattice (down). The bonds J_1 run from south-west to north-east, the bonds J_2 run from south-east to north-west, the bonds J_3 run from south to north.

the model reads

$$H = \sum_{j+l=\text{even}} \left(J_1 s^x_{j,l} s^x_{j+1,l} + J_2 s^y_{j-1,l} s^y_{j,l} + J_3 s^z_{j,l} s^z_{j,l+1} \right); \tag{28}$$

j and l denote the column and row indices of the lattice. We discuss in what follows a fermionic representation for the Kitaev model.[32] Let us perform the Jordan-Wigner transformation

$$s_{j,l}^{+} = a_{j,l}^{\dagger} \exp\left(\mathrm{i}\pi\left(\sum_{i}\sum_{k<l} a_{ik}^{\dagger}a_{ik} + \sum_{i<j} a_{il}^{\dagger}a_{il}\right)\right) \qquad (29)$$

(compare with Eqs. (21), (24)). As a result we find that

$$\begin{aligned}
J_1 s_{j,l}^{x} s_{j+1,l}^{x} &\to \frac{J_1}{4}\left(a_{j,l}^{\dagger}a_{j+1,l}^{\dagger} + a_{j,l}^{\dagger}a_{j+1,l} - a_{j,l}a_{j+1,l}^{\dagger} - a_{j,l}a_{j+1,l}\right),\\
J_2 s_{j-1,l}^{y} s_{j,l}^{y} &\to \frac{J_2}{4}\left(-a_{j-1,l}^{\dagger}a_{j,l}^{\dagger} + a_{j-1,l}^{\dagger}a_{j,l} - a_{j-1,l}a_{j,l}^{\dagger} + a_{j-1,l}a_{j,l}\right),\\
J_3 s_{j,l}^{z} s_{j,l+1}^{z} &\to J_3\left(a_{j,l}^{\dagger}a_{j,l} - \frac{1}{2}\right)\left(a_{j,l+1}^{\dagger}a_{j,l+1} - \frac{1}{2}\right). \qquad (30)
\end{aligned}$$

Next we introduce the following operators

$$\begin{aligned}
c_{j,l} &= a_{j,l}^{\dagger} + a_{j,l}\,, \quad d_{j,l} = \mathrm{i}\left(a_{j,l}^{\dagger} - a_{j,l}\right), \quad j+l = \mathrm{odd};\\
c_{j,l} &= \mathrm{i}\left(a_{j,l}^{\dagger} - a_{j,l}\right), \quad d_{j,l} = a_{j,l}^{\dagger} + a_{j,l}\,, \quad j+l = \mathrm{even}. \qquad (31)
\end{aligned}$$

In terms of these operators the Hamiltonian reads as follows

$$H = \sum_{j+l=\mathrm{even}}\left(-\mathrm{i}\frac{J_1}{4}c_{j,l}c_{j+1,l} + \mathrm{i}\frac{J_2}{4}c_{j-1,l}c_{j,l} + \mathrm{i}\frac{J_3}{4}D_{j,l}c_{j,l}c_{j,l+1}\right). \qquad (32)$$

Since $D_{j,l} = \mathrm{i}d_{j,l}d_{j,l+1}$ are good quantum numbers the Hamiltonian (32) corresponds to a model of spinless fermions with local static Z_2 gauge fields. Thus, Eq. (32) explains a hidden simple structure of the spin model (28).

The generalizations of the Jordan-Wigner transformation for arbitrary spin values were discussed by several authors,[26–28] however, these mappings have not yet provided a substantial break-through for difficult strongly correlated problems.

4. Spin-1/2 isotropic XY chain in a transverse field: dynamic quantities

We start with the simplest spin-1/2 XY model, the transverse XX chain, with the Hamiltonian (15). After performing the Jordan-Wigner transformation we arrive at a tight-binding model for spinless fermions (16) and after performing the Fourier transformation, $c_k = \left(1/\sqrt{N}\right)\sum_{n=1}^{N}\exp(\mathrm{i}kn)\,c_n$ ($k = 2\pi n/N$ if the number of fermions is odd or $k = 2\pi(n+1/2)/N$ if the number of fermions is even, $n = -N/2, -N/2+$

$1, \ldots, N/2-1$ if N is even or $n = -(N-1)/2, -(N-1)/2+1, \ldots, (N-1)/2$ if N is odd), the Hamiltonian (16) becomes diagonal

$$H = \sum_k \Lambda_k \left(c_k^\dagger c_k - \frac{1}{2} \right), \quad \Lambda_k = \Omega + J \cos k \; . \tag{33}$$

As it has been mentioned above, for the analytical calculations discussed below we may consider only periodic boundary conditions for the fermionic Hamiltonian (i.e. $k = 2\pi n/N$ in Eq. (33)).

4.1. *Two-fermion excitations*

We begin with the transverse dynamic structure factor $S_{zz}(k,\omega)$ (4).[33–35] The calculation of the zz time-dependent spin correlation function is straightforward. After exploiting the Jordan-Wigner transformation we have $\langle s_n^z(t) s_{n+l}^z \rangle - \langle s_n^z \rangle \langle s_{n+l}^z \rangle = \langle c_n^\dagger(t) c_n(t) c_{n+l}^\dagger c_{n+l} \rangle - \langle c_n^\dagger c_n \rangle \langle c_{n+l}^\dagger c_{n+l} \rangle$. Here $c_n^\dagger(t) = \left(1/\sqrt{N}\right) \sum_k \exp(\mathrm{i}kn)\, c_k^\dagger(t)$ and $c_k^\dagger(t) = c_k^\dagger \exp(\mathrm{i}\Lambda_k t)$. Next we have to use the Wick-Bloch-de Dominicis theorem, $\langle c_{k_1}^\dagger c_{k_2} c_{k_3}^\dagger c_{k_4} \rangle = \langle c_{k_1}^\dagger c_{k_2} \rangle \langle c_{k_3}^\dagger c_{k_4} \rangle - \langle c_{k_1}^\dagger c_{k_3}^\dagger \rangle \langle c_{k_2} c_{k_4} \rangle + \langle c_{k_1}^\dagger c_{k_4} \rangle \langle c_{k_2} c_{k_3}^\dagger \rangle$, and to calculate the elementary contractions introducing the Fermi function $n_k = 1/\left(1+\exp(\beta\Lambda_k)\right)$, $\langle c_{k_1}^\dagger c_{k_2} \rangle = \delta_{k_1 k_2} n_{k_1}$, $\langle c_{k_1}^\dagger c_{k_2}^\dagger \rangle = 0$. As a result, the final expression for the zz time-dependent spin correlation function reads

$$\begin{aligned}\langle s_n^z(t) s_{n+l}^z \rangle - \langle s^z \rangle^2 = \frac{1}{N^2} \sum_{k_1,k_2} & \exp\left(-\mathrm{i}\left(k_1 - k_2\right) l\right) \\ & \cdot \exp\left(\mathrm{i}\left(\Lambda_{k_1} - \Lambda_{k_2}\right) t\right) n_{k_1} \left(1 - n_{k_2}\right),\end{aligned}$$

$$\langle s^z \rangle = \frac{1}{N} \sum_{n=1}^{N} \langle s_n^z \rangle = -\frac{1}{2N} \sum_k \tanh \frac{\beta \Lambda_k}{2} \, . \tag{34}$$

Plugging Eq. (34) into Eq. (4) we get the desired transverse dynamic structure factor

$$\begin{aligned} S_{zz}(k,\omega) &= \sum_{l=1}^{N} \exp(-\mathrm{i}kl) \int_{-\infty}^{\infty} \mathrm{d}t \exp(\mathrm{i}\omega t) \langle \left(s_n^z(t) - \langle s^z \rangle\right) \left(s_{n+l}^z - \langle s^z \rangle\right) \rangle \\ &= \int_{-\pi}^{\pi} \mathrm{d}k_1 n_{k_1} \left(1 - n_{k_1+k}\right) \delta\left(\omega + \Lambda_{k_1} - \Lambda_{k_1+k}\right) \\ &= \sum_{k^\star} \frac{n_{k^\star}\left(1 - n_{k+k^\star}\right)}{2\left|J \sin\frac{k}{2} \cos\left(\frac{k}{2} + k^\star\right)\right|} \end{aligned} \tag{35}$$

where $-\pi \leq k^\star < \pi$ are the solutions of the equation $\omega = -2J\sin(k/2)\sin(k/2+k^\star)$.

The zz dynamic structure factor (35) is governed exclusively by a two-fermion (one particle and one hole) excitation continuum. The properties of the two-fermion excitation continuum were discussed by G. Müller et al. [34]; we present these results briefly below. The boundaries of the two-fermion continuum in the plane wave-vector k – frequency ω (we assume $\omega \geq 0$, $-\pi \leq k < \pi$) are determined by the equations

$$\omega = -\Lambda_{k_1} + \Lambda_{k_2}, \; k = -k_1 + k_2 (\mod (2\pi)), \; \Lambda_k = \Omega + J\cos k, \quad (36)$$

where $-\pi \leq k_1 < \pi$. Moreover, in the ground state we have to require in addition $n_{k_1} > 0$ and $1 - n_{k_2} > 0$, i.e. $\Lambda_{k_1} \leq 0$ and $\Lambda_{k_2} \geq 0$.

We start with the zero-temperature case. In this case the two-fermion excitation continuum exists as long as $|\Omega| < |J|$. Let us introduce the parameter $\alpha = \arccos(|\Omega|/|J|)$ and the following characteristic lines in the k–ω plane

$$\frac{\omega_1(k)}{|J|} = 2\left|\sin\frac{k}{2}\sin\left(\frac{|k|}{2} - \alpha\right)\right|, \quad (37)$$

$$\frac{\omega_2(k)}{|J|} = 2\left|\sin\frac{k}{2}\sin\left(\frac{|k|}{2} + \alpha\right)\right|, \quad (38)$$

$$\frac{\omega_3(k)}{|J|} = 2\left|\sin\frac{k}{2}\right|. \quad (39)$$

The two-fermion dynamic quantities in the ground state may have non-zero values only within a restricted region of the k–ω plane with the lower boundary $\omega_l(k) = \omega_1(k)$ and the upper boundary $\omega_u(k) = \omega_2(k)$ if $|k| \leq \pi - 2\alpha$ or $\omega_u(k) = \omega_3(k)$ if $\pi - 2\alpha \leq |k|$. Obviously, the two-fermion dynamic quantities may have only three soft modes $k_0 = \{0, \pm 2\alpha\}$. Moreover, there is a middle boundary of the two-fermion excitation continuum $\omega_m(k) = \omega_2(k)$ if $\pi - 2\alpha \leq |k|$ along which the two-fermion dynamic quantities exhibit a jump increasing their values by 2. Finally, the two-fermion dynamic quantities show one-dimensional square-root van Hove divergencies along the curve $\omega_s(k) = \omega_3(k)$. In Fig. 2 we display the characteristic lines (37), (38), (39) which give the boundaries of the two-fermion continuum and potential soft modes and singularities.

As temperature increases the lower boundary is smeared-out and finally disappears, the upper boundary becomes $\omega_3(k)$ along which van Hove singularities occur. In the high-temperature limit the two-fermion dynamic structure factor becomes Ω-independent.

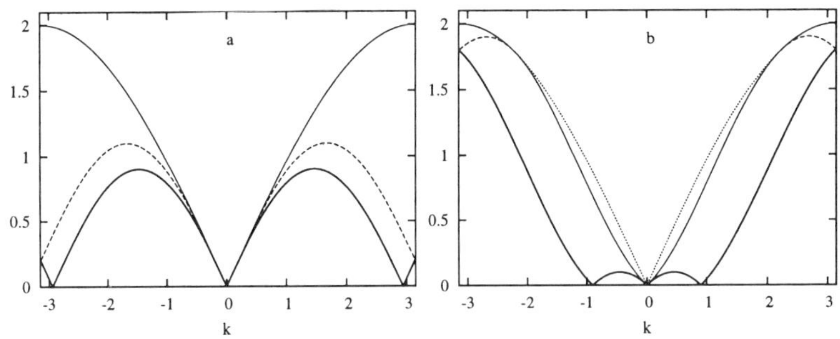

Fig. 2. The two-fermion excitation continuum which governs the ground-state two-fermion dynamic quantities. $|J| = 1$, $|\Omega| = 0.1$ (a), $|\Omega| = 0.9$ (b). We show the lower boundaries (bold lines), the middle boundaries (dashed lines), the upper boundaries (thin lines) and the lines of potential singularities (dotted lines).

In Fig. 3 we display the transverse dynamic structure factor $S_{zz}(k,\omega)$

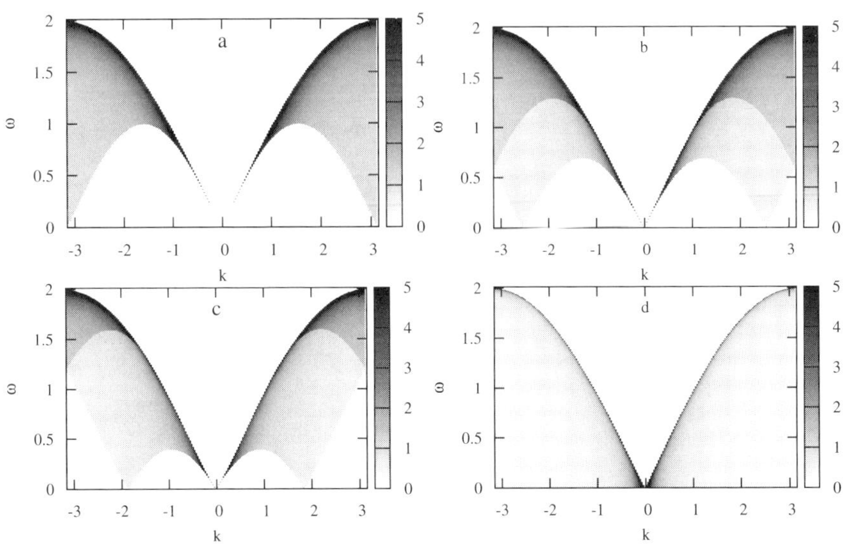

Fig. 3. $S_{zz}(k,\omega)$ (gray-scale plots) for the chain (15) with $J = -1$, $\Omega = 0$ (a), $\Omega = 0.3$ (b), $\Omega = 0.6$ (c) at $T = 0$ and at $T \to \infty$ (d).

(35) at zero temperature (panels a, b, c) and in the high-temperature limit (panel d).

There are other dynamic quantities which probe the two-fermion excitation continuum. Let us consider the dimer operator

$$D_n = s_n^x s_{n+1}^x + s_n^y s_{n+1}^y \to \frac{1}{2}\left(c_n^\dagger c_{n+1} - c_n c_{n+1}^\dagger\right). \tag{40}$$

That operator is related to a perturbation to the Hamiltonian (15) which mimics dimerization, $\epsilon \sum_n \cos(\pi n) D_n$. The dynamics of fluctuations of the dimer operator can be measured experimentally: the corresponding dimer dynamic structure factor is relevant to phonon-assisted optical absorption processes in magnetic-chain compounds.[36]

The calculation of the time-dependent dimer-dimer correlation function repeats all steps discussed above while deriving (34) and ends up with

$$\langle D_n(t) D_{n+l}\rangle - \langle D\rangle^2 = \frac{1}{N^2}\sum_{k_1,k_2}\cos^2\frac{k_1+k_2}{2}\exp\left(-\mathrm{i}\left(k_1-k_2\right)l\right)$$
$$\cdot\exp\left(\mathrm{i}\left(\Lambda_{k_1}-\Lambda_{k_2}\right)t\right)n_{k_1}\left(1-n_{k_2}\right),$$
$$\langle D\rangle = \frac{1}{N}\sum_{n=1}^{N}\langle D_n\rangle = -\frac{1}{2N}\sum_k \cos k \tanh\frac{\beta\Lambda_k}{2}. \tag{41}$$

Inserting Eq. (41) into Eq. (4) we get the dimer dynamic structure factor

$$S_{DD}(k,\omega) = \sum_{l=1}^{N}\exp(-\mathrm{i}kl)\int_{-\infty}^{\infty}\mathrm{d}t\exp(\mathrm{i}\omega t)\langle\left(D_n(t)-\langle D\rangle\right)\left(D_{n+l}-\langle D\rangle\right)\rangle$$
$$= \int_{-\pi}^{\pi}\mathrm{d}k_1\cos^2\left(k_1+\frac{k}{2}\right)n_{k_1}\left(1-n_{k_1+k}\right)\delta\left(\omega+\Lambda_{k_1}-\Lambda_{k_1+k}\right)$$
$$= \sum_{k^\star}\frac{\cos^2\left(\frac{k}{2}+k^\star\right)n_{k^\star}\left(1-n_{k+k^\star}\right)}{2\left|J\sin\frac{k}{2}\cos\left(\frac{k}{2}+k^\star\right)\right|}. \tag{42}$$

We can also calculate the zD and Dz dynamic structure factors

$$S_{zD}(k,\omega) = \exp\left(\mathrm{i}\frac{k}{2}\right)\int_{-\pi}^{\pi}\mathrm{d}k_1\cos\left(k_1+\frac{k}{2}\right)$$
$$\cdot n_{k_1}\left(1-n_{k_1+k}\right)\delta\left(\omega+\Lambda_{k_1}-\Lambda_{k_1+k}\right),$$
$$S_{Dz}(k,\omega) = \left(S_{zD}(k,\omega)\right)^*. \tag{43}$$

In Fig. 4 we display the dynamic structure factors $S_{DD}(k,\omega)$ (42) and $S_{zD}(k,\omega)$ (43) at zero temperature (panels a, c), at low temperature (panel d) and in the high-temperature limit (panel b).

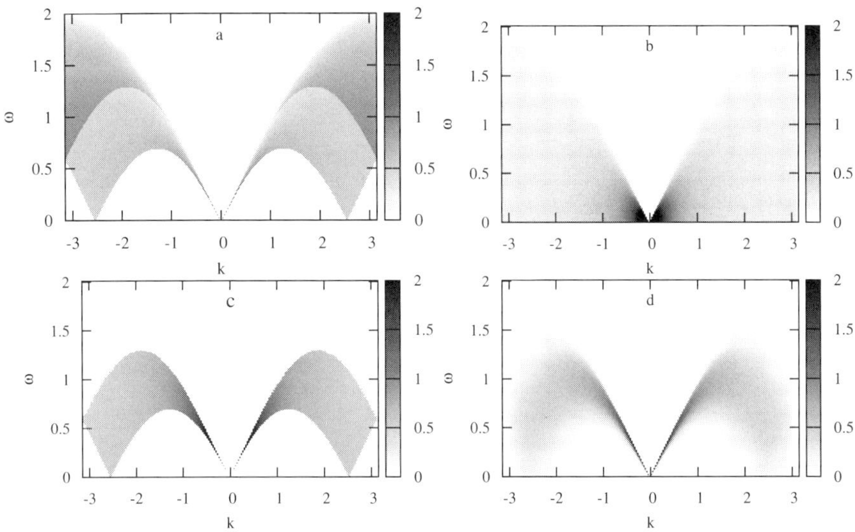

Fig. 4. $S_{DD}(k,\omega)$ (a, b) and $S_{zD}(k,\omega)$ (multiplied by $\exp(-\mathrm{i}k/2)$) (c, d) (gray-scale plots) for the chain (15) with $J=-1$. $S_{DD}(k,\omega)$: $\Omega=0.3$, $T=0$ (a), $T\to\infty$ (b). $S_{zD}(k,\omega)$: $\Omega=0.3$, $T=0$ (c), $\Omega=0.3$, $T=0.1$ (d).

Comparing Eqs. (35), (42), (43) we immediately recognize that all these dynamic quantities are governed exclusively by the two-fermion excitation continuum (for other two-fermion dynamic quantities see below and also Refs. [7,37]) and therefore all of them exhibit generic properties inherent in the two-fermion dynamic quantities. However, they also exhibit some specific properties originating from additional factors in the integrands in Eqs. (35), (42), (43) (e.g. singularities may be suppressed etc., compare Fig. 3b and Fig. 4a). In general, the dynamic structure factor governed by the two-fermion excitation continuum can be written in the following form

$$\begin{aligned}
S_{AB}(k,\omega) &= \int_{-\pi}^{\pi} \mathrm{d}k_1 \mathrm{d}k_2 C_{AB}\left(k_1,k_2\right) \\
&\quad \cdot n_{k_1}\left(1-n_{k_2}\right)\delta\left(\omega+\Lambda_{k_1}-\Lambda_{k_2}\right)\delta_{k+k_1-k_2,0}\,, \\
C_{zz}(k_1,k_2) &= 1\,, \\
C_{DD}(k_1,k_2) &= \cos^2\frac{k_1+k_2}{2}\,, \\
C_{zD}(k_1,k_2) &= \frac{1}{2}\left(\exp\left(-\mathrm{i}k_1\right)+\exp\left(\mathrm{i}k_2\right)\right)\,, \\
C_{Dz}(k_1,k_2) &= \left(C_{zD}(k_1,k_2)\right)^*\,.
\end{aligned} \tag{44}$$

All these quantities show generic properties (spectral boundaries, soft modes, singularity structure) and specific properties controlled by $C_{AB}(k_1, k_2)$.

4.2. *Four-fermion excitations*

We proceed by considering more complicate dynamic quantities. Namely, consider a trimer operator[38]

$$
\begin{aligned}
T_n &= s_n^x s_{n+2}^x + s_n^y s_{n+2}^y \\
&\to \frac{1}{2}\left(c_n^\dagger c_{n+2} - c_n c_{n+2}^\dagger - 2c_n^\dagger c_{n+1}^\dagger c_{n+1} c_{n+2} + 2c_n c_{n+1}^\dagger c_{n+1} c_{n+2}^\dagger\right).
\end{aligned}
\tag{45}
$$

That operator enters as a perturbation to the Hamiltonian (15) which mimics trimerization, $\epsilon \sum_n \cos(2\pi n/3) T_n$. The dynamics of fluctuations of the trimer operator, although it can be analyzed rigorously, is less evident from the experimental point of view. Its importance, however, is justified as a quantity of intermediate complexity between the zz and the xx and xy dynamic quantities.

The calculation of the time-dependent trimer-trimer correlation function contains no new conceptual ideas but is somewhat tedious. The final result for the time-dependent trimer correlation function reads

$$
\begin{aligned}
\langle T_n(t) T_{n+l}\rangle - \langle T\rangle^2 &= \frac{1}{N^2}\sum_{k_1,k_2} C_{TT}^{(2)}(k_1,k_2)\exp\left(-\mathrm{i}\left(k_1-k_2\right)l\right) \\
&\quad\cdot \exp\left(\mathrm{i}\left(\Lambda_{k_1}-\Lambda_{k_2}\right)t\right) n_{k_1}\left(1-n_{k_2}\right) \\
&\quad+\frac{1}{N^4}\sum_{k_1,k_2,k_3,k_4} C_{TT}^{(4)}(k_1,k_2,k_3,k_4) \\
&\quad\cdot \exp\left(-\mathrm{i}\left(k_1+k_2-k_3-k_4\right)l\right) \\
&\quad\cdot \exp\left(\mathrm{i}\left(\Lambda_{k_1}+\Lambda_{k_2}-\Lambda_{k_3}-\Lambda_{k_4}\right)t\right) n_{k_1} n_{k_2}\left(1-n_{k_3}\right)\left(1-n_{k_4}\right), \\
\langle T\rangle &= \frac{1}{N}\sum_{n=1}^{N}\langle T_n\rangle = c_2 + 2c_1^2 - 2c_0 c_2
\end{aligned}
\tag{46}
$$

with

$$
\begin{aligned}
C_{TT}^{(2)}(k_1,k_2) = & \left(1-2c_0\right)^2 \cos^2\left(k_1+k_2\right) \\
& + 4c_1\left(1-2c_0\right)\left(\cos^2\left(k_1+\frac{k_2}{2}\right)+\cos^2\left(\frac{k_1}{2}+k_2\right)\right) \\
& + 4c_1^2\left(\cos^2 k_1+\cos^2 k_2\right) \\
& + 8\left(-c_2+c_1^2+2c_0c_2\right)\cos^2\frac{k_1+k_2}{2} + 8c_1^2\cos^2\frac{k_1-k_2}{2} \\
& + 4c_1\left(1-2c_0-4c_2\right)\left(\cos^2\frac{k_1}{2}+\cos^2\frac{k_2}{2}\right) \\
& + 4c_2-8c_1-8c_1^2+4c_2^2+16c_0c_1-8c_0c_2+16c_1c_2\,,
\end{aligned}
\tag{47}
$$

$$
C_{TT}^{(4)}(k_1,k_2,k_3,k_4) = 16\sin^2\frac{k_1-k_2}{2}\sin^2\frac{k_3-k_4}{2}\cos^2\frac{k_1+k_2+k_3+k_4}{2} \geq 0 \tag{48}
$$

and $c_p = (1/N)\sum_k \cos(pk)\, n_k$. Obviously, the calculation of such an average as $\langle c_{k_1}^\dagger c_{k_2}^\dagger c_{k_3} c_{k_4} c_{k_5}^\dagger c_{k_6}^\dagger c_{k_7} c_{k_8}\rangle$ according to the Wick-Bloch-de Dominicis theorem is rather complicated. Substituting (46) into Eq. (4) we obtain the following result for the trimer dynamic structure factor

$$
\begin{aligned}
S_{TT}(k,\omega) &= \sum_{l=1}^{N} \exp(-\mathrm{i}kl) \int_{-\infty}^{\infty} \mathrm{d}t \exp(\mathrm{i}\omega t) \langle \left(T_n(t)-\langle T\rangle\right)\left(T_{n+l}-\langle T\rangle\right)\rangle \\
&= S_{TT}^{(2)}(k,\omega) + S_{TT}^{(4)}(k,\omega)
\end{aligned}
\tag{49}
$$

with

$$
S_{TT}^{(2)}(k,\omega) = \int_{-\pi}^{\pi} \mathrm{d}k_1 C_{TT}^{(2)}(k_1,k_1+k) n_{k_1}\left(1-n_{k_1+k}\right)\delta\left(\omega+\Lambda_{k_1}-\Lambda_{k_1+k}\right)\,, \tag{50}
$$

$$
\begin{aligned}
S_{TT}^{(4)}(k,\omega) = & \frac{1}{4\pi^2}\int_{-\pi}^{\pi} \mathrm{d}k_1 \mathrm{d}k_2 \mathrm{d}k_3 C_{TT}^{(4)}(k_1,k_2,k_3,k_1+k_2-k_3+k) \\
& \cdot n_{k_1} n_{k_2}\left(1-n_{k_3}\right)\left(1-n_{k_1+k_2-k_3+k}\right) \\
& \cdot \delta\left(\omega+\Lambda_{k_1}+\Lambda_{k_2}-\Lambda_{k_3}-\Lambda_{k_1+k_2-k_3+k}\right).
\end{aligned}
\tag{51}
$$

For further details see Ref. [38].

As can be seen from Eqs. (49), (50), (51) the trimer dynamic structure factor is governed both by the two-fermion (one particle and one hole) excitation continuum discussed above, the term $S_{TT}^{(2)}(k,\omega)$, and by the four-fermion (two particles and two holes) excitation continuum, the term $S_{TT}^{(4)}(k,\omega)$. The four-fermion excitation continuum is determined by the conditions

$$\omega = -\Lambda_{k_1} - \Lambda_{k_2} + \Lambda_{k_3} + \Lambda_{k_4}, \; k = -k_1 - k_2 + k_3 + k_4 \,(\mathrm{mod}(2\pi)), \quad (52)$$

where $-\pi \le k_1, k_2, k_3 < \pi$ and $\Lambda_k = \Omega + J\cos k$. Moreover, in the ground state we have to require in addition $n_{k_1} > 0$, $n_{k_2} > 0$, $1-n_{k_3} > 0$, $1-n_{k_4} > 0$, i.e. $\Lambda_{k_1} \le 0$, $\Lambda_{k_2} \le 0$, $\Lambda_{k_3} \ge 0$, $\Lambda_{k_4} \ge 0$.

We start with the zero-temperature case. The lower boundary is given by one of the following curves

$$\begin{aligned}
\frac{\omega_l^{(1)}(k)}{|J|} &= 2\sin\frac{|k|}{2}\sin\left(\alpha - \frac{|k|}{2}\right),\\
\frac{\omega_l^{(2)}(k)}{|J|} &= 4\cos\frac{k}{4}\cos\left(\alpha + \frac{|k|}{4}\right),\\
\frac{\omega_l^{(3)}(k)}{|J|} &= -2\sin\left(\alpha + \frac{|k|}{2}\right)\sin\left(2\alpha + \frac{|k|}{2}\right),\\
\frac{\omega_l^{(4)}(k)}{|J|} &= -2\sin\left(\alpha - \frac{|k|}{2}\right)\sin\left(2\alpha - \frac{|k|}{2}\right),\\
\frac{\omega_l^{(5)}(k)}{|J|} &= -4\sin\frac{|k|}{4}\sin\left(\alpha - \frac{|k|}{4}\right) \qquad (53)
\end{aligned}$$

depending on the value of Ω, $|\Omega| \le |J|$ and the value of k, $\pi \le k < \pi$ as is shown in the left panel in Fig. 5. The boundary between the region i

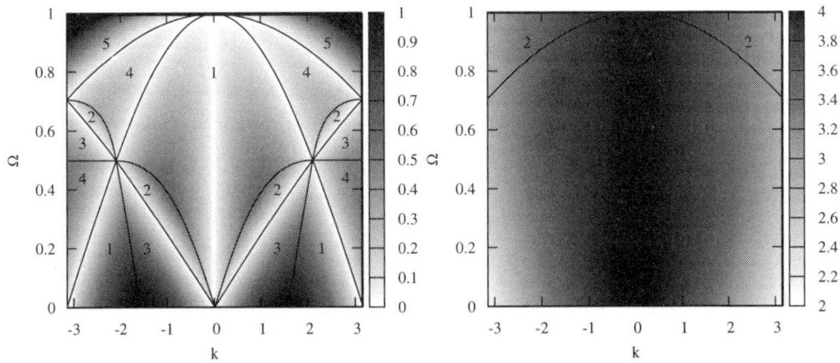

Fig. 5. The lower boundary $\omega_l(k)$ (left panel) and the upper boundary $\omega_u(k)$ (right panel) of the four-fermion excitation continuum in the plane wave-vector k – transverse field Ω, $|J| = 1$.

(where $\omega_l^{(i)}(k)$ is the lower boundary) and the region j (where $\omega_l^{(j)}(k)$ is the lower boundary) (see the left panel in Fig. 5) is given by the formula $|k| = l_{ij}(\alpha)$ where $l_{12}(\alpha) = 4\arctan\left(\left(\tan\alpha - \sqrt{\tan^2\alpha - 3}\right)/3\right)$, $|k| \leq 2\pi/3$; $l_{13}(\alpha) = \pi - \alpha$, $\pi/2 \leq |k| \leq 2\pi/3$; $l_{14}(\alpha) = 2\alpha$; $l_{23}(\alpha) = 2\pi - 4\alpha$; $l_{34}(\alpha) = |k| + \cos\alpha - 1/2$, $2\pi/3 \leq |k| \leq \pi$; $l_{45}(\alpha) = 4\alpha$ (for further details see Ref. [38]). The four-fermion contribution to dynamic quantities may exhibit soft modes $|k_0| = \{0, 2\pi - 4\alpha, 2\alpha, 4\alpha\}$. Next we pass to the upper boundary of the four-fermion excitation continuum which is given by one of the following curves

$$\frac{\omega_u^{(1)}(k)}{|J|} = 4\cos\frac{k}{4}, \tag{54}$$

$$\frac{\omega_u^{(2)}(k)}{|J|} = 4\cos\frac{k}{4}\cos\left(\alpha - \frac{|k|}{4}\right) \tag{55}$$

depending on the value of Ω, $|\Omega| \leq |J|$ and the value of k, $\pi \leq k < \pi$ as is shown in the right panel in Fig. 5. The boundary between the regions 1 and 2 is given by the curve $|k| = 4\alpha$. In Fig. 6 we compare the ground-

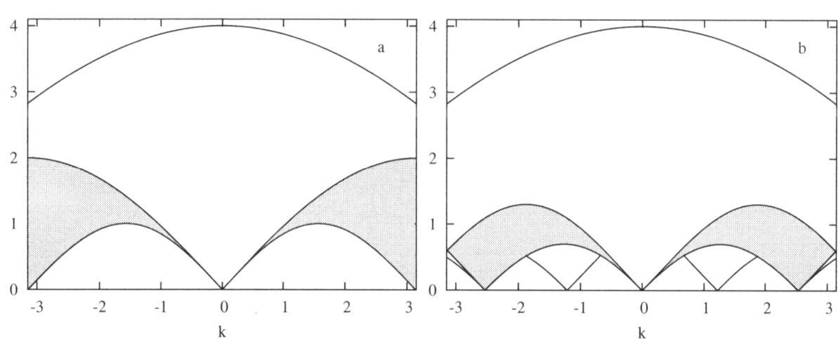

Fig. 6. Lower boundaries and upper boundaries of the two-fermion and four-fermion excitation continua for $|J| = 1$ and $\Omega = 0$ (a) and $\Omega = 0.3$ (b) at $T = 0$. The two-fermion continuum is shown shaded.

state two-fermion and four-fermion excitation continua for two values of the transverse field Ω. The four-fermion excitation continuum always contains the two-fermion excitation continuum. The lower boundaries may coincide (e.g. in the zero-field case the lower boundary is $|J|\sin|k|$ for both continua, panel a in Fig. 6a) whereas the upper boundaries are different.

Next we turn to the van Hove singularities inherent in the four-fermion dynamic quantities. Evidently, the quantity

$$S(k,\omega)=\int_{-\pi}^{\pi}\mathrm{d}k_1\int_{-\pi}^{\pi}\mathrm{d}k_2\int_{-\pi}^{\pi}\mathrm{d}k_3 S(k_1,k_2,k_3,k)$$
$$\cdot\,\delta\left(\omega-\vert J\vert\cos k_1-\vert J\vert\cos k_2+\vert J\vert\cos k_3\right.$$
$$\left.+\vert J\vert\cos\left(k+k_1+k_2-k_3\right)\right) \tag{56}$$

may exhibit van Hove singularities characteristic to the three-dimensional density of states. The lines of potential singularities are as follows

$$\frac{\omega_s^{(1)}(k)}{\vert J\vert}=2\sin\frac{\vert k\vert}{2},$$
$$\frac{\omega_s^{(2)}(k)}{\vert J\vert}=4\sin\frac{\vert k\vert}{4},$$
$$\frac{\omega_s^{(3)}(k)}{\vert J\vert}=4\cos\frac{k}{4}. \tag{57}$$

The four-fermion dynamic quantities may exhibit cusp singularities (akin to density-of-states effects in thee dimensions) along these curves. We illustrate potential singularities in the frequency profiles of $S(k,\omega)$ (56) at different k in Fig. 7.

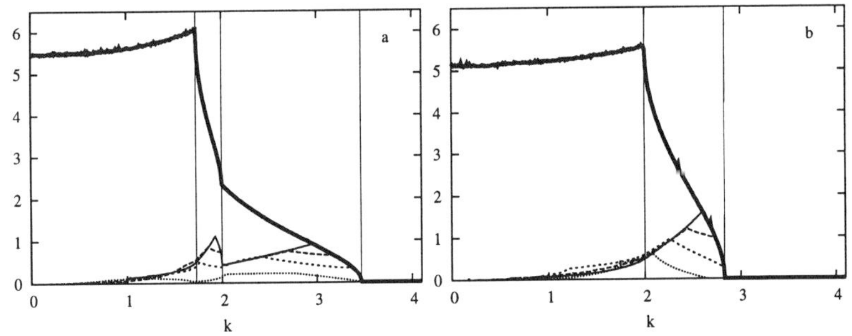

Fig. 7. $S(k,\omega)$ (56) vs ω at $k=2\pi/3$ (a) and $k=\pi$ (b) for $S(k_1,k_2,k_3,k)=1$ (bold curves) and $S(k_1,k_2,k_3,k)=n_{k_1}n_{k_2}\left(1-n_{k_3}\right)\left(1-n_{k_1+k_2-k_3+k}\right)$, $T=0$ for $\Omega=0$ (solid curves), $\Omega=0.3$ (long-dashed curves), $\Omega=0.6$ (short-dashed curves), $\Omega=0.9$ (dotted curves). Vertical lines denote the values of $\omega_s^{(j)}(k)$, $j=1,2,3$ (57).

For nonzero temperatures the lower boundary is smeared out and finally disappears. The upper boundary is given by Eq. (54). In the high-temperature limit the properties of the four-fermion excitation continuum become Ω-independent.

After discussing some generic properties of the four-fermion dynamic quantities (inherent in any four-fermion dynamic quantity) we illustrate some specific properties conditioned by the function $C^{(4)}_{TT}(k_1,k_2,k_3,k_4)$ (48). In Fig. 8 we display the trimer dynamic structure factor (49). The con-

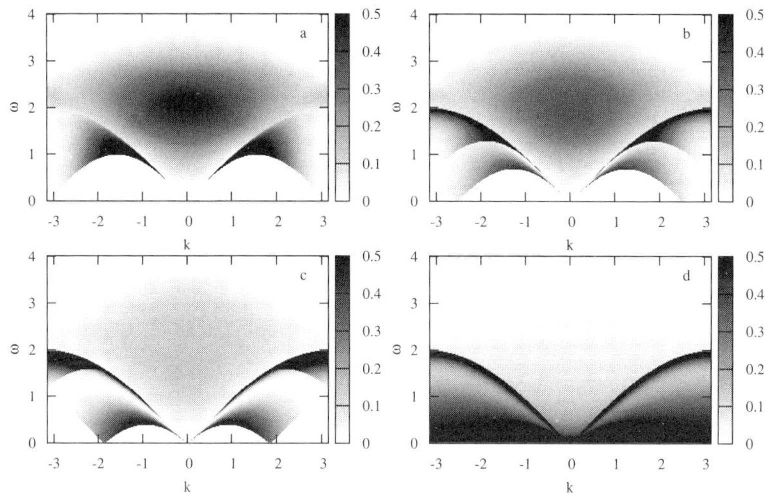

Fig. 8. $S_{TT}(k,\omega)$ (49) (gray-scale plots) of the chain (15) with $J=-1$, $\Omega=0$ (a), $\Omega=0.3$ (b), $\Omega=0.6$ (c) at $T=0$ and at $T\to\infty$ (d).

tributions of the two-fermion excitation continuum and the four-fermion excitation continuum to this quantity can be easily distinguished.

We may formally introduce the polymer operator

$$\mathcal{P}^{(l)}_n = s^x_n s^x_{n+l} + s^y_n s^y_{n+l} \tag{58}$$

(evidently $\mathcal{P}^{(1)}_n = D_n$ and $\mathcal{P}^{(2)}_n = T_n$). Now the dynamic polymer structure factor $S_{\mathcal{P}\mathcal{P}}(k,\omega)$ will involve $2m$-fermion excitations with $m=1,2,\ldots,l$. These quantities are of moderate complexity in comparison with $S_{xx}(k,\omega)$ and $S_{xy}(k,\omega)$ which are enormously complex (see below). We also note that in the limit $l\to\infty$

$$\langle \mathcal{P}^{(l)}_n(t)\mathcal{P}^{(l)}_{n+m}\rangle \stackrel{l\to\infty}{\longrightarrow} 2\langle s^x_n(t)s^x_{n+m}\rangle^2 + 2\langle s^x_n(t)s^y_{n+m}\rangle^2. \tag{59}$$

The last term in Eq. (59) is nonzero only if $\Omega\neq 0$.

In passing, we note that the multimagnon continua of quantum spin chains have been discussed recently on general ground by T. Barnes.[39] Obviously, there is an essential difference in comparison with our case, since the Jordan-Wigner fermions obey the Fermi statistics and this point has important consequences for the four-fermion excitation continuum considered in some detail above.

4.3. *Many-fermion excitations*

We pass to dynamic structure factors which are governed by many-fermion excitations. Let us recall that according to the Jordan-Wigner transformation we have

$$\begin{aligned} s_n^x &= \frac{1}{2}\left(1-2c_1^\dagger c_1\right)\dots\left(1-2c_{n-1}^\dagger c_{n-1}\right)\left(c_n^\dagger + c_n\right) \\ &= \frac{1}{2}\varphi_1^+\varphi_1^-\dots\varphi_{n-1}^+\varphi_{n-1}^-\varphi_n^+, \\ s_n^y &= \frac{1}{2\mathrm{i}}\varphi_1^+\varphi_1^-\dots\varphi_{n-1}^+\varphi_{n-1}^-\varphi_n^-, \\ s_n^z &= -\frac{1}{2}\varphi_n^+\varphi_n^- \end{aligned} \tag{60}$$

where we have introduced the operators $\varphi_m^{\pm} = c_m^\dagger \pm c_m$. Obviously the operators $\varphi_m^{\pm}$ are linear combinations of the operators c_k, $c_k^\dagger$ in terms of which the Hamiltonian is diagonal (see Eq. (33)).

Consider now the xx time-dependent spin correlation function

$$\begin{aligned} 4\langle s_j^x(t)s_{j+n}^x\rangle = &\langle\varphi_1^+(t)\varphi_1^-(t)\dots\varphi_{j-1}^+(t)\varphi_{j-1}^-(t)\varphi_j^+(t) \\ &\cdot\varphi_1^+\varphi_1^-\dots\varphi_{j-1}^+\varphi_{j-1}^-\varphi_j^+\varphi_j^-\varphi_{j+1}^+\varphi_{j+1}^-\dots\varphi_{j+n-1}^+\varphi_{j+n-1}^-\varphi_{j+n}^+\rangle. \end{aligned} \tag{61}$$

It contains a product of $2(2j+n-1)$ $\varphi^{\pm}$ operators (in contrast to $4\langle s_j^z(t)s_{j+n}^z\rangle = \langle\varphi_j^+(t)\varphi_j^-(t)\varphi_{j+n}^+\varphi_{j+n}^-\rangle$ which contains the product of only four $\varphi^{\pm}$ operators). Therefore the calculation of xx and xy dynamic quantities (which are governed by many-fermion excitations) is essentially more complicated.

Exact analytical results for xx and xy dynamic quantities are rather scarce. At the high-temperature limit $T\to\infty$ we know[40,41] that

$$\begin{aligned} 4\langle s_j^x(t)s_{j+n}^x\rangle &= \delta_{n,0}\cos(\Omega t)\exp\left(-\frac{1}{4}J^2t^2\right), \\ 4\langle s_j^x(t)s_{j+n}^y\rangle &= -\delta_{n,0}\sin(\Omega t)\exp\left(-\frac{1}{4}J^2t^2\right). \end{aligned} \tag{62}$$

At zero temperature the xx and xy time-dependent correlation functions are extremely simple only when $|\Omega|>|J|$.[42] Consider for example the case $\Omega>|J|$ when the ground state is completely polarized $|\mathrm{GS}_s\rangle = \prod_{n=1}^N|\downarrow_n\rangle$

(in spin language) or completely empty $c_k|\mathrm{GS}_c\rangle = 0$ (in fermionic language). Owing to the simplicity of the ground state $s_m^+|\mathrm{GS}_s\rangle = c_m^\dagger|\mathrm{GS}_c\rangle = \left(1/\sqrt{N}\right)\sum_k \exp(\mathrm{i}km)\, c_k^\dagger|\mathrm{GS}_c\rangle$, $s_m^-|\mathrm{GS}_s\rangle = 0$. Therefore

$$\begin{aligned}4\langle s_j^x(t)s_{j+n}^x\rangle &= \langle \mathrm{GS}_s|s_j^-(t)s_{j+n}^+|\mathrm{GS}_s\rangle \\ &= \frac{1}{N}\sum_k \exp\left(\mathrm{i}kn - \mathrm{i}\left(\Omega + J\cos k\right)t\right), \\ 4\langle s_j^x(t)s_{j+n}^y\rangle &= -\mathrm{i}\langle \mathrm{GS}_s|s_j^-(t)s_{j+n}^+|\mathrm{GS}_s\rangle = -4\mathrm{i}\langle s_j^x(t)s_{j+n}^x\rangle\end{aligned} \tag{63}$$

and as a result

$$S_{xx}(k,\omega) = \mathrm{i}S_{xy}(k,\omega) = \frac{\pi}{2}\delta\left(\omega - \Omega - J\cos k\right). \tag{64}$$

Many results at $T = 0$ refer to the asymptotic behavior of the xx or xy time-dependent spin correlation functions.[43,44] From the paper by A. R. Its et al. we know the long-time asymptotic behavior at nonzero temperatures

$$\begin{aligned}\langle s_j^+(t)s_{j+n}^-\rangle &\sim \begin{cases}\exp\left(f(n,0)\right), & \frac{n}{Jt} > 1, \\ t^{2\left(\nu_-^2+\nu_+^2\right)}\exp\left(f(n,t)\right), & \frac{n}{Jt} < 1,\end{cases} \\ f(n,t) &= \frac{1}{2\pi}\int_{-\pi}^{\pi}\mathrm{d}p|n + Jt\sin p|\ln\left|\tanh\frac{\beta\left(\Omega - J\cos p\right)}{2}\right|, \\ \nu_\pm &= \frac{1}{2\pi}\ln\left|\tanh\frac{\beta\left(\Omega \mp J\sqrt{1-\left(\frac{n}{Jt}\right)^2}\right)}{2}\right|.\end{aligned} \tag{65}$$

On the other hand, we can obtain the xx and xy dynamic quantities numerically.[45–51] Consider the slightly more complicated inhomogeneous spin-1/2 anisotropic XY chain in a transverse field with the Hamiltonian

$$\begin{aligned}H &= \sum_{j=1}^{N}\Omega_j s_j^z + \sum_{j=1}^{N-1}\left(J_j^{xx}s_j^x s_{j+1}^x + J_j^{xy}s_j^x s_{j+1}^y + J_j^{yx}s_j^y s_{j+1}^x + J_j^{yy}s_j^y s_{j+1}^y\right) \\ &\to -\frac{1}{2}\sum_{j=1}^{N}\Omega_j + \sum_{i,j=1}^{N}\left(c_i^\dagger A_{ij}c_j + \frac{1}{2}\left(c_i^\dagger B_{ij}c_j^\dagger - c_iB_{ij}^*c_j\right)\right)\end{aligned} \tag{66}$$

where

$$\begin{aligned}
A_{ij} &= \Omega_i\delta_{ij} + J_i^{+-}\delta_{j,i+1} + J_{i-1}^{-+}\delta_{j,i-1} = A_{ji}^*\,,\\
B_{ij} &= J_i^{++}\delta_{j,i+1} - J_{i-1}^{++}\delta_{j,i-1} = -B_{ji}\,,\\
J_j^{+-} &= \frac{1}{4}\left(J_j^{xx} + J_j^{yy} + \mathrm{i}\left(J_j^{xy} - J_j^{yx}\right)\right) = \left(J_j^{-+}\right)^*\,,\\
J_j^{++} &= \frac{1}{4}\left(J_j^{xx} - J_j^{yy} - \mathrm{i}\left(J_j^{xy} + J_j^{yx}\right)\right) = \left(J_j^{--}\right)^*\,.
\end{aligned} \tag{67}$$

To diagonalize a form bilinear in Fermi operators like (66) we perform the linear canonical transformation

$$\eta_k = \sum_{n=1}^{N}\left(g_{kn}c_n + h_{kn}c_n^\dagger\right), \quad \eta_k^\dagger = \sum_{n=1}^{N}\left(g_{kn}^*c_n^\dagger + h_{kn}^*c_n\right). \tag{68}$$

The resulting Hamiltonian reads as follows

$$\begin{aligned}
H &= \sum_{k=1}^{N}\Lambda_k\left(\eta_k^\dagger\eta_k - \frac{1}{2}\right),\\
\left\{\eta_{k'},\eta_{k''}^\dagger\right\} &= \delta_{k'k''}, \quad \left\{\eta_{k'},\eta_{k''}\right\} = \left\{\eta_{k'}^\dagger,\eta_{k''}^\dagger\right\} = 0
\end{aligned} \tag{69}$$

if the coefficients g_{kn}, h_{kn} satisfy the set of equations

$$\begin{aligned}
&\left(\mathbf{g}_k\ \mathbf{h}_k\right)\mathbf{M} = \Lambda_k\left(\mathbf{g}_k\ \mathbf{h}_k\right),\\
&\mathbf{g}_k = \left(g_{k1}\ \ldots\ g_{kN}\right), \quad \mathbf{h}_k = \left(h_{k1}\ \ldots\ h_{kN}\right), \quad \mathbf{M} = \begin{pmatrix}\mathbf{A} & \mathbf{B}\\ -\mathbf{B}^* & -\mathbf{A}^*\end{pmatrix}.
\end{aligned} \tag{70}$$

Further it may be convenient to introduce the linear combinations $\Phi_{kn} = g_{kn} + h_{kn}$ and $\Psi_{kn} = g_{kn} - h_{kn}$ which enter the relations

$$\begin{aligned}
\varphi_j^+ &= c_j^\dagger + c_j = \sum_{p=1}^{N}\left(\Phi_{pj}\eta_p^\dagger + \Phi_{pj}^*\eta_p\right),\\
\varphi_j^- &= c_j^\dagger - c_j = \sum_{p=1}^{N}\left(\Psi_{pj}\eta_p^\dagger - \Psi_{pj}^*\eta_p\right).
\end{aligned} \tag{71}$$

We calculate the time-dependent spin correlation functions using the Wick-Bloch-de Dominicis theorem. For example,

$$\begin{aligned}
4\langle s_n^z(t)s_{n+m}^z\rangle &= \langle\varphi_n^+(t)\varphi_n^-(t)\varphi_{n+m}^+\varphi_{n+m}^-\rangle\\
&= \langle\varphi_n^+\varphi_n^-\rangle\langle\varphi_{n+m}^+\varphi_{n+m}^-\rangle\\
&\quad -\langle\varphi_n^+(t)\varphi_{n+m}^+\rangle\langle\varphi_n^-(t)\varphi_{n+m}^-\rangle + \langle\varphi_n^+(t)\varphi_{n+m}^-\rangle\langle\varphi_n^-(t)\varphi_{n+m}^+\rangle\,.
\end{aligned} \tag{72}$$

The r.h.s. of Eq. (72) may be compactly written as the Pfaffian of the 4×4 antisymmetric matrix

$$4\langle s_n^z(t)s_{n+m}^z\rangle = \mathrm{Pf}\begin{pmatrix} 0 & \langle\varphi_n^+\varphi_n^-\rangle & \langle\varphi_n^+(t)\varphi_{n+m}^+\rangle & \langle\varphi_n^+(t)\varphi_{n+m}^-\rangle \\ -\langle\varphi_n^+\varphi_n^-\rangle & 0 & \langle\varphi_n^-(t)\varphi_{n+m}^+\rangle & \langle\varphi_n^-(t)\varphi_{n+m}^-\rangle \\ -\langle\varphi_n^+(t)\varphi_{n+m}^+\rangle & -\langle\varphi_n^-(t)\varphi_{n+m}^+\rangle & 0 & \langle\varphi_{n+m}^+\varphi_{n+m}^-\rangle \\ -\langle\varphi_n^+(t)\varphi_{n+m}^-\rangle & -\langle\varphi_n^-(t)\varphi_{n+m}^-\rangle & -\langle\varphi_{n+m}^+\varphi_{n+m}^-\rangle & 0 \end{pmatrix}. \tag{73}$$

Similarly (see Eq. (61)), for the more complicated xx time-dependent spin correlation function we have

$$4\langle s_n^x(t)s_{n+m}^x\rangle = \mathrm{Pf}\begin{pmatrix} 0 & \langle\varphi_1^+\varphi_1^-\rangle & \langle\varphi_1^+\varphi_2^+\rangle & \cdots & \langle\varphi_1^+(t)\varphi_{n+m}^+\rangle \\ -\langle\varphi_1^+\varphi_1^-\rangle & 0 & \langle\varphi_1^-\varphi_2^+\rangle & \cdots & \langle\varphi_1^-(t)\varphi_{n+m}^+\rangle \\ \vdots & \vdots & \vdots & \cdots & \vdots \\ -\langle\varphi_1^+(t)\varphi_{n+m}^+\rangle & -\langle\varphi_1^-(t)\varphi_{n+m}^+\rangle & -\langle\varphi_2^+(t)\varphi_{n+m}^+\rangle & \cdots & 0 \end{pmatrix}, \tag{74}$$

i.e. $\langle s_j^x(t)s_{j+n}^x\rangle$ can be written as a Pfaffian of a $2(2j+n-1)\times 2(2j+n-1)$ antisymmetric matrix. The elementary contractions involved in (73), (74) read

$$\begin{aligned} \langle\varphi_j^+(t)\varphi_m^+\rangle &= \sum_{p=1}^{N}\left(\Phi_{pj}\Phi_{pm}^*F(\Lambda_p) + \Phi_{pj}^*\Phi_{pm}F(-\Lambda_p)\right), \\ \langle\varphi_j^+(t)\varphi_m^-\rangle &= \sum_{p=1}^{N}\left(-\Phi_{pj}\Psi_{pm}^*F(\Lambda_p) + \Phi_{pj}^*\Psi_{pm}F(-\Lambda_p)\right), \\ \langle\varphi_j^-(t)\varphi_m^+\rangle &= \sum_{p=1}^{N}\left(\Psi_{pj}\Phi_{pm}^*F(\Lambda_p) - \Psi_{pj}^*\Phi_{pm}F(-\Lambda_p)\right), \\ \langle\varphi_j^-(t)\varphi_m^-\rangle &= -\sum_{p=1}^{N}\left(\Psi_{pj}\Psi_{pm}^*F(\Lambda_p) + \Psi_{pj}^*\Psi_{pm}F(-\Lambda_p)\right), \\ F(\Lambda_p) &= \frac{\exp\left(\mathrm{i}\Lambda_p t\right)}{1+\exp\left(\beta\Lambda_p\right)}. \end{aligned} \tag{75}$$

It is worthwhile to recall some properties of the Pfaffians which are used for calculating them. In the first numerical studies the authors used the relation

$$(\mathrm{Pf}\mathbf{A})^2 = \det\mathbf{A} \tag{76}$$

and computed numerically the determinants which gave the values of Pfaffians according to (76). On the other hand, the Pfaffian may be computed directly[46] noting that

$$\mathrm{Pf}\left(\mathbf{U}^T\mathbf{A}\mathbf{U}\right) = \det\mathbf{U}\ \mathrm{Pf}\mathbf{A} \tag{77}$$

and that

$$\mathrm{Pf}\begin{pmatrix} 0 & R_{12} & 0 & 0 & \dots 0 \\ -R_{12} & 0 & 0 & 0 & \dots 0 \\ 0 & 0 & 0 & R_{34} & \dots 0 \\ 0 & 0 & -R_{34} & 0 & \dots 0 \\ \vdots & \vdots & \vdots & \vdots & \dots \vdots \\ 0 & 0 & 0 & 0 & \dots 0 \end{pmatrix} = R_{12}R_{34}\dots . \tag{78}$$

We use the approach described to calculate the xx and xy dynamic structure factors for the spin-1/2 transverse XX chain numerically.[52] To estimate the quality of the numerical procedure we compare our numerical findings with exact analytical results in the high-temperature limit and with exact asymptotics for finite temperatures in Fig. 9. Knowing the

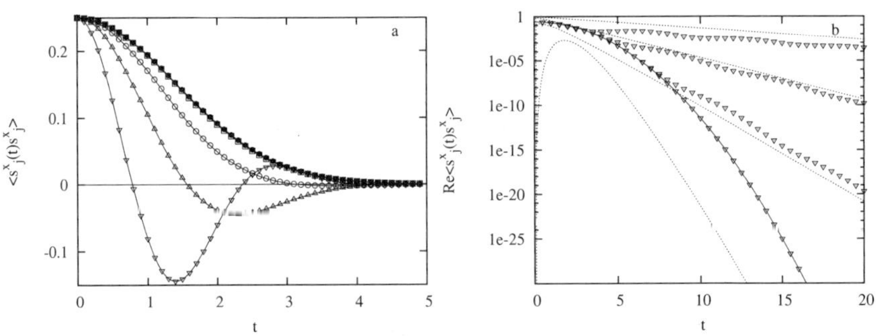

Fig. 9. Panel a: Time-dependence of the autocorrelation function $\langle s_j^x(t)s_j^x\rangle$, $j = 51$ at infinite temperature obtained numerically (symbols) and analytically (see Eq. (62)) (solid curves). $\Omega = 2$, 1 (downward and upward triangles), $\Omega = 0.5$ (open circles), $\Omega = 0.1$ (squares), $\Omega = 0$ (filled circles). Panel b: Time-dependence of the real part of the autocorrelation function $\langle s_j^x(t)s_j^x\rangle$, $j = 51$ at $\Omega = 0$ for various temperatures obtained numerically (symbols) in comparison with asymptotics (65). $\beta = 5$, 1, 0.1, 0.00001 (from top to bottom). The exact analytical result for $\beta = 0$ is also shown (the lowest curve). Evidently only the slopes of the asymptotics should be compared with the numerical results.

time-dependent correlation functions we obtain the corresponding dynamic structure factors according to

$$S_{xx}(k,\omega) = \sum_{n=0,\pm1,\ldots} \exp\left(-\mathrm{i}kn\right) 2\Re \left(\int_0^\infty \mathrm{d}t \exp\left(\mathrm{i}\left(\omega + \mathrm{i}\epsilon\right)t\right) \langle s_j^x(t) s_{j+n}^x \rangle \right),$$
$$S_{xy}(k,\omega) = \sum_{n=0,\pm1,\ldots} \exp\left(-\mathrm{i}kn\right) 2\mathrm{i}\Im \left(\int_0^\infty \mathrm{d}t \exp\left(\mathrm{i}\left(\omega + \mathrm{i}\epsilon\right)t\right) \langle s_j^x(t) s_{j+n}^y \rangle \right) \quad (79)$$

with $\epsilon \to +0$. In practice we consider chains of $N = 400$ sites, take $j = 41, 51, 61$, n up to 50 or up to 100, and set $\epsilon = 0 \ldots 0.001 \ldots 0.1$ (see Ref. [52]). The results of our calculations for $S_{xx}(k,\omega)$ and $S_{xy}(k,\omega)$ are illustrated in Figs. 10, 11.

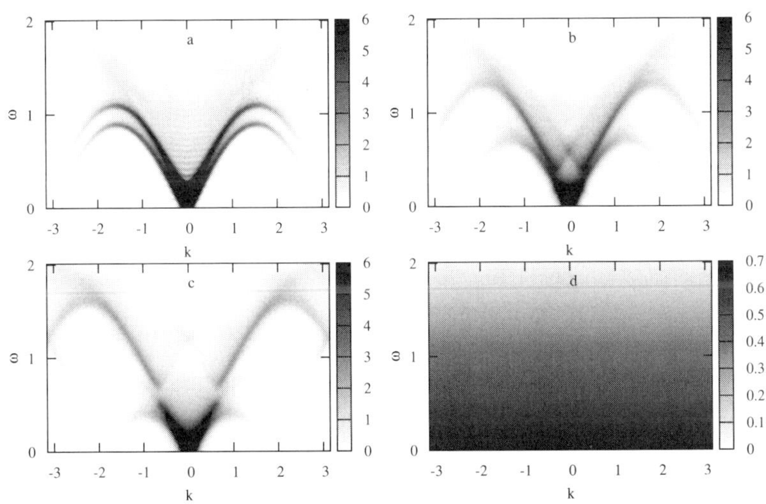

Fig. 10. $S_{xx}(k,\omega)$ (gray-scale plots) for the chain (15) with $J = -1$, for $\Omega = 0.0001$ (a), $\Omega = 0.3$ (b), $\Omega = 0.6$ (c) at low temperature $\beta = 20$ and for $\Omega = 0.6$ in the high-temperature limit $\beta = 0$ (d).

Let us recall, the transverse dynamic structure factor $S_{zz}(k,\omega)$ probes two-particle excitations, i.e. it is governed by the excitations which are composed of two Jordan-Wigner spinless fermions. The two-fermion excitation continuum has a sharp upper frequency cutoff at which $S_{zz}(k,\omega)$ may diverge. At $T = 0$ it has also a sharp lower frequency cutoff which touches $\omega = 0$ at k_0 (soft modes). $S_{zz}(k,\omega)$ is almost structureless (apart

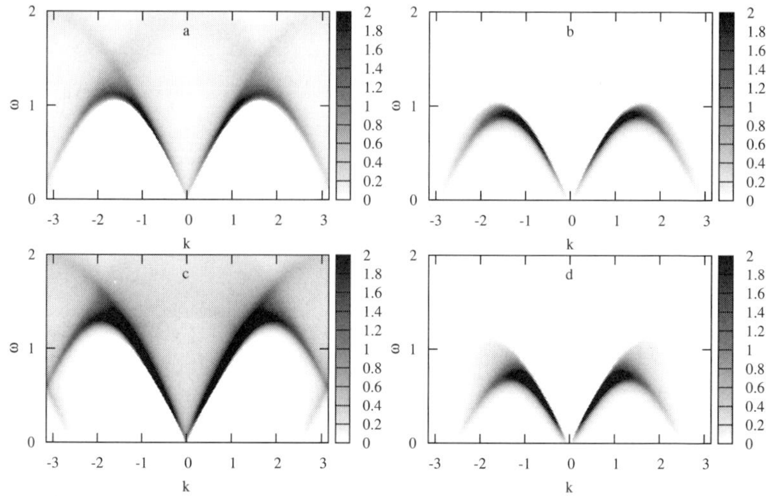

Fig. 11. $\mathrm{i}\,S_{xy}(k,\omega)$ (gray-scale plots) for the chain (15) with $J=-1$ for $\Omega=0.1$ (a, b), $\Omega=0.3$ (c, d) at low temperature $\beta=20$. Positive parts are shown in panels a and c, negative parts are shown in panels b and d.

from upper boundary singularities) and exists only for $|\Omega|<|J|$. In the high-temperature limit $T\to\infty$ $S_{zz}(k,\omega)$ becomes Ω-independent. All these features are nicely seen in Fig. 3.

In contrast, the dynamic structure factors $S_{xx}(k,\omega)$ and $S_{xy}(k,\omega)$ are many-particle quantities in terms of the Jordan-Wigner spinless fermions. The frequency range of these quantities is not *a priori* restricted, however, in the low-temperature limit $T\to 0$ $S_{xx}(k,\omega)$ and $S_{xy}(k,\omega)$ are rather small (but nonzero) outside the two-fermion excitation continuum. These quantities show washed-out excitation branches roughly following the boundary of the two-fermion excitation continuum. [Although the results presented in Figs. 10, 11 refer to the case $J<0$ (the ferromagnetic sign of the XX exchange interaction) the results for $J>0$ (the antiferromagnetic sign of the XX exchange interaction) follow by symmetry. In fact, while changing the sign of XX exchange interaction, $+J\to -J$, we get $S_{xx}(k,\omega)$, $S_{xy}(k,\omega)$ given by Eq. (79) in which the wave-vector is changed $k\to k\mp\pi$.] From the exact calculation in the strong-field zero-temperature limit (64) we know that $S_{xx}(k,\omega)$, $S_{xy}(k,\omega)$ are proportional to $\delta\left(\omega-\Lambda_k\right)$, $\Lambda_k=\Omega+J\cos k$. In the high-temperature limit $T\to\infty$ $S_{xx}(k,\omega)$ and $S_{xy}(k,\omega)$ become k-independent, but depend on Ω. All the features described can be seen in Figs. 10, 11.

It is instructive to compare our precise numerical findings in the low-temperature limit with the results for the ground-state dynamic structure factors obtained by bosonization.[18–20] In the case $\Omega = 0$ within the framework of the bosonization approach we have

$$S_{\alpha\alpha}(k,\omega) \sim \frac{\theta\left(\omega - |vk|\right)}{\left(\omega^2 - (vk)^2\right)^{1-\frac{\eta_\alpha}{2}}}, \quad \alpha = x, z. \tag{80}$$

Here $v = J$ is the velocity and $\eta_x = 1/2$, $\eta_z = 2$ are the exponents which describe correctly the singularity at the lower continuum boundary $\omega_l(k) = |J \sin k| \to |Jk|$ as $k \to 0$ or $k \to \pm\pi$. In the case of nonzero transverse fields $\Omega \neq 0$ the values of the Fermi momentum and Fermi velocity are changed. In Fig. 12 we compare the predictions of the bosonization approach (80)

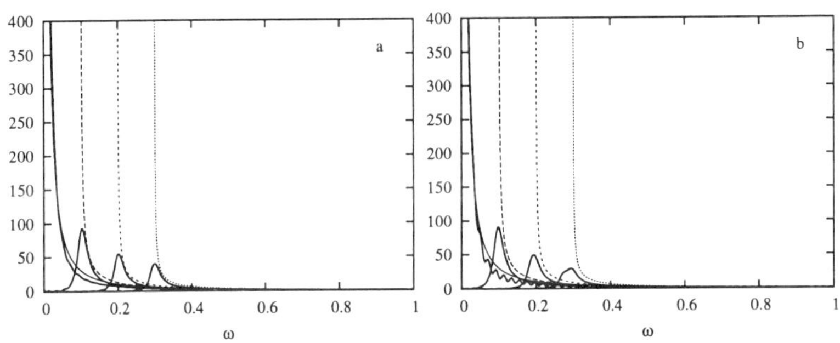

Fig. 12. $S_{xx}(k,\omega)$ for the chain (15) with $J = -1$; frequency profiles at $k = 0, 0.1, 0.2, 0.3$ (from left to right) at $\Omega = 0$ (panel a) and $\Omega = 0.3$ (panel b). Bosonization results which follow from Eq. (80) are shown by thin lines ($v = 1$ for $\Omega = 0$ and $v = 0.9539\ldots$ for $\Omega = 0.3$); numerical results at low temperature $\beta = 100$ are shown by solid lines.

with the numerical results at low temperatures.

To summarize, in this section we have discussed dynamic properties of the spin-1/2 transverse XX chain within the Jordan-Wigner fermionization approach. Within this scheme the spin Hamiltonian corresponds to the Hamiltonian of noninteracting spinless fermions. The transverse dynamic structure factor corresponds to the fermionic density dynamic structure factor and probes the two-fermion excitation continuum. There are more dynamic structure factors which probe the two-fermion excitation continuum, e.g., the dimer dynamic structure factor. All two-fermion dynamic quantities have common features (spectral boundaries, potential soft modes and singularities) and specific features. There are also dynamic quantities which probe the four-fermion excitation continuum; as an example of such a quantity we have discussed the trimer dynamic structure factor. Remarkably, the dynamic structure factors which are associated with the dynamics of fluctuations of the x or y spin components (in contrast to the transverse dynamic structure factor which is associated with dynamics of fluctuations of the z spin component) are enormously complex within the Jordan-Wigner description since they probe many-fermion excitations. Nevertheless the two-fermion excitation continuum is still important for these dynamic quantities at low temperatures. As we have observed in our numerics, most of the spectral weight is concentrated along the boundaries of the two-fermion excitation continuum (it was also noted earlier for the XXZ Heisenberg chain[53]). This is not the case in the high-temperature limit when these dynamic structure factors show Gaussian ridges. In the next two sections we shall follow to what extent our observations survive for more complicated spin-1/2 XY chains.

5. Dimerized spin-1/2 isotropic XY chain in a transverse field

Now we pass to the dimerized spin-1/2 XX chain in a transverse field. The Hamiltonian of the model reads

$$
\begin{aligned}
H &= \sum_n J\left(1-(-1)^n\delta\right)\left(s_n^x s_{n+1}^x + s_n^y s_{n+1}^y\right) + \Omega\sum_n s_n^z \\
&\to \sum_n \frac{J}{2}\left(1-(-1)^n\delta\right)\left(c_n^\dagger c_{n+1} - c_n c_{n+1}^\dagger\right) + \Omega\sum_n\left(c_n^\dagger c_n - \frac{1}{2}\right),
\end{aligned}
\tag{81}
$$

where δ is the dimerization parameter ($0 < \delta < 1$). After performing consequently the Fourier transformation, $c_n^\dagger = \left(1/\sqrt{N}\right)\sum_k \exp(\mathrm{i}kn)\, c_k^\dagger$, $k = 2\pi p/N$, $p = -N/2, \ldots, N/2-1$ (N is even), and the Bogolyubov transformation, $c_k = u_{k+\pi}\eta_k + \mathrm{i}v_k\eta_{k+\pi}$, $u_k = \left(1/\sqrt{2}\right)\sqrt{1+|\cos k|/\epsilon_k}$, $v_k = \operatorname{sgn}(\sin(2k))\left(1/\sqrt{2}\right)\sqrt{1-|\cos k|/\epsilon_k}$, $\epsilon_k = \sqrt{\cos^2 k + \delta^2\sin^2 k}$, the Hamiltonian becomes diagonal, $H = \sum_k \Lambda_k\left(\eta_k^\dagger\eta_k - 1/2\right)$ with the elementary excitation energy $\Lambda_k = \Omega + \lambda_k$, $\lambda_k = \operatorname{sgn}(\cos k)J\epsilon_k$ (for further details see Refs. [35,54]).

The calculation of the transverse dynamic structure factor follows the scheme described in Sec. 4 and ends up with the following result

$$S_{zz}(k,\omega)$$
$$= \int_{-\pi}^{\pi} \mathrm{d}k_1 \left((u_{k_1}u_{k_1+k} + v_{k_1}v_{k_1+k})^2 n_{k_1} (1 - n_{k_1+k}) \delta (\omega + \lambda_{k_1} - \lambda_{k_1+k}) \right.$$
$$\left. + (u_{k_1}v_{k_1+k} - v_{k_1}u_{k_1+k})^2 n_{k_1} (1 - n_{k_1+k+\pi}) \delta (\omega + \lambda_{k_1} - \lambda_{k_1+k+\pi}) \right) \tag{82}$$

(see also Refs. [35,54–56]).

Again for the xx and xy dynamic structure factors exact analytical results are rather scarce. In the high-temperature limit only the autocorrelation functions survive[41]

$$\langle s_j^x(t)s_j^x\rangle = \frac{1}{4}\Re Z_j(t), \quad \langle s_j^x(t)s_j^y\rangle = \frac{1}{4}\Im Z_j(t),$$
$$Z_j(t) = \frac{\Theta_1\left(J_+t, \frac{J_-}{J_+}\right)}{\Theta_1\left(0, \frac{J_-}{J_+}\right)} \frac{H_1\left(J_+t, \frac{J_-}{J_+}\right)}{H_1\left(0, \frac{J_-}{J_+}\right)}$$
$$\cdot \exp\left(-\mathrm{i}\Omega t - \left(1 - \frac{\mathcal{E}\left(\frac{J_-}{J_+}\right)}{\mathcal{K}\left(\frac{J_-}{J_+}\right)} \right) J_+^2 t^2 \right). \tag{83}$$

Here the Jacobian theta and eta functions are defined as follows

$$\Theta_1(u,k) = \sum_{n=-\infty}^{\infty} q^{n^2} \exp(2n\mathrm{i}z),$$
$$H_1(u,k) = \sum_{n=-\infty}^{\infty} q^{\left(n+\frac{1}{2}\right)^2} \exp((2n+1)\mathrm{i}z),$$
$$q = \exp\left(-\pi \frac{\mathcal{K}\left(\sqrt{1-k^2}\right)}{\mathcal{K}(k)} \right), \quad z = \frac{\pi u}{2\mathcal{K}(k)} \tag{84}$$

and the complete elliptic integrals of the 1st and the 2nd kinds are given by

$$\mathcal{K}(k) = \int_0^{\frac{\pi}{2}} \frac{\mathrm{d}\theta}{\sqrt{1-k^2\sin^2\theta}}, \quad \mathcal{E}(k) = \int_0^{\frac{\pi}{2}} \mathrm{d}\theta \sqrt{1-k^2\sin^2\theta}\,. \tag{85}$$

Moreover, $J_\pm^2 = J^2(1\pm\delta)^2/4$.

In the strong-field limit $|\Omega| > |J|$ at $T=0$ we can repeat the calculation of the previous section to find, for example, for the xx time-dependent correlation function and the corresponding dynamic structure factor the following result

$$4\langle s_j^x(t)s_{j+n}^x\rangle = \frac{1}{N}\sum_k \exp(\mathrm{i}kn)\left(u_k^2\exp(-\mathrm{i}\Lambda_k t)+v_k^2\exp(-\mathrm{i}\Lambda_{k+\pi}t)\right.$$
$$\left.-\mathrm{i}(-1)^{j+n}u_kv_k\left(\exp(-\mathrm{i}\Lambda_k t)-\exp(-\mathrm{i}\Lambda_{k+\pi}t)\right)\right), \quad (86)$$

$$S_{xx}(k,\omega) = \frac{\pi}{2}\left(u_k^2\delta(\omega-\Lambda_k)+v_k^2\delta(\omega-\Lambda_{k+\pi})\right). \quad (87)$$

For arbitrary values of temperature and transverse field the xx and xy dynamic structure can be obtained numerically.[54]

Let us discuss the dynamic quantities of the dimerized transverse XX chain. We begin with the transverse dynamic structure factor which can be written as (compare with Eq. (44))

$$S_{zz}(k,\omega) = \int_{-\pi}^{\pi}\mathrm{d}k_1\mathrm{d}k_2 C^{(1)}(k_1,k_2)$$
$$\cdot n_{k_1}(1-n_{k_2})\delta(\omega+\lambda_{k_1}-\lambda_{k_2})\delta_{k+k_1-k_2,0}$$
$$+\int_{-\pi}^{\pi}\mathrm{d}k_1\mathrm{d}k_2 C^{(2)}(k_1,k_2)$$
$$\cdot n_{k_1}(1-n_{k_2})\delta(\omega+\lambda_{k_1}-\lambda_{k_2})\delta_{k+k_1-k_2+\pi,0},$$
$$C^{(1)}(k_1,k_2) = \left(u_{k_1}u_{k_2}+v_{k_1}v_{k_2}\right)^2,$$
$$C^{(2)}(k_1,k_2) = \left(u_{k_1}v_{k_2}-v_{k_1}u_{k_2}\right)^2. \quad (88)$$

As can be seen from Eq. (88) the transverse dynamic structure probes two-fermion excitations. $S_{zz}(k,\omega)$ may have nonzero value within a restricted region of the k–ω plane when there is such a wave-vector k_1, $-\pi\le k_1<\pi$ that $\omega=-\lambda_{k_1}+\lambda_{k_1+k}$ or $\omega=-\lambda_{k_1}+\lambda_{k_1+k+\pi}$. Moreover, at zero temperature there are additional restrictions arising from the Fermi functions. The lines of potential singularities follow from the analysis of the equations $\mathrm{d}\lambda_{k_1}/\mathrm{d}k_1-\mathrm{d}\lambda_{k_1+k}/\mathrm{d}k_1=0$ and $\mathrm{d}\lambda_{k_1}/\mathrm{d}k_1-\mathrm{d}\lambda_{k_1+k+\pi}/\mathrm{d}k_1=0$. The characteristic lines in the k–ω plane which determine the behavior of the transverse dynamic structure factor were reported for the first time by J. H. Taylor and G. Müller.[35]

In Fig. 13 we show the region of the k–ω plane in which the two-fermion dynamic quantity may have nonzero values. In Fig. 14 we show the transverse dynamic structure factor at different temperatures. Comparing Fig. 13b and Fig. 14 one can see the van Hove singularities and the effects of the Fermi functions and the $C^{(1)}$- and $C^{(2)}$-functions. In Figs. 15, 16 we

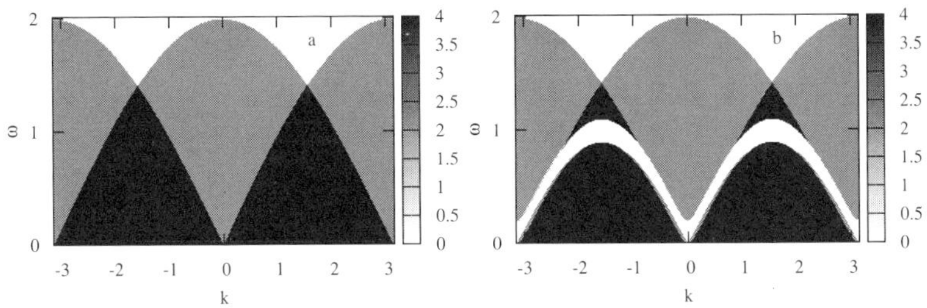

Fig. 13. Location of the roots of equations $\omega = -\lambda_{k_1} + \lambda_{k_1+k}$ and $\omega = -\lambda_{k_1} + \lambda_{k_1+k+\pi}$ ($-\pi \leq k_1 < \pi$) in the k–ω plane for $\delta = 0$ (panel a) and $\delta = 0.1$ (panel b): light region: no roots, light gray region: two roots, dark gray region: four roots.

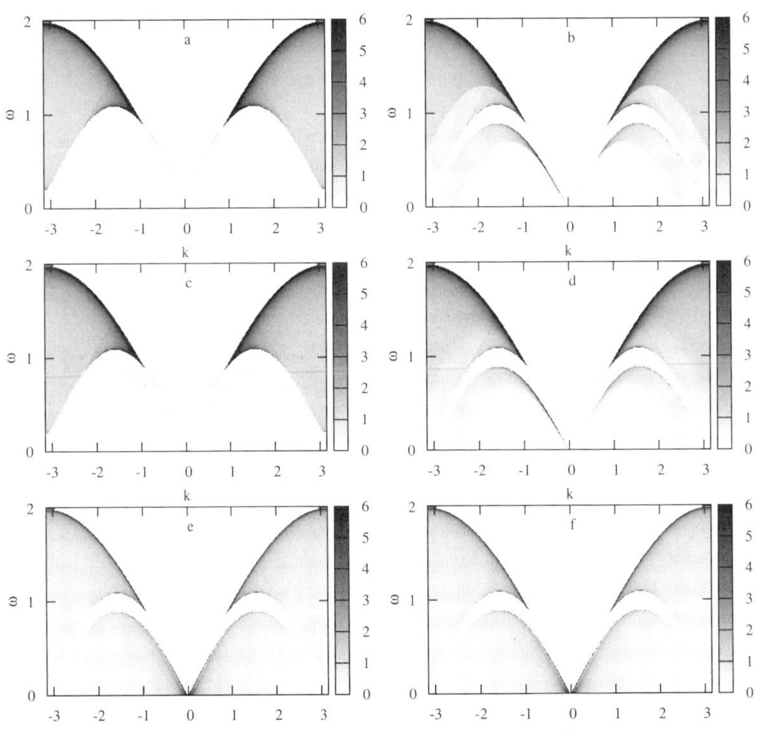

Fig. 14. $S_{zz}(k,\omega)$ (gray-scale plots) for the chain (81) with $J = -1$, $\delta = 0.1$ at different temperatures $\beta = \infty$ (a, b), $\beta = 20$ (c, d), $\beta = 1$ (e, f) for $\Omega = 0$ (left panels a, c, e) and $\Omega = 0.3$ (right panels b, d, f). Note that the results at $\beta = 1$ for $\Omega = 0$ and $\Omega = 0.3$ (panels e and f) are practically indistinguishable.

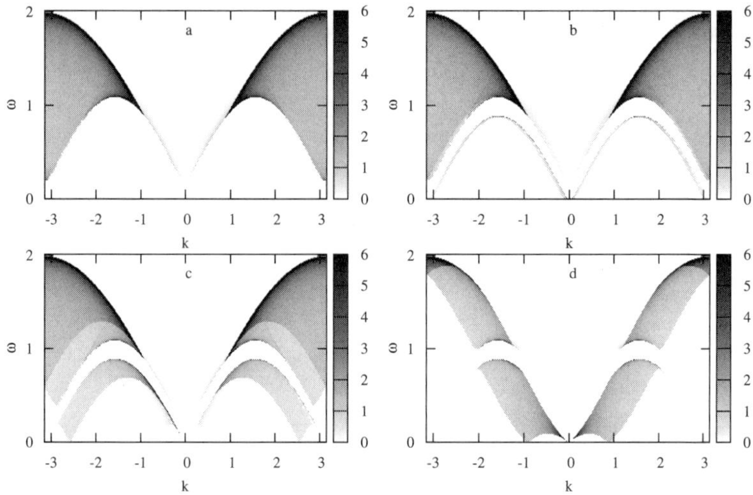

Fig. 15. $S_{zz}(k,\omega)$ (gray-scale plots) for the chain (81) with $J = -1$, $\delta = 0.1$ at zero temperature $\beta = \infty$ and different values of the transverse field $\Omega = 0.1$ (a), $\Omega = 0.11$ (b), $\Omega = 0.3$ (c) and $\Omega = 0.9$ (d).

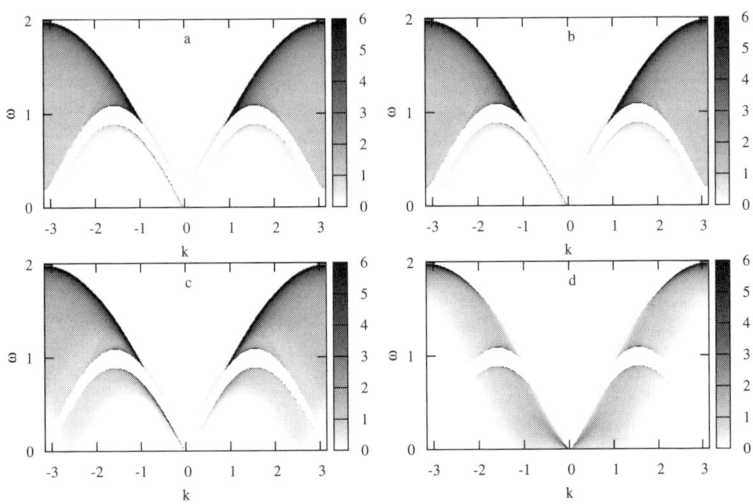

Fig. 16. The same as in Fig. 15 for $\beta = 20$.

show $S_{zz}(k,\omega)$ at various values of the transverse field for two temperatures $\beta = \infty$ and $\beta = 20$.

Next we pass to the xx dynamic structure factor obtained numerically. Typically we consider chains of $N = 400$ sites assume in (4) $n = 41$, l up to 50, consider t up to $t_c = 200$ and take $\epsilon = 0.001$.[54] In Fig. 17 we show

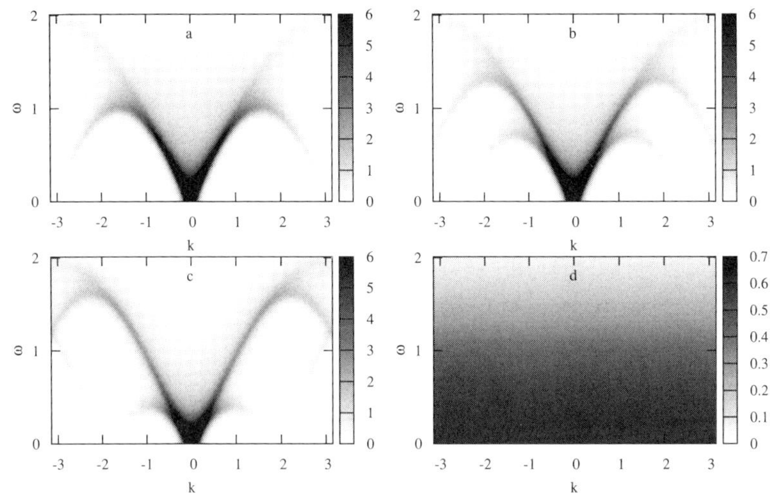

Fig. 17. $S_{xx}(k,\omega)$ (gray-scale plots) for the chain (81) with $J = -1$, $\delta = 0.1$ at low temperature $\beta = 20$ and for $\Omega = 0.1$ (a), $\Omega = 0.3$ (b), $\Omega = 0.6$ (c) and in the high-temperature limit $\beta = 0$ for $\Omega = 0.6$ (d).

$S_{xx}(k,\omega)$ of the dimerized transverse XX chain at low temperatures for different values of the transverse field.

In contrast to the zz dynamic structure factor which is a two-fermion dynamic quantity, the xx dynamic structure factor is a many-particle dynamic quantity within the Jordan-Wigner fermionization approach. Therefore, nonzero values of $S_{xx}(k,\omega)$ far above the two-fermion excitation continua may be expected. However, as can be seen in Fig. 17 the opposite is true: At low-temperatures $S_{xx}(k,\omega)$ shows several well-defined excitation branches which follow roughly the boundaries of the two-fermion excitation continua. Although we can describe the low-energy physics also using the bosonization treatment, high-frequency features cannot be reproduced within such an approach.

Finally we note that the dimerized XX chain does not show bound-state branches; within the fermionization approach this may be related to the absence of interactions between fermions. In contrast, a particle-hole bound state can be observed in the dimerized Heisenberg chain.[57]

6. Spin-1/2 XY chains with the Dzyaloshinskii-Moriya interaction

In this section we examine the effect of the Dzyaloshinskii-Moriya interaction (actually, the z component of the vector of the Dzyaloshinskii-Moriya interaction, see Eq. (18)). The Hamiltonian of the transverse XX chain with the Dzyaloshinskii-Moriya interaction reads

$$H=\sum_n\left(J\left(s_n^x s_{n+1}^x+s_n^y s_{n+1}^y\right)+D\left(s_n^x s_{n+1}^y-s_n^y s_{n+1}^x\right)\right)-h\sum_n s_n^z. \quad (89)$$

Interestingly, the Dzyaloshinskii-Moriya interaction can be eliminated from the Hamiltonian (89) resulting in renormalization of the isotropic XY exchange interaction.[58] To see this, consider the following spin axes rotation

$$\begin{aligned}
s_n^x &\to \tilde{s}_n^x = s_n^x\cos\phi_n+s_n^y\sin\phi_n,\\
s_n^y &\to \tilde{s}_n^y = -s_n^x\sin\phi_n+s_n^y\cos\phi_n,\\
s_n^z &\to \tilde{s}_n^z = s_n^z,\\
&\phi_n=(n-1)\varphi,\quad \tan\varphi=\frac{D}{J}.
\end{aligned} \quad (90)$$

After such a unitary transformation the Hamiltonian (89) becomes

$$\begin{aligned}
H&=\sum_n\tilde{J}\left(\tilde{s}_n^x\tilde{s}_{n+1}^x+\tilde{s}_n^y\tilde{s}_{n+1}^y\right)-h\sum_n\tilde{s}_n^z,\\
\tilde{J}&=\mathrm{sgn}(J)\sqrt{J^2+D^2}.
\end{aligned} \quad (91)$$

Using the unitary transformation (90) we can examine the effect of the Dzyaloshinskii-Moriya interaction using the dynamic quantities of the transverse XX chain without the Dzyaloshinskii-Moriya interaction discussed already in Sec. 4. First of all we note that the zz dynamic structure factor is given by Eq. (35), however, with $\Lambda_k=-h+\tilde{J}\cos k$. The formulas determining the two-fermion excitation continua boundaries are still given by Eqs. (37), (38), (39) but with $\tilde{J}$ instead of J on the l.h.s. of these equations and in the definition of the parameter α.

Exploiting (90) we find the following relations between the xx and xy dynamic structure factors of the model (89) (l.h.s. of Eq. (92)) and the dynamic structure factors of the model (91) (r.h.s. of Eq. (92))[59]

$$\begin{aligned}
S_{xx}(k,\omega)&=\frac{1}{2}\left(S_{xx}(k-\varphi,\omega)|_{\tilde{J}}+S_{xx}(k+\varphi,\omega)|_{\tilde{J}}\right.\\
&\quad\left.+\mathrm{i}\left(S_{xy}(k-\varphi,\omega)|_{\tilde{J}}-S_{xy}(k+\varphi,\omega)|_{\tilde{J}}\right)\right),\\
S_{xy}(k,\omega)&=\frac{1}{2}\left(S_{xy}(k-\varphi,\omega)|_{\tilde{J}}+S_{xy}(k+\varphi,\omega)|_{\tilde{J}}\right.\\
&\quad\left.-\mathrm{i}\left(S_{xx}(k-\varphi,\omega)|_{\tilde{J}}-S_{xx}(k+\varphi,\omega)|_{\tilde{J}}\right)\right).
\end{aligned} \quad (92)$$

Therefore, using Eq. (62) we obtain for the model (89)

$$S_{xx}(k,\omega) = \frac{\sqrt{\pi}}{4\tilde{J}}\left(\exp\left(-\frac{(\omega+h)^2}{\tilde{J}^2}\right) + \exp\left(-\frac{(\omega-h)^2}{\tilde{J}^2}\right)\right),$$

$$\mathrm{i}S_{xy}(k,\omega) = \frac{\sqrt{\pi}}{4\tilde{J}}\left(\exp\left(-\frac{(\omega+h)^2}{\tilde{J}^2}\right) - \exp\left(-\frac{(\omega-h)^2}{\tilde{J}^2}\right)\right), \quad (93)$$

i.e. in the high-temperature limit $S_{xx}(k,\omega)$ and $S_{xy}(k,\omega)$ are k-independent and display a single Gaussian ridge at $\omega = |h|$.

In the zero-temperature and strong-field limit ($T = 0$, $|h| > \sqrt{J^2 + D^2}$) according to (92) and (64) we find

$$\begin{aligned} S_{xx}(k,\omega) &= -\operatorname{sgn}(h)\mathrm{i}S_{xy}(k,\omega) \\ &= \frac{\pi}{2}\delta\left(\omega - |h| - \tilde{J}\cos\left(k + \operatorname{sgn}(h)\varphi\right)\right). \end{aligned} \quad (94)$$

For arbitrary values of temperature and transverse field we use Eq. (92) and numerical results for the xx and xy dynamic structure factors of the transverse XX chain (91) (see Sec. 4) to reveal the effect of the Dzyaloshinskii-Moriya interaction. Some of our findings are plotted in Fig. 18 where we show $S_{xx}(k,\omega)$ for $D = 0$ (left panels) and $D \neq 0$ (right panels) at different values of the transverse field h.

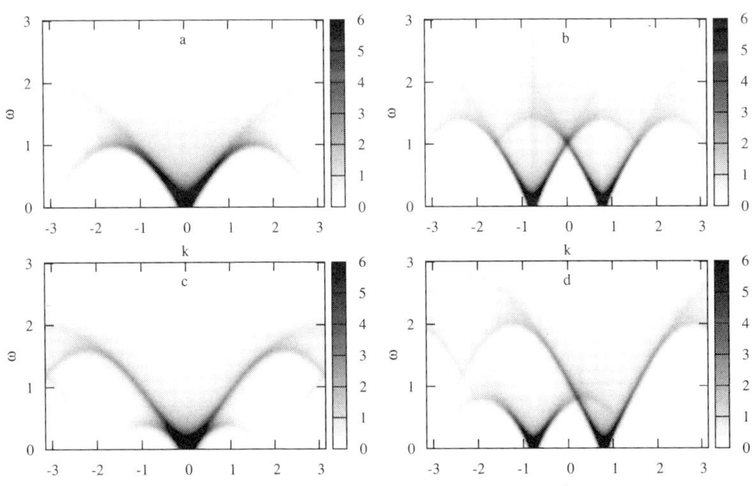

Fig. 18. $S_{xx}(k,\omega)$ (gray-scale plots) for the model (89) with $J = -1$, $D = 0$ (left panels a, c) and $D = 1$ (right panels b, d) for $h = 0$ (upper panels a, b) and $h = -0.6$ (lower panels c, d) at low temperature $\beta = 20$.

We recall that in the low-temperature limit when $J < 0$ and $D = 0$ $S_{xx}(k,\omega)$ and $S_{xy}(k,\omega)$ are concentrated in the k–ω plane along the curves (37), (38), (39) which determine the boundaries of the two-fermion excitation continuum $\omega_l(k)$, $\omega_m(k)$ and $\omega_u(k)$ (see Sec. 4, Figs. 10, 11). [For the antiferromagnetic sign of XX exchange interaction $J > 0$ these dynamic quantities are concentrated along the curves $\omega_l(k \pm \pi)$, $\omega_m(k \pm \pi)$ and $\omega_u(k \pm \pi)$ as it follows from simple symmetry arguments.] In the case when the Dzyaloshinskii-Moriya interaction is present, $D \neq 0$, the two-fermion excitation continuum splits into two continua (see Fig. 18), the 'left' one with the boundaries $\omega_l(k - \varphi)$, $\omega_m(k - \varphi)$ and $\omega_u(k - \varphi)$ and the 'right' one with the boundaries $\omega_l(k + \varphi)$, $\omega_m(k + \varphi)$ and $\omega_u(k + \varphi)$. (The 'left' and the 'right' continua are connected by symmetry operation.) The larger D is the larger is the splitting controlled by $\varphi = \arctan(D/J)$. At fixed $D \neq 0$ and $h = 0$ the spectral weight is equally distributed between the left and the right continua (panel b in Fig. 18). While $|h|$ increases from 0 to $\sqrt{J^2 + D^2}$ the spectral weight moves from one continuum to another continuum (panel d in Fig. 18).

We note in passing that the discussed peculiarities of the xx dynamic structure factor may be used for an unambiguous determination of the Dzyaloshinskii-Moriya interaction in chain compounds for example, in resonance experiments.[50,59–61]

In the case of the anisotropic XY chain the Dzyaloshinskii-Moriya interaction cannot be eliminated by the transformation (90). Now we face the Hamiltonian

$$\begin{aligned} H &= \sum_n \left(J^x s_n^x s_{n+1}^x + J^y s_n^y s_{n+1}^y + D\left(s_n^x s_{n+1}^y - s_n^y s_{n+1}^x\right)\right) + \Omega \sum_n s_n^z \\ &\to \sum_n \left(\frac{J + \mathrm{i}D}{2} c_n^\dagger c_{n+1} - \frac{J - \mathrm{i}D}{2} c_n c_{n+1}^\dagger + \frac{\gamma}{2}\left(c_n^\dagger c_{n+1}^\dagger - c_n c_{n+1}\right)\right. \\ &\quad \left. + \Omega\left(c_n^\dagger c_n - \frac{1}{2}\right)\right) \end{aligned} \tag{95}$$

with $J = (J^x + J^y)/2$ and $\gamma = (J^x - J^y)/2$. This Hamiltonian can be put into a diagonal form by performing the Fourier transformation, $c_k = \left(1/\sqrt{N}\right)\sum_n \exp(\mathrm{i}kn)\, c_n$, $c_n = \left(1/\sqrt{N}\right)\sum_k \exp(-\mathrm{i}kn)\, c_k$, and the Bogolyubov transformation, $c_k = -\mathrm{i}u_k\beta_k + v_k\beta_{-k}^\dagger$, $\beta_k = \mathrm{i}u_k c_k + v_k c_{-k}^\dagger$, $u_k = \mathrm{sgn}(\gamma \sin k)\left(1/\sqrt{2}\right)\sqrt{1 + (\Omega + J\cos k)/\lambda_k}$, $v_k = \left(1/\sqrt{2}\right)\sqrt{1 - (\Omega + J\cos k)/\lambda_k}$, $\lambda_k = \sqrt{(\Omega + J\cos k)^2 + \gamma^2 \sin^2 k}$. The final result reads $H = \sum_k \Lambda_k \left(\beta_k^\dagger \beta_k - 1/2\right)$, $\Lambda_k = D\sin k + \lambda_k$. We notice that the elementary excitation energy spectrum is gapless when $\Omega^2 \leq J^2 + D^2 - \gamma^2$ and $\gamma^2 \leq D^2$ or when $\Omega^2 = J^2$ and $\gamma^2 > D^2$.

The calculation of the transverse dynamic structure factor follows the lines explained in some detail in Sec. 4 and ends up with[62]

$$S_{zz}(k,\omega) = \sum_{j=1}^{3} S_{zz}^{(j)}(k,\omega),$$

$$S_{zz}^{(j)}(k,\omega) = \int_{-\pi}^{\pi} \mathrm{d}k_1 B^{(j)}(k_1,k) C^{(j)}(k_1,k) \delta\left(\omega - E^{(j)}(k_1,k)\right),$$

$$B^{(1)}(k_1,k) = B^{(3)}(k_1,k) = \frac{1 - f(k_1,k)}{4}, \quad B^{(2)}(k_1,k) = \frac{1 + f(k_1,k)}{2},$$

$$f(k_1,k) = \frac{\left(\Omega + J\cos\left(k_1 - \frac{k}{2}\right)\right)\left(\Omega + J\cos\left(k_1 + \frac{k}{2}\right)\right) - \gamma^2 \sin\left(k_1 - \frac{k}{2}\right)\sin\left(k_1 + \frac{k}{2}\right)}{\lambda_{k_1 - \frac{k}{2}}\lambda_{k_1 + \frac{k}{2}}},$$

$$\begin{aligned}
C^{(1)}(k_1,k) &= \left(1 - n_{k_1 + \frac{k}{2}}\right)\left(1 - n_{-k_1 + \frac{k}{2}}\right), \\
C^{(2)}(k_1,k) &= \left(1 - n_{k_1 + \frac{k}{2}}\right) n_{k_1 - \frac{k}{2}}, \\
C^{(3)}(k_1,k) &= n_{k_1 - \frac{k}{2}} n_{-k_1 - \frac{k}{2}}, \\
E^{(1)}(k_1,k) &= \Lambda_{k_1 + \frac{k}{2}} + \Lambda_{-k_1 + \frac{k}{2}}, \\
E^{(2)}(k_1,k) &= \Lambda_{k_1 + \frac{k}{2}} - \Lambda_{k_1 - \frac{k}{2}}, \\
E^{(3)}(k_1,k) &= -\Lambda_{k_1 - \frac{k}{2}} - \Lambda_{-k_1 - \frac{k}{2}}.
\end{aligned} \tag{96}$$

The transverse dynamic factor, as it follows from Eq. (96), is shown in panel c in Fig. 19 for a typical set of parameters.

From Eq. (96) we see that the transverse dynamic structure factor is governed by three two-fermion excitation continua. Let us discuss some properties of these continua (see Fig. 20 where we show two-fermion excitation continua for a specific set of parameters $J = -1$, $\gamma = 0.5$, $D = 1$, $\Omega = 0.5$). We begin with the high-temperature limit when the Fermi factors are not essential (left panels in Fig. 20). The two-fermion dynamic structure factor may have nonzero values in the k–ω plane if the equation $\omega - E^{(j)}(k_1,k) = 0$ has at least one solution $k_1^\star$, $-\pi \leq k_1^\star < \pi$. Next, the

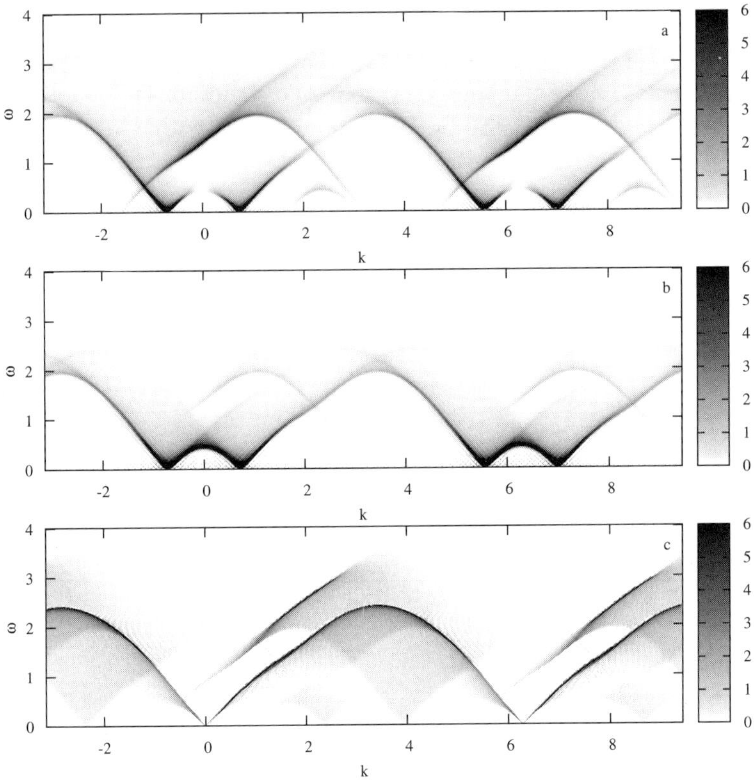

Fig. 19. $S_{xx}(k,\omega)$ (a), $S_{yy}(k,\omega)$ (b), $S_{zz}(k,\omega)$ (c) for the spin chain (95) with $J=-1$, $\gamma=0.5$, $D=1$, $\Omega=0.5$ at low temperature $\beta=50$. Note that these quantities are shown for k that varies from $-\pi$ to 3π.

lower and the upper boundaries are given by

$$\omega_l^{(j)}(k) = \min_{-\pi\le k_1<\pi}\left\{0, E^{(j)}(k_1,k)\right\},$$
$$\omega_u^{(j)}(k) = \max_{-\pi\le k_1<\pi}\left\{E^{(j)}(k_1,k)\right\}. \tag{97}$$

The two-fermion dynamic quantities may exhibit van Hove singularities along the line $\omega_s^{(j)}(k)=E^{(j)}(k_1,k)$ where k_1 satisfies the equation

$$\frac{\partial}{\partial k_1}E^{(j)}(k_1,k)=0. \tag{98}$$

If for the solution of Eq. (98) we also have $\partial^2E^{(j)}(k_1,k)/\partial k_1^2\neq 0$ the two-fermion dynamic structure factor shows the well-known square-root

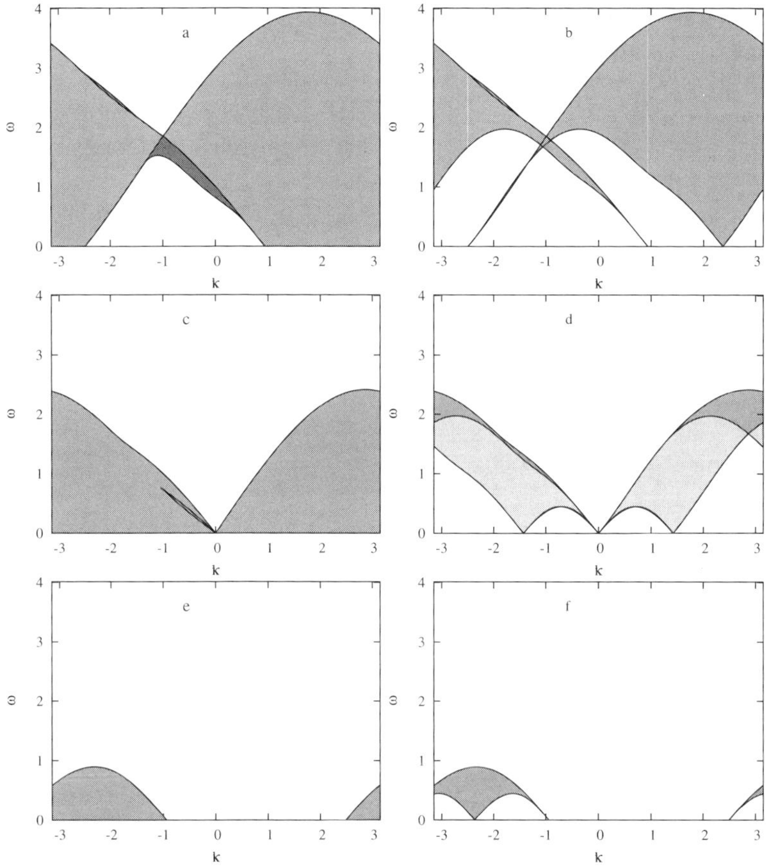

Fig. 20. Two-fermion excitation continua ($j = 1$ (a, b), $j = 2$ (c, d), $j = 3$ (e, f)) for $J = -1$, $\gamma = 0.5$, $D = 1$, $\Omega = 0.5$. Left panels: $T \to \infty$; right panels: $T = 0$.

singularity. However, it may happen that $\partial^2 E^{(j)}(k_1, k)/\partial k_1^2 = 0$ but $\partial^3 E^{(j)}(k_1, k)/\partial k_1^3 \neq 0$. Then a van Hove singularity is characterized by the exponent 2/3. That is really the case, for example, for $J = 1$, $\gamma = 0.5$, $D = 1$, $\Omega = 0.5$ for $j = 2$ at $k = 1.0784\ldots$. We have $\partial E^{(2)}(k_1, k)/\partial k_1 = \partial^2 E^{(2)}(k_1, k)/\partial k_1^2 = 0$, $\partial^3 E^{(2)}(k_1, k)/\partial k_1^3 \neq 0$ at $k_1 = k_1^\star = 2.1648\ldots$. Therefore in the ϵ-vicinity of $\omega = 0.7859\ldots$ the two-fermion dynamic structure factor should be proportional to $|\epsilon|^{-2/3}$. We mention that the singularity with this exponent is also present for $D = 0$.[62]

In the zero-temperature case the effect of the Fermi functions involved in the $C^{(j)}$-functions (see Eq. (96)) becomes important (right panels in

Fig. 20). As a result the region of possible values of k_1 is contracted. Further details can be found in Ref. [62].

Finally, we mention the role of the $B^{(j)}$-functions (see Eq. (96)) for the two-fermion dynamic structure factors which are responsible for some specific features of the transverse dynamic structure factor (compare panels b, d, f in Fig. 20 and panel c in Fig. 19).

We pass to the xx and yy dynamic structure factors (see panels a and b in Fig. 19). These dynamic structure factors are many-fermion dynamic quantities and although they are not restricted to some region in the k–ω plane, they are rather small outside the two-fermion excitation continua (compare panels a and b with panel c in Fig. 19). In the low-temperature limit the xx and yy dynamic structure factors show several washed-out excitation branches which are in correspondence with characteristic lines of the two-fermion excitation continua. In the high-temperature limit $S_{xx}(k,\omega)$ and $S_{yy}(k,\omega)$ become k-independent.

It should be stressed that the constant frequency or constant wave-vector scans of the dynamic structure factors clearly manifest the presence of the Dzyaloshinskii-Moriya interaction and some easily recognized features of these quantities can be used for determining the Dzyaloshinskii-Moriya interaction.

7. Square-lattice spin-1/2 isotropic XY model

Let us discuss what kind of results for spin models can be obtained in two dimensions after applying the Jordan-Wigner transformation (Sec. 3). We consider the spin-1/2 isotropic XY model on a spatially anisotropic square lattice with the Hamiltonian

$$H = \sum_{i=0}^{\infty}\sum_{j=0}^{\infty}\left(\frac{J}{2}\left(s_{i,j}^{+}s_{i+1,j}^{-}+s_{i,j}^{-}s_{i+1,j}^{+}\right)\right.$$
$$\left.+\frac{J_{\perp}}{2}\left(s_{i,j}^{+}s_{i,j+1}^{-}+s_{i,j}^{-}s_{i,j+1}^{+}\right)\right), \tag{99}$$

where J and $J_{\perp}$ are the XX exchange interactions in the horizontal and vertical directions. Our aim is to calculate the transverse dynamic structure factor

$$S_{zz}(\mathbf{k},\omega) = \sum_{p=0}^{\infty}\sum_{q=0}^{\infty}\exp\left(\mathrm{i}\left(k_x p + k_y q\right)\right)\int_{-\infty}^{\infty}\mathrm{d}t\exp\left(\mathrm{i}\omega t\right)$$
$$\cdot\left(\langle s_{n,m}^{z}(t)s_{n+p,m+q}^{z}\rangle - \langle s_{n,m}^{z}\rangle\langle s_{n+p,m+q}^{z}\rangle\right). \tag{100}$$

We notice that the transverse dynamic structure factor $S_{zz}(\mathbf{k},\omega)$ (100) for the spin model (99) is related to the density-density dynamic structure factor of hard-core bosons on a square-lattice.[63]

We apply the two-dimensional Jordan-Wigner transformation (21), (23) to the spin Hamiltonian (99). Moreover, we adopt the mean-field approach for the phase factors and change the gauge leaving the flux per elementary plaquette Φ_0 to be equal to π. As a result we arrive at the Hamiltonian like (27), i.e.

$$H = \sum_{i=0}^{\infty}\sum_{j=0}^{\infty}\left(\frac{J}{2}(-1)^{i+j}\left(d_{i,j}^{\dagger}d_{i+1,j} - d_{i,j}d_{i+1,j}^{\dagger}\right) + \frac{J_{\perp}}{2}\left(d_{i,j}^{\dagger}d_{i,j+1} - d_{i,j}d_{i,j+1}^{\dagger}\right)\right). \tag{101}$$

The Hamiltonian (101) contains the correct results in the one-dimensional limit when either $J_{\perp}=0$ or $J=0$ (in the former case to recover the one-dimensional Hamiltonian (16) (with $\Omega=0$) one has to perform in addition a gauge transformation $d_{i,j}^{\dagger} = \exp(\mathrm{i}\pi\psi_i)\,f_{i,j}^{\dagger}$, $\psi_0=0$, $\psi_{i+1}=\psi_i+i$). The Hamiltonian (101) can be diagonalized by performing 1) the Fourier transformation, $d_{i,j} = \left(1/\sqrt{N_xN_y}\right)\sum_{k_x,k_y}\exp\left(\mathrm{i}\left(k_xi+k_yj\right)\right)d_{k_x,k_y}$, $k_\alpha = 2\pi n_\alpha/N_\alpha$, $n_\alpha = -N_\alpha/2, -N_\alpha/2+1,\ldots,N_\alpha/2-1$, $\alpha=x,y$, $N_x=N_y=\sqrt{N}\to\infty$ is even, which yields

$$\begin{aligned}
H &= \frac{1}{2}\sum_{\mathbf{k}}|E_{\mathbf{k}}|\left(\cos\gamma_{\mathbf{k}}\left(b_{\mathbf{k}}^{\dagger}b_{\mathbf{k}} - a_{\mathbf{k}}^{\dagger}a_{\mathbf{k}}\right) + \mathrm{i}\sin\gamma_{\mathbf{k}}\left(b_{\mathbf{k}}^{\dagger}a_{\mathbf{k}} - a_{\mathbf{k}}^{\dagger}b_{\mathbf{k}}\right)\right),\\
|E_{\mathbf{k}}| &= \sqrt{J_{\perp}^2\cos^2 k_y + J^2\sin^2 k_x},\\
\cos\gamma_{\mathbf{k}} &= \frac{J_{\perp}\cos k_y}{|E_{\mathbf{k}}|},\quad \sin\gamma_{\mathbf{k}} = \frac{J\sin k_x}{|E_{\mathbf{k}}|}
\end{aligned} \tag{102}$$

with $b_{\mathbf{k}} = d_{k_x,k_y}$ and $a_{\mathbf{k}} = d_{k_x\pm\pi,k_y\pm\pi}$ and 2) the Bogolyubov transformation, $\alpha_{\mathbf{k}} = \cos(\gamma_{\mathbf{k}}/2)\,b_{\mathbf{k}} + \mathrm{i}\sin(\gamma_{\mathbf{k}}/2)\,a_{\mathbf{k}}$, $\beta_{\mathbf{k}} = \sin(\gamma_{\mathbf{k}}/2)\,b_{\mathbf{k}} - \mathrm{i}\cos(\gamma_{\mathbf{k}}/2)\,a_{\mathbf{k}}$, which yields

$$\begin{aligned}
H &= {\sum_{\mathbf{k}}}^{\prime}\Lambda_{\mathbf{k}}\left(\alpha_{\mathbf{k}}^{\dagger}\alpha_{\mathbf{k}} - \beta_{\mathbf{k}}^{\dagger}\beta_{\mathbf{k}}\right),\\
\Lambda_{\mathbf{k}} &= |E_{\mathbf{k}}| = \sqrt{J_{\perp}^2\cos^2 k_y + J^2\sin^2 k_x} \geq 0
\end{aligned} \tag{103}$$

(the prime near the sum in Eq. (103) means that $\mathbf{k}$ varies in the thermodynamic limit in the region $-\pi\leq k_y\leq\pi$, $-\pi+|k_y|\leq k_x\leq\pi-|k_y|$).

The calculation of the transverse dynamic structure factor repeats the steps elaborated in some detail for the one-dimensional case. First, we use the Wick-Bloch-de Dominicis theorem to obtain the zz time-dependent spin correlation function

$$
\begin{aligned}
&\langle s^z_{n,m}(t)s^z_{n+p,m+q}\rangle - \langle s^z_{n,m}\rangle\langle s^z_{n+p,m+q}\rangle \\
&\qquad = \langle d^\dagger_{n,m}(t)d_{n+p,m+q}\rangle\langle d_{n,m}(t)d^\dagger_{n+p,m+q}\rangle,
\end{aligned}
$$

$$
\begin{aligned}
&\langle d^\dagger_{n,m}(t)d_{n+p,m+q}\rangle = \frac{1}{2N}\sum_{\mathbf{k}}\exp\left(\mathrm{i}\left(k_x p + k_y q\right)\right) \\
&\cdot\left(\left(\cos^2\frac{\gamma_{\mathbf{k}}}{2} - \mathrm{i}\left(-1\right)^{n+m}\left(\left(-1\right)^{p+q}-1\right)\cos\frac{\gamma_{\mathbf{k}}}{2}\sin\frac{\gamma_{\mathbf{k}}}{2} + \left(-1\right)^{p+q}\sin^2\frac{\gamma_{\mathbf{k}}}{2}\right)\right. \\
&\cdot n_{\mathbf{k}}\exp\left(\mathrm{i}\Lambda_{\mathbf{k}}t\right) \\
&+\left(\sin^2\frac{\gamma_{\mathbf{k}}}{2} + \mathrm{i}\left(-1\right)^{n+m}\left(\left(-1\right)^{p+q}-1\right)\cos\frac{\gamma_{\mathbf{k}}}{2}\sin\frac{\gamma_{\mathbf{k}}}{2} + \left(-1\right)^{p+q}\cos^2\frac{\gamma_{\mathbf{k}}}{2}\right) \\
&\left.\cdot\left(1-n_{\mathbf{k}}\right)\exp\left(-\mathrm{i}\Lambda_{\mathbf{k}}t\right)\right), \\
&\langle d_{n,m}(t)d^\dagger_{n+p,m+q}\rangle = \frac{1}{2N}\sum_{\mathbf{k}}\exp\left(-\mathrm{i}\left(k_x p + k_y q\right)\right) \\
&\cdot\left(\left(\cos^2\frac{\gamma_{\mathbf{k}}}{2} + \mathrm{i}\left(-1\right)^{n+m}\left(\left(-1\right)^{p+q}-1\right)\cos\frac{\gamma_{\mathbf{k}}}{2}\sin\frac{\gamma_{\mathbf{k}}}{2} + \left(-1\right)^{p+q}\sin^2\frac{\gamma_{\mathbf{k}}}{2}\right)\right. \\
&\cdot\left(1-n_{\mathbf{k}}\right)\exp\left(-\mathrm{i}\Lambda_{\mathbf{k}}t\right) \\
&+\left(\sin^2\frac{\gamma_{\mathbf{k}}}{2} - \mathrm{i}\left(-1\right)^{n+m}\left(\left(-1\right)^{p+q}-1\right)\cos\frac{\gamma_{\mathbf{k}}}{2}\sin\frac{\gamma_{\mathbf{k}}}{2} + \left(-1\right)^{p+q}\cos^2\frac{\gamma_{\mathbf{k}}}{2}\right) \\
&\left.\cdot n_{\mathbf{k}}\exp\left(\mathrm{i}\Lambda_{\mathbf{k}}t\right)\right). \qquad (104)
\end{aligned}
$$

Then we plug Eq. (104) into Eq. (100) to obtain the following expression for the zz dynamic structure factor

$$
\begin{aligned}
S_{zz}(\mathbf{k},\omega) = \pi &\int_{-\pi}^{\pi}\frac{\mathrm{d}k_{1y}}{2\pi}\int_{-\pi}^{\pi}\frac{\mathrm{d}k_{1x}}{2\pi} \\
&\cdot\left(\cos^2\frac{\gamma_{\mathbf{k}_1+\mathbf{k}}-\gamma_{\mathbf{k}_1}}{2}\left(n_{\mathbf{k}_1}\left(1-n_{\mathbf{k}_1+\mathbf{k}}\right)\delta\left(\omega+\Lambda_{\mathbf{k}_1}-\Lambda_{\mathbf{k}_1+\mathbf{k}}\right)\right.\right. \\
&+\left.\left(1-n_{\mathbf{k}_1}\right)n_{\mathbf{k}_1+\mathbf{k}}\delta\left(\omega-\Lambda_{\mathbf{k}_1}+\Lambda_{\mathbf{k}_1+\mathbf{k}}\right)\right) \\
&+\sin^2\frac{\gamma_{\mathbf{k}_1+\mathbf{k}}-\gamma_{\mathbf{k}_1}}{2}\left(n_{\mathbf{k}_1}n_{\mathbf{k}_1+\mathbf{k}}\delta\left(\omega+\Lambda_{\mathbf{k}_1}+\Lambda_{\mathbf{k}_1+\mathbf{k}}\right)\right. \\
&+\left.\left.\left(1-n_{\mathbf{k}_1}\right)\left(1-n_{\mathbf{k}_1+\mathbf{k}}\right)\delta\left(\omega-\Lambda_{\mathbf{k}_1}-\Lambda_{\mathbf{k}_1+\mathbf{k}}\right)\right)\right). \qquad (105)
\end{aligned}
$$

One can easily convince oneself that Eq. (105) contains the correct result in the one-dimensional limit (35) (with $\Omega = 0$) when either $J_\perp = 0$ or $J = 0$. In the two-dimensional case Eq. (105) is an approximate formula for the transverse dynamic structure factor of the spin-1/2 isotropic XY model on a spatially anisotropic square lattice.

Let us discuss the dynamic quantity obtained in some detail.[64] In Fig. 21 we show gray-scale plots for $S_{zz}(\mathbf{k},\omega)$ and in Fig. 22 we show frequency

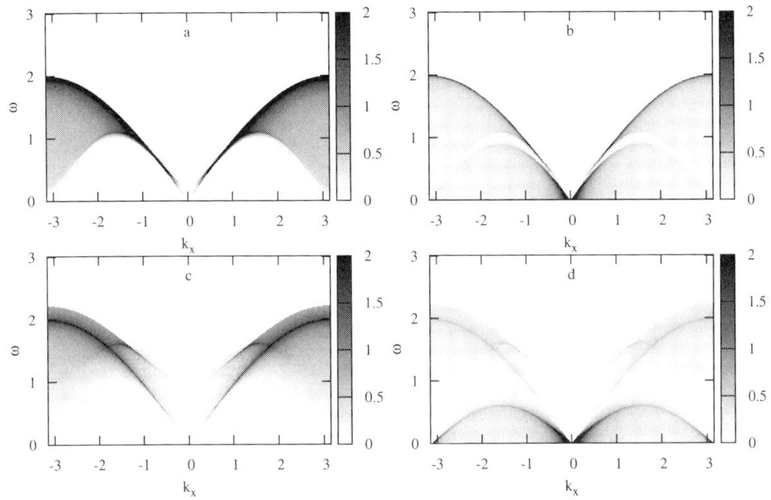

Fig. 21. The zz dynamic structure factor $S_{zz}(k_x, 0, \omega)$ (gray-scale plots) for the square-lattice $s = 1/2$ XX model (99) as it follows from Eq. (105) at $T = 0$ (left column) and at $T = 10$ (right column). $J = -1$, $J_\perp = -0.1$ (a, b), $J_\perp = -0.5$ (c, d).

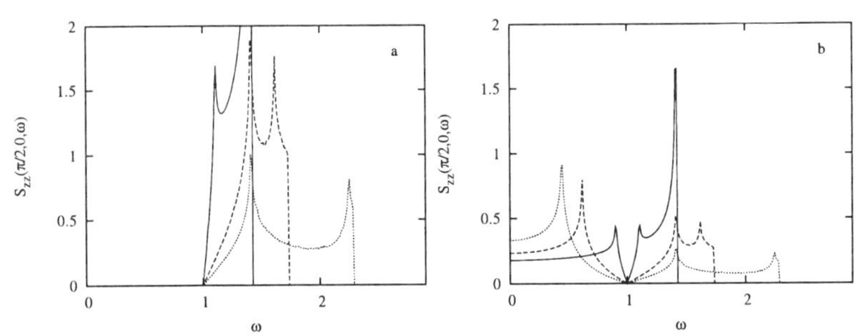

Fig. 22. Frequency dependence of the zz dynamic structure factor (105) for momentum transfer along the chain $k_x = \pi/2$ as the interchain interaction changes ($J = -1$, $J_\perp = -0.1$ (solid curves), $J = -1$, $J_\perp = -0.5$ (dashed curves), $J = -1$, $J_\perp = -0.9$ (dotted curves)) at zero temperature $T = 0$ (a) and high temperature $T = 10$ (b).

profiles of $S_{zz}(\mathbf{k},\omega)$ for a representative set of parameters. Formula (105) implies the interpretation of $S_{zz}(\mathbf{k},\omega)$ as a two-fermion excitation quantity. As can be seen from Figs. 21, 22 $S_{zz}(\mathbf{k},\omega)$ exhibits several washed-out excitation branches which can be generated by two spinless fermions in accordance with (105). We begin with the low-temperature limit when only the fourth term in Eq. (105) (which contains $(1-n_{\mathbf{k}_1})(1-n_{\mathbf{k}_1+\mathbf{k}})$) survives. Consider, for example, two fermions with $\mathbf{k}_1=(0,\pi/2)-\mathbf{k}/2$ and $\mathbf{k}_1+\mathbf{k}=(0,\pi/2)+\mathbf{k}/2$ with the energy of the pair

$$\omega_{\mathbf{k}}=2\sqrt{J^2\sin^2\frac{k_x}{2}+J_\perp^2\sin^2\frac{k_y}{2}} \tag{106}$$

or two fermions with $\mathbf{k}_1=(\pi/2,0)-\mathbf{k}/2$ and $\mathbf{k}_1+\mathbf{k}=(\pi/2,0)+\mathbf{k}/2$ with the energy of the pair

$$\omega_{\mathbf{k}}=2\sqrt{J^2\cos^2\frac{k_x}{2}+J_\perp^2\cos^2\frac{k_y}{2}}. \tag{107}$$

These modes are the well-known spin waves[65] clearly present at low temperatures (panels a and c in Fig. 21). Further, one can recognize the high frequency modes $[\mathbf{k}_1=-\mathbf{k}/2,\ \mathbf{k}_1=(\pi/2,\pi/2)-\mathbf{k}/2]$ with the dispersion relations

$$\omega_{\mathbf{k}}=2\sqrt{J^2\sin^2\frac{k_x}{2}+J_\perp^2\cos^2\frac{k_y}{2}}, \tag{108}$$

$$\omega_{\mathbf{k}}=2\sqrt{J^2\cos^2\frac{k_x}{2}+J_\perp^2\sin^2\frac{k_y}{2}}. \tag{109}$$

Another set of high-frequency modes $[\mathbf{k}_1=0,\ \mathbf{k}_1=(\pi/2,\pi/2)]$ have the dispersion relations

$$\omega_{\mathbf{k}}=J_\perp+\sqrt{J^2\sin^2k_x+J_\perp^2\cos^2k_y}, \tag{110}$$

$$\omega_{\mathbf{k}}=J+\sqrt{J^2\cos^2k_x+J_\perp^2\sin^2\,k_y}. \tag{111}$$

The low-frequency mode $[\mathbf{k}_1=(0,\pi/2)]$ with the dispersion relation

$$\omega_{\mathbf{k}}=\sqrt{J^2\sin^2k_x+J_\perp^2\sin^2k_y} \tag{112}$$

(it is composed of two fermions, the energy of one of which equals zero) forms the low-frequency cutoff at zero temperature. Comparing left and right panels in Figs. 21 and 22 one can also see the modes which become visible only as temperature increases (at zero temperature they are forbidden because of the Fermi factors in Eq. (105)). Putting $k_{1x}=-k_x$,

$k_{1y} = \pi/2 - k_x$ for $k_y = 0$ and $k_{1x} = k_y$, $k_{1y} = \pi/2 - k_y$ for $k_x = 0$ we get

$$\omega_{k_x,0} = \left(\sqrt{J^2 + J_\perp^2} - J_\perp\right) |\sin k_x|,$$
$$\omega_{0,k_y} = \left(\sqrt{J^2 + J_\perp^2} - J\right) |\sin k_y|. \tag{113}$$

This excitation branch contains most of the spectral weight at high temperatures (see panel d in Fig. 21 and panel b in Fig. 22).

The established modes (106) – (113) manifest themselves as peaks, cusps or cutoffs in the frequency or wave-vector profiles of $S_{zz}(\mathbf{k},\omega)$. The frequency profiles depicted in Fig. 22 may be almost symmetric or asymmetric, they may resemble δ-peaks or result from two coalesced peaks, they may gradually disappear or may be abruptly cut off.

It is worthwhile to mention here some experimental studies on dynamic properties of two-dimensional quantum spin models, in particular, the neutron scattering experiments on Cs_2CuCl_4[66] (for a theory of dynamic correlations in the spin-liquid phase in Cs_2CuCl_4 see Ref. [67]). Cs_2CuCl_4 is a two-dimensional low-exchange quantum magnet. It has a layered crystal structure; in each layer the exchange paths form a triangle lattice with nonequivalent interactions along chains $J = 0.374(5)$ meV and along zig-zag bonds $J' = 0.34(3)J$. The interlayer coupling is small $J'' = 0.045(5)J$ and it stabilizes the long-range magnetic order below $T_{\rm N} = 0.62(1)$ K. The neutron scattering measurements in the spin-liquid phase (i.e. above $T_{\rm N}$ but below J, J' when the two-dimensional magnetic layers are decoupled) clearly indicate that the dynamic correlations are dominated by highly dispersive excitation continua which is a characteristic signature of fractionalization of spin-1 spin waves into pairs of deconfined spin-1/2 spinons. Linear spin-wave theory including one- and two-magnon processes cannot describe the continuum scattering. The proposed theories[67] are based either on a quasi-one-dimensional approach (that immediately introduces spinon language) or on the explicitly two-dimensional resonating-valence-bond picture.

As a final remark we recall that Eq. (105) contains the exact result (35) in the one-dimensional limit. On the other hand, Eq. (105) gives an approximate result in the two-dimensional case because of the mean-field description of the phase factors which arise after fermionization. The adopted mean-field treatment neglects a complicated interaction between spinless fermions. In the case of the XXZ Heisenberg model the interaction between fermions is present even within the adopted mean-field procedure due to the interaction between z spin components. The quartic terms in the fermionic Hamiltonian may be treated after making further approximation (see references cited in Ref. [29] and also Ref. [68]).

8. Conclusions

The Jordan-Wigner transformation which realizes a spin-to-fermion mapping was suggested as a rigorous framework for the description of spin-1/2 XY chains in the early 1960s. In general, the Jordan-Wigner fermionization permits to map a system of interacting spins $s = 1/2$ onto a system of spinless fermions. It may happen that the spinless fermions are noninteracting. In this case this approach reveals an exactly solvable spin model. However, even for exactly solvable spin models not all 'simple' quantities of interest in spin language remain simple in fermionic language. For example, the z spin component attached to the site j, s_j^z, becomes the product of two Fermi operators attached to this site, $c_j^\dagger c_j - 1/2$. In contrast, the local spin operators s_j^x, s_j^y, $s_j^\pm$ become nonlocal objects in fermionic description involving a string of sites $1, 2, \ldots, j$ (see Eqs. (12), (13)). This leads to some complications in studying the dynamics of fluctuations of these operators: the dynamics of fluctuations of operators which seem to be rather simple in spin language may be governed by many-particle correlations in fermionic language. As we have discussed in sections 4, 5, 6, the Jordan-Wigner fermionization approach permits to establish a number of rigorous results for the dynamics of spin-1/2 XY chains. Especially easy are the cases of two- and four-fermion dynamic quantities which are amenable mostly to analytical calculations. The case of many-fermion dynamic quantities is more complicated, however, these quantities can be examined numerically at very high precision.

For more realistic spin-1/2 XXZ Heisenberg chains the Jordan-Wigner fermionization approach leads to a system of interacting spinless fermions. The simplest way to proceed in this case is to apply Hartree-Fock-like approximations.[17] If we are interested in low-energy physics only it might be helpful to apply the bosonization approach.[18–20]

The results for one-dimensional quantum spin systems obtained using the Jordan-Wigner fermionization can be compared with the outcomes of alternative approaches: field-theoretic bosonization techniques[18–20] valid in the low-energy limit (see Fig. 12), Bethe ansatz calculations (for calculation of dynamic structure factors of spin-1/2 XXZ chains see Refs. [69,70]) or exact diagonalization computations which, however, are restricted to small finite systems.

For two-dimensional quantum spin models achievements are rather modest. In this case the Jordan-Wigner fermionization approach can provide an approximate theory; the simplest one treats in the mean-field spirit the phase factors which arise after fermionization.

We end up with a brief comment about the experimental relevance of some of the dynamic quantities calculated. They may be used for interpretation of the energy absorption in electron spin resonance (ESR) experiments.[60] Consider an ESR experiment in which the static magnetic field is directed along the z axis and the electromagnetic wave with the polarization in $\alpha \perp z$ direction (say $\alpha = x$) are applied to a magnetic system which is described as a spin-1/2 XY chain (ESR experiment in the standard Faraday configuration). In such an ESR experiment one measures the intensity of the radiation absorption $I(\omega)$ as a function of frequency $\omega > 0$ of the electromagnetic wave. Within the linear response theory the absorption intensity is written as

$$I(\omega) \propto \omega \Im \chi_{\alpha\alpha}(0, \omega), \tag{114}$$

where $\Im \chi_{\alpha\alpha}(0, \omega)$ is the imaginary part of the $\alpha\alpha$ component of the dynamic susceptibility $\chi_{\alpha\alpha}(k, \omega)$ at zero wave-vector $k = 0$. We notice that

$$\Im \chi_{\alpha\alpha}(0, \omega) = \frac{1 - \exp(-\beta\omega)}{2} S_{\alpha\alpha}(0, \omega), \tag{115}$$

where the dynamic structure factor is defined by Eq. (4). Thus, the peculiarities of the dynamic structure factor $S_{\alpha\alpha}(0, \omega)$ caused, e.g., by the XY exchange interaction anisotropy, Dzyaloshinskii-Moriya interaction or dimerization should manifest themselves in ESR experiments. The time-dependent spin correlation functions taken at the same site or at the neighboring sites manifest themselves in the spin-lattice relaxation rate $1/T_1$ measured by nuclear magnetic resonance (NMR).[61]

The activity in the field of the Jordan-Wigner fermionization approach has expanded much over the last few decades. Despite some limitations, the Jordan-Wigner fermionization approach has a wide range of applicability. Particularly attractive is that it allows one to handle complicated problems of low-dimensional quantum spin systems armed with relatively simple tools. It thus seems quite likely that it will continue to be used successfully in the coming years.

Acknowledgments

The author would like to thank T. Krokhmalskii, T. Verkholyak, J. Stolze, G. Müller and H. Büttner in collaboration with whom the study of the dynamics was performed. He is grateful to T. Krokhmalskii for preparing all figures for the paper, many interesting conversations and helpful comments and suggestions. He thanks J. Stolze and T. Verkholyak for a critical reading of the manuscript. NATO support is acknowledged (the grant reference number CBP.NUKR.CLG 982540, project "Dynamic Probes of Low-Dimensional Quantum Magnets"). The author acknowledges kind hospitality of the Organizers of the 43rd Karpacz Winter School of Theoretical Physics "Condensed Matter Physics in the Prime of XXI Century: Phenomena, Materials, Ideas, Methods" in Lądek Zdrój in February 2007.

References

1. E. Ising, Z. Phys. **31**, 253 (1925).
2. L. Onsager, Phys. Rev. **65**, 117 (1944).
3. R.J. Baxter, *Exactly Solved Models in Statistical Mechanics* (Academic Press, London, 1982).
4. H. Bethe, Z. Phys. **71**, 205 (1931).
5. M. Karbach and G. Müller, arXiv:cond-mat/9809162;
   M. Karbach, K. Hu and G. Müller, arXiv:cond-mat/9809163;
   M. Karbach, K. Hu and G. Müller, arXiv:cond-mat/0008018.
6. M. Kenzelmann, R. Coldea, D.A. Tennant, D. Visser, M. Hofmann, P. Smeibidl and Z. Tylczynski, Phys. Rev. **B65**, 144432 (2002).
7. J.-S. Caux, F.H.L. Essler and U. Löw, Phys. Rev. **B68**, 134431 (2003);
   D.V. Dmitriev and V.Ya. Krivnov, Phys. Rev. **B70**, 144414 (2004);
   D.V. Dmitriev and V.Ya. Krivnov, Pis'ma Zh. Eksp. Teor. Fiz. **80**, 349 (2004).
8. M. D'Iorio, R.L. Armstrong and D.R. Taylor, Phys. Rev. **B27**, 1664 (1983);
   M. D'Iorio, U. Glaus and E. Stoll, Solid State Commun. **47**, 313 (1983).
9. P. Jordan and E. Wigner, Z. Phys. **47**, 631 (1928).
10. Th.J. Siskens and P. Mazur, Physica **A71**, 560 (1974).
11. E. Lieb, T. Schultz and D. Mattis, Ann. Phys. (N.Y.) **16**, 407 (1961).
12. S. Katsura, Phys. Rev. **127**, 1508 (1962);
    S. Katsura, Phys. Rev. **129**, 2835 (1963).
13. V.M. Kontorovich and V.M. Tsukernik, Sov. Phys. JETP **25**, 960 (1967);
    Th. J. Siskens, H.W. Capel and K.J.F. Gaemers, Physica **A79**, 259 (1975).
14. M. Suzuki, Prog. Theor. Phys. **46**, 1337 (1971);
    D. Gottlieb and J. Rössler, Phys. Rev. **B60**, 9232 (1999);
    O. Derzhko, J. Richter and V. Derzhko, Ann. Phys. (Leipzig) **8**, SI-49 (1999);
    I. Titvinidze and G.I. Japaridze, Eur. Phys. J. **B32**, 383 (2003).
15. J.P. de Lima and L.L. Gonçalves, J. Magn. Magn. Mater. **206**, 135 (1999);
    O. Derzhko, J. Richter and O. Zaburannyi, Physica **A285**, 495 (2000);
    O. Derzhko, J. Richter, T. Krokhmalskii and O. Zaburannyi, Phys. Rev. **E69**, 066112 (2004);

J.P. de Lima, L.L. Gonçalves and T.F.A. Alves, Phys. Rev. **B75**, 214406 (2007).
16. H. Nishimori, Phys. Lett. **A100**, 239 (1984);
O. Derzhko and J. Richter, Phys. Rev. **B55**, 14298 (1997);
O. Derzhko and J. Richter, Phys. Rev. **B59**, 100 (1999).
17. L.N. Bulaevskii, Zh. Eksp. Theor. Fiz. **44**, 1008 (1963) [Sov. Phys. JETP **17**, 684 (1963)].
18. A. Luther and I. Peschel, Phys. Rev. **B12**, 3908 (1975).
19. H.J. Schulz, Phys. Rev. **B34**, 6372 (1986).
20. I. Affleck, Field theory methods and quantum critical phenomena. In: *Fields, Strings and Critical Phenomena*, Eds. E. Brézin and J. Zinn-Justin (Elsevier, Amsterdam, 1989) pp. 563-640;
J. von Delft and H. Schoeller, Ann. Phys. (Leipzig) **7**, 225 (1998);
S. Rao and D. Sen, arXiv:cond-mat/0005492.
21. H.H. Fu, K.L. Yao and Z.L. Liu, Phys. Rev. **B73**, 104454 (2006).
22. E. Fradkin, Phys. Rev. Lett. **63**, 322 (1989);
E. Fradkin, *Field Theories of Condensed Matter Systems* (Addison-Wesley, 1991).
23. Y.R. Wang, Phys. Rev. **B43**, 3786 (1991);
Y.R. Wang, Phys. Rev. **B43**, 13774 (1991);
Y.R. Wang, Phys. Rev. **B46**, 151 (1992).
24. M. Azzouz, Phys. Rev. **B48**, 6136 (1993);
M. Azzouz, L. Chen and S. Moukouri, Phys. Rev. **B50**, 6233 (1994);
M. Azzouz and C. Bourbonnais, Phys. Rev. **B53**, 5090 (1996);
M. Azzouz, B. Dumoulin and A. Benyoussef, Phys. Rev. **B55**, R11957 (1997).
25. D.C. Cabra and G.L. Rossini, Phys. Rev. **B69**, 184425 (2004);
C.A. Lames, D.C. Cabra, M.D. Grynberg and G.L. Rossini, Phys. Rev. **B74**, 224435 (2006).
26. C.D. Batista and G. Ortiz, Phys. Rev. Lett. **86**, 1082 (2001).
27. S.V. Dobrov, J. Phys. **A36**, L503 (2003).
28. M.N. Kiselev, D.N. Aristov and K. Kikoin, Phys. Rev. **B71**, 092404 (2005).
29. O. Derzhko, Journal of Physical Studies (L'viv) **5**, 49 (2001).
30. G. Misguich, Th. Jolicoeur and S.M. Girvin, Phys. Rev. Lett. **87**, 097203 (2001);
D.C. Cabra, M.D. Grynberg, P.C.W. Holdsworth and P. Pujol, Phys. Rev. **B65**, 094418 (2002);
M.-C. Chang and M.-F. Yang, Phys. Rev. **B66**, 184416 (2002).
31. A. Kitaev, Ann. Phys. (N.Y.) **321**, 2 (2006).
32. X.-Y. Feng, G.-M. Zhang and T. Xiang, Phys. Rev. Lett. **98**, 087204 (2007).
33. Th. Niemejer, Physica **36**, 377 (1967).
34. G. Müller, H. Thomas, H. Beck and J. Bonner, Phys. Rev. **B24**, 1429 (1981);
G. Müller, H. Thomas, M.W. Puga and H. Beck, J. Phys. **C14**, 3399 (1981).
35. J.H. Taylor and G. Müller, Physica **A130**, 1 (1985).
36. J. Lorenzana and G.A. Sawatzky, Phys. Rev. Lett. **74**, 1867 (1995);

J. Lorenzana and G.A. Sawatzky, Phys. Rev. **B52**, 9576 (1995);
H. Suzuura, H. Yasuhara, A. Furusaki, N. Nagaosa and Y. Tokura, Phys. Rev. Lett. **76**, 2579 (1996);
Yongmin Yu, G. Müller and V.S. Viswanath, Phys. Rev. **B54**, 9242 (1996);
J. Lorenzana and R. Eder, Phys. Rev. **B55**, R3358 (1997);
R. Werner, Phys. Rev. **B63**, 174416 (2001).

37. H.G. Vaidya and C.A. Tracy, Physica **A92**, 1 (1978).
38. O. Derzhko, T. Krokhmalskii, J. Stolze and G. Müller, Phys. Rev. **B71**, 104432 (2005);
O. Derzhko, T. Krokhmalskii, J. Stolze and G. Müller, Physica **B378-380**, 445 (2006).
39. T. Barnes, Phys. Rev. **B67**, 024412 (2003).
40. U. Brandt and K. Jacoby, Z. Phys. **B25**, 181 (1976);
U. Brandt and K. Jacoby, Z. Phys. **B26**, 245 (1977).
41. H.W. Capel and J.H.H. Perk, Physica **A87**, 211 (1977);
J.H.H. Perk and H.W. Capel, Physica **A100**, 1 (1980).
42. H.B. Cruz and L.L. Gonçalves, J. Phys. **C14**, 2785 (1981).
43. B.M. McCoy, E. Barouch and D.B. Abraham, Phys. Rev. **A4**, 2331 (1971);
J. Lajzerowicz and P. Pfeuty, Phys. Rev. **B11**, 4560 (1975);
B.M. McCoy, J.H.H. Perk and R.E. Shrock, Nucl. Phys. **B220**, 35 (1983);
B.M. McCoy, J.H.H. Perk and R.E. Shrock, Nucl. Phys. **B220**, 269 (1983);
G. Müller and R.E. Shrock, Phys. Rev. **B29**, 288 (1984).
44. F. Colomo, A.G. Izergin, V.E. Korepin and V. Tognetti, Phys. Lett. **A169**, 243 (1992);
A.R. Its, A.G. Izergin, V.E. Korepin and N.A. Slavnov, Phys. Rev. Lett. **70**, 1704 (1993);
A.R. Its, A.G. Izergin, V.E. Korepin and N.A. Slavnov, Phys. Rev. Lett. **70**, 2357 (1993);
F. Colomo, A.G. Izergin and V. Tognetti, J. Phys. **A30**, 361 (1997).
45. H. Asakawa, Physica **A233**, 39 (1996).
46. O. Derzhko and T. Krokhmalskii, Phys. Rev. **B56**, 11659 (1997);
O. Derzhko and T. Krokhmalskii, Phys. Stat. Sol. (b) **208**, 221 (1998).
47. A.P. Young and H. Rieger, Phys. Rev. **B53**, 8486 (1996);
S. Sachdev and A.P. Young, Phys. Rev. Lett. **78**, 2220 (1997);
A.P. Young, Phys. Rev. **B56**, 11691 (1997).
48. J. Stolze, A. Nöppert and G. Müller, Phys. Rev. **B52**, 4319 (1995).
49. J. Stolze and M. Vogel, Phys. Rev. **B61**, 4026 (2000).
50. Y. Maeda and M. Oshikawa, Phys. Rev. **B67**, 224424 (2003).
51. X. Jia and S. Chakravarty, Phys. Rev. **B74**, 172414 (2006).
52. O. Derzhko, T. Krokhmalskii and J. Stolze, J. Phys. **A33**, 3063 (2000).
53. K. Fabricius, U. Löw and J. Stolze, Phys. Rev. **B55**, 5833 (1997).
54. O. Derzhko, T. Krokhmalskii and J. Stolze, J. Phys. **A35**, 3573 (2002).
55. K. Kawasaki, N. Maya, A. Kouzuki and K. Nakamura, J. Phys. Soc. Jpn. **66**, 839 (1997).
56. A. Kouzuki, K. Kawasaki and K. Nakamura, Phys. Rev. **B60**, 12874 (1999).
57. W. Yu and S. Haas, Phys. Rev. **B62**, 344 (2000).

58. M. Oshikawa and I. Affleck, Phys. Rev. Lett. **79**, 2883 (1997);
D.N. Aristov and S.V. Maleyev, Phys. Rev. **B62**, R751 (2000);
O. Derzhko, J. Richter and O. Zaburannyi, J. Phys.: Condens. Matter **12**, 8661 (2000).
59. O. Derzhko and T. Verkholyak J. Phys. Soc. Jpn. **75**, 104711 (2006).
60. M. Oshikawa and I. Affleck, Phys. Rev. **B65**, 134410 (2002).
61. J. Sirker, Phys. Rev. **B73**, 224424 (2006).
62. O. Derzhko, T. Verkholyak, T. Krokhmalskii and H. Büttner, Phys. Rev. **B73**, 214407 (2006);
O. Derzhko, T. Verkholyak, T. Krokhmalskii and H. Büttner, Physica **B378-380**, 443 (2006).
63. T. Matsubara and H. Matsuda, Prog. Theor. Phys. **16**, 569 (1956);
H. Matsuda and T. Matsubara, Prog. Theor. Phys. **17**, 19 (1957);
H. Matsuda, Prog. Theor. Phys. **18**, 357 (1957);
T. Morita, Prog. Theor. Phys. **18**, 462 (1957).
64. O. Derzhko and T. Krokhmalskii, Physica **B337**, 357 (2003);
O. Derzhko and T. Krokhmalskii, Acta Physica Polonica **B32**, 3421 (2001);
O. Derzhko and T. Krokhmalskii, J. Magn. Magn. Mater. **232-245**, 778 (2002);
O. Derzhko and T. Krokhmalskii, Czech. J. Phys. **55**, 601 (2005).
65. G. Gomez-Santos and J.D. Joannopoulos, Phys. Rev. **B36**, 8707 (1987).
66. R. Coldea, D.A. Tennant, R.A. Cowley, D.F. McMorrow, B. Dorner and Z. Tylczynski, Phys. Rev. Lett. **79**, 151 (1997);
R. Coldea, D.A. Tennant, A.M. Tsvelik and Z. Tylczynski, Phys. Rev. Lett. **86**, 1335 (2001);
R. Coldea, D. A. Tennant, K. Habicht, P. Smeibidl, C. Wolters and Z. Tylczynski, Phys. Rev. Lett. **88**, 137203 (2002);
R. Coldea, D.A. Tennant and Z. Tylczynski, Phys. Rev. **B68**, 134424 (2003).
67. M. Bocquet, F.H.L. Essler, A.M. Tsvelik and A.O. Gogolin, Phys. Rev. **B64**, 094425 (2001);
M. Bocquet, Phys. Rev. **B65**, 184415 (2002);
C.H. Chung, J.B. Marston and R.H. McKenzie, J. Phys.: Condens. Matter **13**, 5159 (2001);
C.-H. Chung, K. Voelker and Y.B. Kim, Phys. Rev. **B68**, 094412 (2003);
Yi Zhou and X.-G. Wen, arXiv:cond-mat/0210662.
68. A. Lopez, A.G. Rojo and E. Fradkin, Phys. Rev. **B49**, 15139 (1994).
69. M. Karbach, G. Müller, A.H. Bougourzi, A. Fledderjohann and K.-H. Mütter, Phys. Rev. **B55**, 12510 (1997);
M. Karbach, D. Biegel and G. Müller, Phys. Rev. **B66**, 054405 (2002).
70. J.-S. Caux and J.M. Maillet, Phys. Rev. Lett. **95**, 077201 (2005);
J.-S. Caux, R. Hagemans and J.M. Maillet, J. Stat. Mech.: Theor. Exp., P09003 (2005);
J.-S. Caux and R. Hagemans, J. Stat. Mech.: Theor. Exp., P12013 (2006).

QUANTUM COMPUTING WITH ELECTRICAL CIRCUITS: HAMILTONIAN CONSTRUCTION FOR BASIC QUBIT-RESONATOR MODELS

M.R. GELLER

Department of Physics and Astronomy
University of Georgia, Athens, Georgia 30602, USA

Recent experiments motivated by applications to quantum information processing are probing a new and fascinating regime of electrical engineering—that of quantum electrical circuits—where macroscopic collective variables such as polarization charge and electric current exhibit quantum coherence. Here I discuss the problem of constructing a quantum mechanical Hamiltonian for the low-frequency modes of such a circuit, focusing on the case of a superconducting qubit coupled to a harmonic oscillator or resonator, an architecture that is being pursued by several experimental groups.

1. Quantum gate design

In the quantum circuit model of quantum information processing, an arbitrary unitary transformation on N qubits can be decomposed into a sequence of certain universal two-qubit logical operations acting on pairs of qubits, combined with arbitrary single-qubit rotations.[1] The purpose of quantum gate design is to develop experimental protocols or "machine language code" to implement these elementary operations.

For quantum information processing architectures based on superconducting circuits,[2,3] the first step is to construct an effective Hamiltonian for the system. Whereas the fully microscopic Hamiltonian for the electronic and ionic degrees of freedom in the conductors forming the circuit is known, at least in principle, the Hamiltonian of interest here describes only the relevant low-energy modes of that circuit.

A rigorous construction might involve making a canonical transformation from the microscopic quantum degrees of freedom to a set of collective modes. Here I follow a simpler and more intuitive phenomenological quantization method, whereby a classical description based on Kirkoff's laws is derived first, and then later canonically quantized. It is important to realize that such an approach is not based on first principles and must be confirmed experimentally.

2. The phase qubit

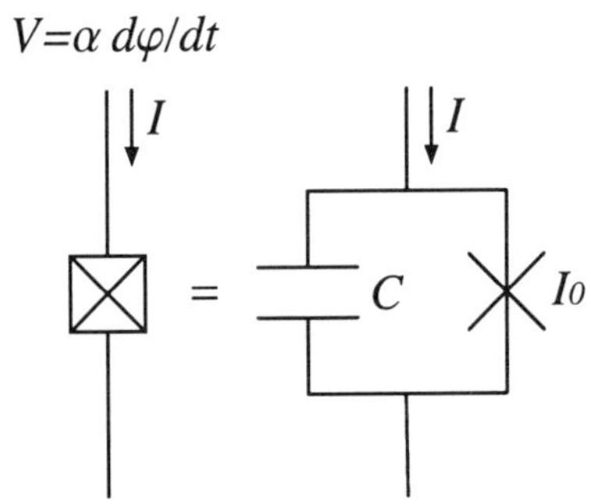

Fig. 1. Circuit model for a current-biased JJ, neglecting dissipation. Here $\alpha \equiv \hbar/2e$.

The primitive building block for any superconducting qubit is the Josephson junction (JJ) shown in Fig. 1. The low-energy dynamics of this system is governed by the phase difference φ between the condensate wave functions or order parameters on the two sides of the insulating barrier. The phase difference is an operator canonically conjugate to the Cooper-pair number difference N, according to[a]

$$[\varphi, N] = i\,. \tag{1}$$

The low-energy eigenstates $\psi_m(\varphi)$ of the JJ can be regarded as probability-amplitude distributions in φ. As will be explained below, the potential energy $U(\varphi)$ of the JJ is manipulated by applying a bias current I to the junction, providing an external control of the quantum states $\psi_m(\varphi)$, including the qubit energy-level spacing ϵ. The crossed box in Fig. 1 represents a "real" JJ. The cross alone represents a nonlinear element that satisfies the Josephson equations[b]

$$I = I_0 \sin\varphi \qquad \text{and} \qquad V = \alpha\dot{\varphi}\,, \tag{2}$$

[a]We define the momentum P to be canonically conjugate to φ, and $N \equiv P/\hbar$. In the phase representation, $N = -i\frac{\partial}{\partial\varphi}$.
[b]$\alpha \equiv \hbar/2e$.

with critical current I_0. The capacitor accounts for junction charging.[c] A single JJ is characterized by two energy scales, the Josephson coupling energy

$$E_{\rm J} \equiv \frac{\hbar I_0}{2e}, \tag{3}$$

where e is the magnitude of the electron charge, and the Cooper-pair charging energy

$$E_{\rm c} \equiv \frac{(2e)^2}{2C}, \tag{4}$$

with C the junction capacitance. For example,

$$E_{\rm J} = 2.05\,{\rm meV} \times I_0[\mu{\rm A}] \quad \text{and} \quad E_{\rm c} = \frac{320\,{\rm neV}}{C[{\rm pF}]}, \tag{5}$$

where $I_0[\mu\text{A}]$ and $C[\text{pF}]$ are the critical current and junction capacitance in microamperes and picofarads, respectively. In the regimes of interest to quantum computation, $E_{\rm J}$ and $E_{\rm c}$ are assumed to be larger than the thermal energy $k_{\rm B}T$ but smaller than the superconducting energy gap $\Delta_{\rm sc}$, which is about $180\,\mu$eV in Al. The relative size of $E_{\rm J}$ and $E_{\rm c}$ vary, depending on the specific qubit implementation.

The basic phase qubit considered here consists of a JJ with an external current bias, and is shown in Fig. 2. The classical Lagrangian for this circuit is

$$L_{\rm JJ} = \frac{1}{2} M \dot{\varphi}^2 - U, \qquad M \equiv \frac{\hbar^2}{2E_{\rm c}}. \tag{6}$$

Here

$$U \equiv -E_{\rm J}\big(\cos\varphi + s\,\varphi\big), \quad \text{with} \quad s \equiv \frac{I}{I_0}, \tag{7}$$

is the effective potential energy of the JJ, shown in Fig. 3. Note that the "mass" M in (6) actually has dimensions of mass × length^2. The form (6) results from equating the sum of the currents flowing through the capacitor and ideal Josephson element to I. The phase qubit implementation uses $E_{\rm J} \gg E_{\rm c}$.

According to the Josephson equations, the classical canonical momentum $P = \frac{\partial L}{\partial \dot{\varphi}}$ is proportional to the charge Q or to the number of Cooper

---

[c]This provides a simple mean-field treatment of the *inter*-condensate electron-electron interaction neglected in the standard tunneling Hamiltonian formalism on which the Josephson equations are based.

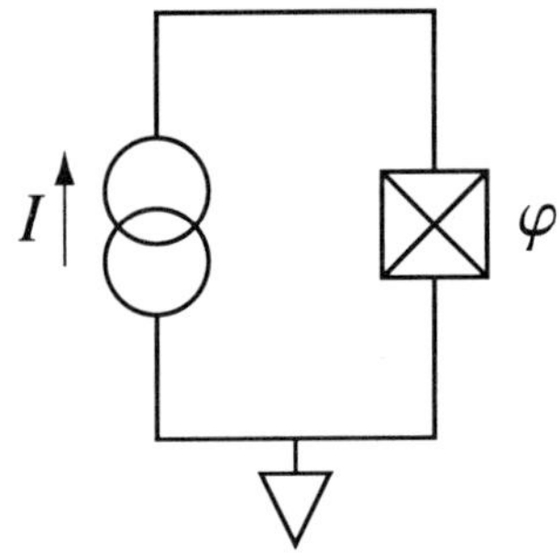

Fig. 2. Basic phase qubit circuit.

pairs Q/2e on the capacitor according to $P = \hbar Q/2e$. The quantum Hamiltonian can then be written as

$$H_{\rm JJ} = E_{\rm c} N^2 + U \,, \tag{8}$$

where φ and N are operators satisfying (1). Because U depends on s, which itself depends on time, $H_{\rm JJ}$ is generally time-dependent. The low lying stationary states when $s < 1$ are shown in Fig. 4. The two lowest eigenstates $|0\rangle$ and $|1\rangle$ are used to make a qubit. ϵ is the level spacing and ΔU is the height of the barrier.

A useful "spin $\frac{1}{2}$" form of the phase qubit Hamiltonian follows by projecting (8) to the qubit subspace. There are two natural ways of doing this. The first is to use the basis of the s-dependent eigenstates, in which case

$$H = -\frac{\hbar\omega_{\rm p}}{2}\,\sigma^z \,, \tag{9}$$

where

$$\omega_{\rm p} \equiv \omega_{\rm p0}(1-s^2)^{\frac{1}{4}} \quad \text{and} \quad \omega_{\rm p0} \equiv \sqrt{2E_{\rm c}E_{\rm J}}/\hbar \,. \tag{10}$$

The s-dependent eigenstates are called instantaneous eigenstates, because s is usually changing with time. The time-dependent Schrödinger equation in this basis contains additional terms coming from the time-dependence of the basis states themselves, which can be calculated in closed form in the harmonic limit.[4] These additional terms account for all nonadiabatic effects.

The second spin form uses a basis of eigenstates with a fixed value of bias, s_0. In this case

$$H = -\frac{\hbar\omega_{\rm p}(s_0)}{2}\,\sigma^z - \frac{E_{\rm J}\ell}{\sqrt{2}}(s-s_0)\,\sigma^x \,, \tag{11}$$

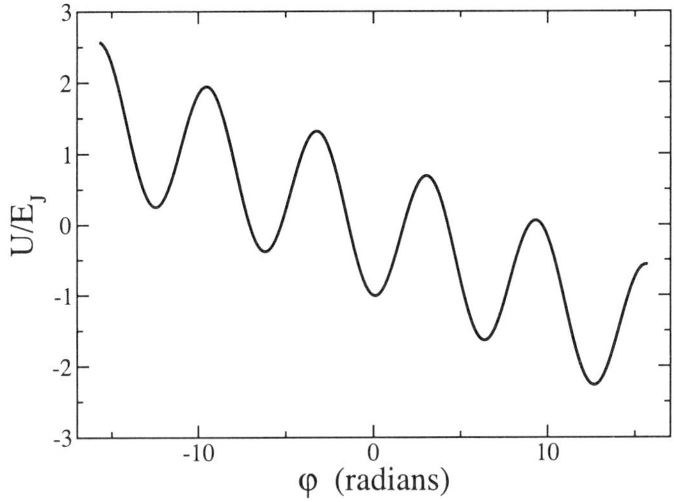

Fig. 3. Effective potential for a current-biased JJ. The slope of the cosine potential is s. The potential is harmonic for the qubit states unless s is very close to 1.

where

$$\ell \equiv \ell_0 (1 - s_0)^{-\frac{1}{8}} \quad \text{and} \quad \ell_0 \equiv \left(\frac{2E_c}{E_J} \right)^{\frac{1}{4}} . \tag{12}$$

This form is restricted to $|s - s_0| \ll 1$, but it is very useful for describing rf pulses.

The angle ℓ characterizes the width of the eigenstates in φ. For example, in the s_0-eigenstate basis (and with s_0 in the harmonic regime), we have

$$\varphi = \varphi_{01} \sigma^x + \arcsin(s_0)\, I, \quad \text{with} \quad \varphi_{mm'} \equiv \langle m|\varphi|m'\rangle . \tag{13}$$

Here $\varphi_{mm'}$ is an effective dipole moment (with dimensions of angle, not length), and $\varphi_{01} = \ell/\sqrt{2}$.

3. Qubit-oscillator models

Circuit diagrams for an rf squid capacitively coupled to parallel and series LC oscillators are shown in Figs. 5 and 6. Φ_x is an external flux bias, and φ is the phase difference across the JJ (the phase of the ungrounded superconductor relative to the grounded side is φ). Quantization of the total magnetic flux Φ in the squid loop leads to the condition (in cgs units)

$$\Phi = \Phi_x - cLI = \Phi_{sc} \frac{\varphi}{2\pi} , \qquad \Phi_{sc} \equiv \frac{hc}{2e} , \tag{14}$$

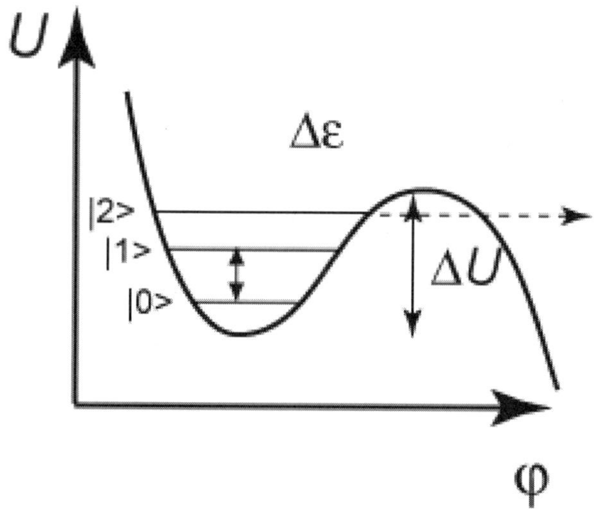

Fig. 4. Effective potential in the anharmonic regime, with s very close to 1. State preparation and readout are carried out in this regime.

where I is the current flowing downward through the Josephson junction, related to φ by

$$\alpha C\ddot{\varphi} + I_0 \sin\varphi = I \,. \tag{15}$$

Here C and I_0 are the usual JJ capacitance and critical-current parameters, and

$$\alpha \equiv \frac{\hbar}{2e} \,. \tag{16}$$

The minus sign in (14) reflects the diamagnetic (for $0 < \varphi < \pi$) screening by the superconducting loop. The quantization condition (14) assumes an isolated squid (specifically, that no current is being provided by the coupling capacitor). In Figs. 5 and 6 the voltage across the JJ is

$$V = \alpha\dot{\varphi} \,. \tag{17}$$

3.1. *JJ coupled to parallel LC oscillator*

Referring to Fig. 5, the equations of motion for φ and q are

$$\alpha^2 (C + C_{\rm int})\ddot{\varphi} + E_{\rm J}\sin\varphi + \frac{\alpha\Phi_{\rm sc}}{2\pi cL}\varphi - \frac{\alpha\Phi_{\rm ext}}{cL} = \alpha\frac{C_{\rm int}}{C_{\rm osc}}\dot{q} \tag{18}$$

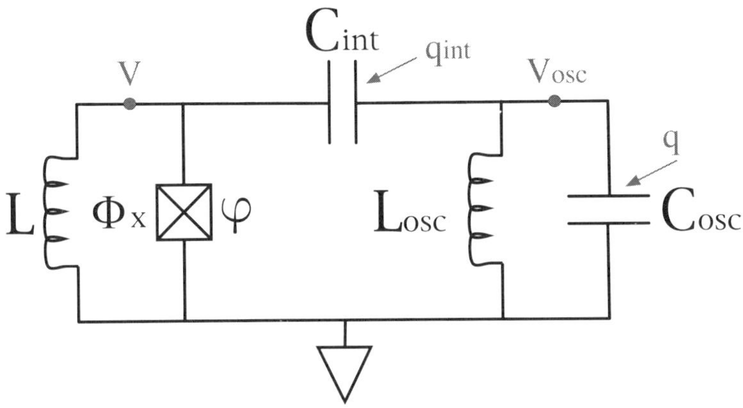

Fig. 5. Circuit model for a superconducting qubit coupled to a parallel LC oscillator.

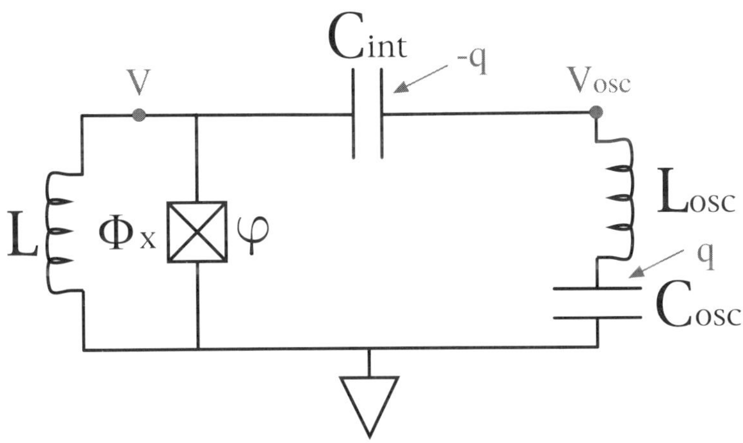

Fig. 6. Circuit model for qubit coupled to series LC oscillator.

and

$$L_{\text{osc}}\left(1+\frac{C_{\text{int}}}{C_{\text{osc}}}\right)\ddot{q}+\frac{q}{C_{\text{osc}}}=\alpha L_{\text{osc}}C_{\text{int}}\dddot{\varphi}\,. \tag{19}$$

Surprisingly, it is not possible to find a Lagrangian (local in time and a polynomial in $\frac{d^m\varphi}{dt^m}$ and $\frac{d^nq}{dt^n}$) that gives these equations of motion. To proceed, we make a transformation from q to a dimensionless node-flux variable ϕ, defined as

$$\phi(t)\equiv\frac{1}{\alpha C_{\text{osc}}}\int_{-\infty}^{t}q(t')\,dt'\,, \tag{20}$$

use $q = \alpha C_{\rm osc}\dot{\phi}$, and integrate the equation resulting from (19) over time. This leads to the coupled equations

$$\alpha^2(C + C_{\rm int})\ddot{\varphi} + E_{\rm J}\sin\varphi + \frac{\alpha^2}{L}\varphi - \frac{2\pi\alpha^2 x}{L} = \alpha^2 C_{\rm int}\ddot{\phi} \tag{21}$$

and

$$\alpha^2(C_{\rm osc} + C_{\rm int})\ddot{\phi} + \frac{\alpha^2}{L_{\rm osc}}\phi = \alpha^2 C_{\rm int}\ddot{\varphi} + {\rm const}\,, \tag{22}$$

where

$$x \equiv \frac{\Phi_{\rm x}}{\Phi_{\rm sc}} \tag{23}$$

is the dimensionless flux bias. The integration constant in (22) acts as an applied static force and can be dropped (corresponding to a shift in ϕ). Note the symmetry in the cross-coupling terms on the right-hand-sides of (21) and (22).

A Lagrangian leading to (21) and (22) is

$$L = \frac{\alpha^2 C}{2}\dot{\varphi}^2 + E_{\rm J}\cos\varphi - \frac{\alpha^2}{2L}(\varphi - 2\pi x)^2 + \frac{\alpha^2 C_{\rm osc}}{2}\dot{\phi}^2 - \frac{\alpha^2}{2L_{\rm osc}}\phi^2 - \alpha^2 C_{\rm int}\,\dot{\varphi}\dot{\phi}\,. \tag{24}$$

The simple capacitance renormalizations $C \to C + C_{\rm int}$ and $C_{\rm osc} \to C_{\rm osc} + C_{\rm int}$ present in (21) and (22) have been ignored here but can easily be accounted for below. The velocity-velocity coupling in (24) will lead to a $\sigma_{\rm J}^y \sigma_{\rm osc}^y$ interaction term in the Hamiltonian.

The canonical momenta are

$$p_\varphi = \alpha^2 C\dot{\varphi} - \alpha^2 C_{\rm int}\dot{\phi} \quad \text{and} \quad p_\phi = \alpha^2 C_{\rm osc}\dot{\phi} - \alpha^2 C_{\rm int}\dot{\varphi}\,. \tag{25}$$

The velocities in terms of these momenta are

$$\dot{\varphi} = \frac{C_{\rm osc}\,p_\varphi + C_{\rm int}\,p_\phi}{\alpha^2(CC_{\rm osc} - C_{\rm int}^2)} \quad \text{and} \quad \dot{\phi} = \frac{C_{\rm int}\,p_\varphi + C\,p_\phi}{\alpha^2(CC_{\rm osc} - C_{\rm int}^2)}\,. \tag{26}$$

Quantization then leads to

$$H = p_\varphi\dot{\varphi} + p_\phi\dot{\phi} - L = H_\varphi + H_\phi + \delta H\,, \tag{27}$$

where $[\varphi, p_\varphi] = i\hbar$, $[\phi, p_\phi] = i\hbar$, and

$$\delta H = \frac{C_{\rm int}}{\alpha^2(CC_{\rm osc} - C_{\rm int}^2)}\,p_\varphi\,p_\phi\,. \tag{28}$$

Typically $C_{\rm int} \ll \sqrt{CC_{\rm osc}}$, allowing the $C_{\rm int}^2$ in the denominator to be dropped. Furthermore,

$$H_\varphi = E_{\rm c}N^2 - E_{\rm J}\cos\varphi + \frac{\alpha^2}{2L}(\varphi - 2\pi x)^2 \tag{29}$$

and

$$H_\phi = \frac{p_\phi^2}{2\alpha^2 C_{\rm osc}} + \frac{\alpha^2 \phi^2}{2L_{\rm osc}} \,. \tag{30}$$

The kinetic energy in H_ϕ is electrical in origin and the potential energy is magnetic. The strength of quantum fluctuations is characterized by the dimensionless quantity

$$\ell_\phi \equiv \sqrt{\frac{\hbar}{\alpha^2 C_{\rm osc}\, \omega_{\rm osc}}} \,. \tag{31}$$

Finally, I simplify (28) by projecting the squid and oscillator into their $\{|0\rangle, |1\rangle\}$ subspaces. Then

$$p_\varphi \rightarrow \frac{\hbar\, \varphi_{01} \epsilon}{2E_{\rm c}}\, \sigma^y \,, \tag{32}$$

where $\varphi_{01} \equiv \langle 0|\varphi|1\rangle$ is the JJ dipole moment and $\epsilon \equiv \epsilon_1 - \epsilon_0$ is the qubit level spacing (both calculated in the absence of coupling to the oscillator). To obtain this result I have used the identity $[\varphi, H_\varphi] = 2iE_{\rm c}N$, allowing us to relate momentum and dipole matrix elements. The oscillator momentum operator projects similarly,

$$p_\phi \rightarrow \frac{\hbar}{\sqrt{2}\, \ell_\phi}\, \sigma^y \,. \tag{33}$$

Then we obtain [dropping the $C_{\rm int}^2$ in the denominator of (28)]

$$\delta H = g \sigma_{\rm J}^y \sigma_{\rm osc}^y, \qquad \text{(parallel circuit)}\,, \tag{34}$$

where

$$g = \frac{\hbar^2 C_{\rm int} \varphi_{01} \epsilon}{2\sqrt{2}\, \alpha^2 C C_{\rm osc} E_{\rm c} \ell_\phi} = \frac{1}{2} \left(\frac{C_{\rm int}}{C_{\rm osc}} \right) \left(\frac{\sqrt{2}\, \varphi_{01}}{\ell_\phi} \right) \epsilon \,. \tag{35}$$

Note that in the harmonic junction limit, $\varphi_{01} = \ell_\varphi / \sqrt{2}$, with ℓ_φ the width of the wave functions in the junction.

3.2. *JJ coupled to series LC oscillator*

Referring to Fig. 6, the equations of motion are

$$\alpha^2 C \ddot{\varphi} + E_{\rm J} \sin\varphi + \frac{\alpha^2}{L}\varphi - \frac{2\pi\alpha^2 x}{L} = -\alpha \dot{q} \tag{36}$$

and, assuming $C_{\rm int} \neq 0$,

$$L_{\rm osc} \ddot{q} + \frac{q}{C'_{\rm osc}} = \alpha \dot{\varphi} \,, \tag{37}$$

where $C'_{\rm osc} \equiv (C_{\rm osc}^{-1} + C_{\rm int}^{-1})^{-1}$. Note that the capacitance $C_{\rm int}$ does not enter the cross-coupling terms on the right-hand-sides of (36) and (37). This is an indication that the qubit-oscillator coupling in this system is nonperturbative: The limit $C_{\rm int} \rightarrow 0$ differs from the case $C_{\rm int} = 0$, and there is no small parameter associated with the interaction. A Lagrangian leading to (36) and (37) is[d]

$$L = \frac{\alpha^2 C}{2} \dot{\varphi}^2 + E_{\rm J} \cos\varphi - \frac{\alpha^2}{2L} (\varphi - 2\pi x)^2 + \frac{L_{\rm osc}}{2} \dot{q}^2 - \frac{1}{2C'_{\rm osc}} q^2 + \alpha \dot{\varphi} q \, . \tag{38}$$

The canonical momenta are

$$p_\varphi = \alpha^2 C \dot{\varphi} + \alpha q \quad \text{and} \quad p_q = L_{\rm osc} \dot{q} \, , \tag{39}$$

leading to the quantum Hamiltonian

$$H = \frac{(p_\varphi - \alpha q)^2}{2\alpha^2 C} - E_{\rm J} \cos\varphi + \frac{\alpha^2}{2L} (\varphi - 2\pi x)^2 + \frac{p_q^2}{2L_{\rm osc}} + \frac{q^2}{2C'_{\rm osc}} \, . \tag{40}$$

The squid sees the oscillator as a source of vector potential $A \propto \alpha q$, whose time derivative describes an effective electric field. Noting that the "diamagnetic" A^2 term serves to further decrease the oscillator capacitance, I obtain

$$H = H_\varphi + H_q + \delta H \, , \tag{41}$$

where H_q is the oscillator Hamiltonian

$$H_q = \frac{p_q^2}{2L_{\rm osc}} + \frac{q^2}{2C'_{\rm osc}} = \hbar\omega_{\rm osc} \big(a^\dagger a + \tfrac{1}{2} \big) , \qquad \omega_{\rm osc} = \sqrt{\frac{1}{L_{\rm osc} C'_{\rm osc}}} \, , \tag{42}$$

where

$$C'_{\rm osc} \equiv \big(C_{\rm osc}^{-1} + C^{-1} + C_{\rm int}^{-1} \big)^{-1} , \tag{43}$$

and

$$\delta H = - \frac{p_\varphi \, q}{\alpha C} \tag{44}$$

is the qubit-oscillator interaction.

In the $\{|0\rangle, |1\rangle\}$ subspace of the series oscillator,

$$q \rightarrow \frac{\ell_q}{\sqrt{2}} \, \sigma^x \, , \tag{45}$$

[d] An alternative form for L has an interaction term $\delta L = -\alpha \, \varphi \dot{q}$.

where

$$\ell_q \equiv \sqrt{\frac{\hbar}{L_{\rm osc}\,\omega_{\rm osc}}}\,. \tag{46}$$

Then we have

$$\delta H = -g\sigma_{\rm J}^y \sigma_{\rm osc}^x, \qquad (\text{series circuit})\,, \tag{47}$$

where

$$g = \frac{\varphi_{01}\ell_q}{2\sqrt{2}\,e}\,\epsilon\,. \tag{48}$$

Note that there is no factor of $C_{\rm int}/C_{\rm osc}$ here. The coupling constant (48) is small (much less than the qubit level spacing ϵ) only if the quantum fluctuations in both the squid and oscillator are small.

3.3. *Relation to capacitively coupled qubits*

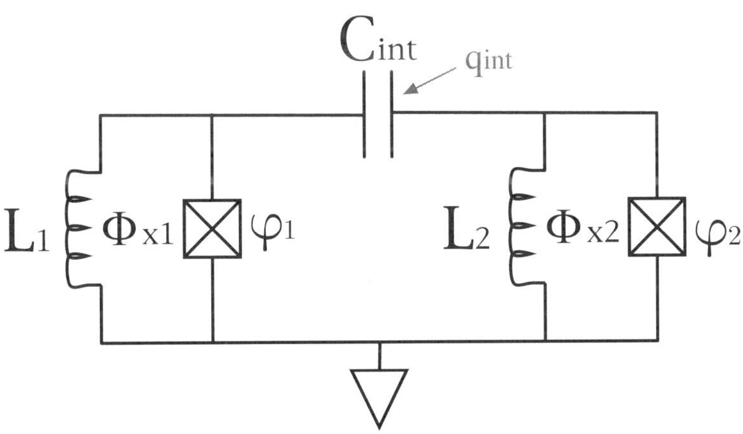

Fig. 7. Capacitively coupled qubits.

It is useful to compare the result for the qubit–parallel-LC system to a pair of capacitively coupled qubits. Referring to Fig. 7, the equations of motion are

$$\alpha^2 C_1' \ddot{\varphi}_1 + E_{\rm J1}\sin\varphi_1 + \frac{\alpha^2}{L_1}(\varphi_1 - 2\pi x_1) = \alpha^2 C_{\rm int}\ddot{\varphi}_2\,, \tag{49}$$

$$\alpha^2 C_2' \ddot{\varphi}_2 + E_{\rm J2}\sin\varphi_2 + \frac{\alpha^2}{L_2}(\varphi_1 - 2\pi x_2) = \alpha^2 C_{\rm int}\ddot{\varphi}_1\,, \tag{50}$$

where

$$C_i' \equiv C_i + C_{\rm int}, \qquad (i = 1, 2) \,. \tag{51}$$

The Lagrangian for the coupled system is

$$L = \sum_i \left[\frac{\alpha^2 C_i'}{2} \dot{\varphi}_i^2 + E_{{\rm J}i} \cos\varphi_i - \frac{\alpha^2}{2L_i} \big(\varphi_i - 2\pi x_i\big)^2 \right] - \alpha C_{\rm int} \dot{\varphi}_1 \dot{\varphi}_2 \,, \tag{52}$$

leading to the Hamiltonian

$$H = \frac{C_2' p_1^2 + C_1' p_2^2 + 2C_{\rm int} p_1 p_2}{2\alpha^2 (C_1' C_2' - C_{\rm int}^2)} \,. \tag{53}$$

In the 2-qubit subspace the interaction Hamiltonian is $\delta H = g \sigma_1^y \sigma_2^y$, where

$$\begin{aligned} g &= \frac{C_{\rm int} \varphi_{01}^{(1)} \varphi_{01}^{(2)} \epsilon^{(1)} \epsilon^{(2)}}{4e^2} \\ &= \frac{1}{2} \frac{C_{\rm int}}{\sqrt{C_1 C_2}} \left[\left(\frac{\sqrt{2} \varphi_{01}^{(1)}}{\ell^{(1)}} \right) \left(\frac{\sqrt{2} \varphi_{01}^{(2)}}{\ell^{(2)}} \right) \sqrt{\frac{\epsilon^{(1)}}{\hbar\omega^{(1)}}} \sqrt{\frac{\epsilon^{(2)}}{\hbar\omega^{(2)}}} \right] \sqrt{\epsilon^{(1)} \epsilon^{(2)}} \,. \end{aligned} \tag{54}$$

Here ω is the classical oscillation frequency of the JJ, and $\ell \equiv \sqrt{2E_{\rm c}/\hbar\omega}$ is the associated wave function width. The factor in square brackets is unity for JJs in the harmonic limit. If we now assume identical junctions, with the second biased in the harmonic regime so that it is similar to an oscillator, then after some rearrangement we obtain

$$g = \frac{1}{2} \left(\frac{C_{\rm int}}{C} \right) \left(\frac{\sqrt{2} \varphi_{01}^{(1)}}{\ell^{(2)}} \right) \epsilon^{(1)} \,, \tag{55}$$

which corresponds precisely to (35).

4. Qubit coupled to electromagnetic resonator

I now consider an rf squid coupled to a coplanar waveguide resonator. The charge qubit case has been addressed by Blais *et al.*[5] A simplified form of the system layout is shown in Fig. 8. The squid has been discussed in Sec. 3. The coplanar waveguide resonator consists of a conducting strip of length d and width w, capacitively coupled to rf transmission lines. Fig. 9 shows a hybrid circuit model for the system, where the resonator is described at the level of microscopic electrodynamics and the squid is in the usual lumped circuit limit. The geometry considered also allows for the position x_0 of the qubit along the resonator to vary; $x_0 = 0$ is the case shown in Fig. 8.

The system Hamiltonian is derived in two different ways. The simplest is to treat the resonator in the continuum limit, the approach followed in

Sec. 4.2. However, for numerical simulations the discrete LC ladder model of the resonator used in Sec. 4.3 is preferable. Both models lead to the Hamiltonian and coupling constant given below in (56) and (57).

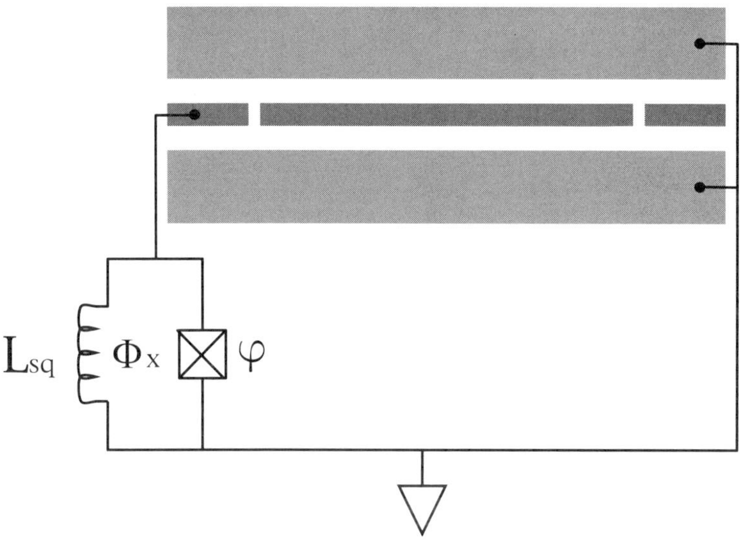

Fig. 8. Superconducting qubit coupled to coplanar waveguide resonator. The inner regions represent the resonator and capacitively coupled rf transmission lines. The wide outer regions are ground planes.

4.1. *Summary of results and mapping to qubit-oscillator*

I will show below that after projection into the qubit subspace of the squid and the vacuum and 1-photon subspace of the fundamental mode of the resonator, the interaction Hamiltonian for the system shown in Fig. 9 is

$$\delta H = g\,\sigma^y_\varphi \sigma^y_\phi\,, \tag{56}$$

where

$$g \equiv \cos\left(\tfrac{\pi x_0}{d}\right) \frac{C_{\rm int}\varphi_{01}\sqrt{\hbar\omega_{\rm res}}}{2e\sqrt{Cd}}\ \epsilon\,. \tag{57}$$

The subscripts φ and ϕ refer to the JJ phase and oscillator node-flux degrees of freedom, respectively (it will be necessary to distinguish between matrix representations of the oscillator variables written in different bases). In addition, $C_{\rm int}$ is the coupling capacitance, φ_{01} is the squid dipole moment,

$$\omega_{\rm res} = \frac{\pi v}{d} \tag{58}$$

is the angular frequency of the fundamental mode of the resonator, written in terms of the transmission line wave speed

$$v \equiv \sqrt{\frac{1}{\mathcal{L}\mathcal{C}}}, \tag{59}$$

$\mathcal{L}$ and $\mathcal{C}$ are the inductance and capacitance per unit length of the coplanar waveguide, d is the resonator length, and ϵ is the qubit energy level spacing. On resonance we have $\epsilon = \hbar\omega_{\rm res}$.

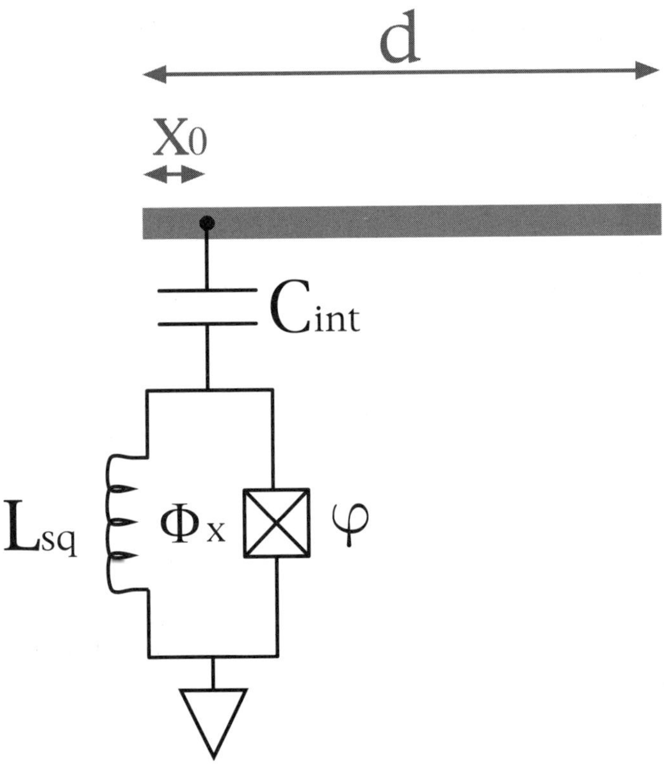

Fig. 9. Hybrid circuit model used to construct Hamiltonian. The qubit is coupled via a capacitance $C_{\rm int}$ to a resonator of length d at a distance x_0 from the end; the layout of Fig. 8 corresponds to $x_0 = 0$. The width of the resonator is w. The ground planes are not shown and the figure is not to scale. Coupling to transmission lines is ignored. The diameter of the wire connecting $C_{\rm int}$ to the resonator also enters the model and is denoted by b.

The coupling constant quoted in (57) assumes two conditions on the allowed values of $C_{\rm int}$. First, $C_{\rm int}$ must be much smaller than the JJ capacitance C. In particular, the calculation is done to leading nontrivial order in the parameter $C_{\rm int}/C$. This is the usual condition for weak coupling, and it is easily satisfied experimentally. The second condition on $C_{\rm int}$ is more restrictive and arises because of the modification of the resonator modes themselves by the attached squid. This modification depends on both $C_{\rm int}$ and on the size of the attachment point of the lumped part of the circuit to the microscopic continuous part, and is denoted by b in Fig. 9. In the design of Fig. 8, b is just the resonator width w. The condition that the qubit couples to modes of the *isolated* resonator requires that $C_{\rm int}$ be much smaller than

$$C^* \equiv \mathcal{C} b, \tag{60}$$

which can be interpreted as the capacitance "under" the attachment wire. If $C_{\rm int}$ is not much smaller than C^*, then the resonator modes the qubit couple to are themselves nontrivially modified by the coupling to the squid, and the coupling constant (57) is modified.

The Hamiltonian (56) can be mapped to a qubit coupled to a single parallel LC oscillator. To do this, define an effective oscillator inductance and capacitance

$$L_{\rm eff} \equiv \frac{2}{\pi^2} \mathcal{L} d \quad \text{and} \quad C_{\rm eff} \equiv \frac{1}{2} \mathcal{C} d . \tag{61}$$

Note that the oscillator frequency

$$\sqrt{\frac{1}{L_{\rm eff} C_{\rm eff}}} \tag{62}$$

implied by these effective quantities is equal to the actual fundamental mode frequency (58), as expected. In terms of $L_{\rm eff}$ and $C_{\rm eff}$ we can write (57) as

$$g = \frac{\cos(\frac{\pi x_0}{d})}{2} \left(\frac{C_{\rm int}}{C_{\rm eff}} \right) \left(\frac{\sqrt{2}\, \varphi_{01}}{\ell_{\rm eff}} \right) \epsilon, \quad \text{with} \quad \ell_{\rm eff} \equiv \sqrt{\frac{\hbar}{\alpha^2 C_{\rm eff} \omega_{\rm res}}} = \sqrt{\frac{(\frac{4e^2}{C_{\rm eff}})}{\hbar \omega_{\rm res}}} . \tag{63}$$

When $x_0 = 0$, this expression has precisely the form for coupling to a parallel LC oscillator with inductance $L_{\rm eff}$ and capacitance $C_{\rm eff}$.

The second expression for $\ell_{\rm eff}$ in (63) emphasizes that it is a dimensionless measure of the electric field energy in an LC oscillator. In the quantum description of a squid coupled to an parallel LC oscillator, the relevant oscillator degree of freedom is a node-flux variable, and in the node-flux representation the kinetic energy term in the Hamiltonian is electric in origin. Thus, $\ell_{\rm eff}$ is also a dimensionless measure of the quantum zero-point motion in the fundamental mode of the resonator.

4.2. *Continuum resonator model*

The resonator lies on the x axis with its left end at the origin. Referring to Fig. 9, let $\rho(x,t)$, $I(x,t)$, and $V(x,t)$ be the charge per unit length, the current in the x direction, and the electric potential on the resonator, and let $\mathcal{L}$ and $\mathcal{C}$ be the inductance and capacitance per unit length of the coplanar waveguide.

The equation of motion for an infinite waveguide follows from the inductance equation

$$\partial_x V + \mathcal{L}\, \partial_t I = 0\,, \tag{64}$$

the capacitance equation

$$\rho = \mathcal{C}\, V\,, \tag{65}$$

and the continuity equation

$$\partial_t \rho + \partial_x I = 0\,. \tag{66}$$

These lead to the wave equation

$$\left(\partial_t^2 - v^2 \partial_x^2\right)\rho = 0\,, \tag{67}$$

with velocity given in (59). The potential V and current I satisfy identical wave equations, but these will not be needed here.

A finite segment of waveguide—a resonator—satisfies the wave equation (67) together with the boundary conditions that $I = 0$ at the ends. Using (64), we see that these boundary conditions require $\partial_x \rho$ (or $\partial_x V$) to vanish at the ends, leading to charge (or voltage) antinodes there. I also assume that the resonator carries no net charge, so that

$$\int_0^d dx\, \rho = 0\,. \tag{68}$$

The charge density eigenmodes are

$$f_n(x) \equiv \sqrt{\frac{2}{d}}\, \cos\!\left(\frac{n\pi x}{d}\right), \qquad n = 1, 2, 3, \dots\,, \tag{69}$$

the $n = 0$ mode excluded because of the charge neutrality condition (68). These satisfy the orthonormality condition

$$\int_0^d dx\; f_n\, f_{n'} = \delta_{nn'}\,. \tag{70}$$

The mode angular frequencies are

$$\omega_n = \frac{n\pi v}{d}\,. \tag{71}$$

Below we will primarily be interested in the fundamental mode, $n = 1$, with the frequency given in (58).

To derive a Hamiltonian for the system shown in Fig. 9, it will be necessary to account for the finite width b of the wire connecting the resonator to the coupling capacitor. In the actual device, b is equal to the width w of the waveguide, but in future designs these may differ. The finite width of the wire smears the squid-resonator interaction over a region of size b. I account for this by introducing a broadened delta function $\Delta(x)$ of width b, satisfying

$$\int_0^d dx\, \Delta(x - x_0) = 1, \qquad b < x_0 < d - b\,. \tag{72}$$

The actual shape of Δ is determined by the microscopic current density at the squid-resonator junction. However, for definiteness I assume a square shape

$$\Delta(x) \equiv \begin{cases} \frac{1}{b} & |x| \leq \frac{b}{2} \\ 0 & |x| > \frac{b}{2} \end{cases}. \tag{73}$$

Because the wavelengths of the modes of interest here are much larger than b, the detailed shape of $\Delta(x)$ should be irrelevant as long as $\Delta(x)$ is everywhere finite. In particular, it is not possible to take the $b \rightarrow 0$ limit, where $\Delta(x)$ becomes a delta function, as $\delta(0)$ diverges.

To find the equations of motion for the system of Fig. 9, let $q_{\rm int}$ be the charge induced on the upper (resonator side) plate of the coupling capacitor. We take the fundamental degrees of freedom of the circuit to be the JJ coordinate φ and the resonator density field $\rho(x)$, suppressing the time argument in all quantities when not necessary. In terms of these degrees of freedom,

$$q_{\rm int} = C_{\rm int} \left[\frac{\bar{\rho}(x_0)}{\mathcal{C}} - \alpha \dot{\varphi} \right], \tag{74}$$

where

$$\bar{f}(x) \equiv \int dx'\, \Delta(x' - x)\, f(x') \tag{75}$$

denotes the average of a quantity $f(x)$ over a width b, and

$$\alpha \equiv \frac{\hbar}{2e}\,. \tag{76}$$

The equation of motion for φ is

$$\alpha^2 C \ddot{\varphi} + E_{\rm J} \sin\varphi + \frac{\alpha^2}{L_{\rm sq}} \left(\varphi - 2\pi \frac{\Phi_{\rm x}}{\Phi_{\rm sc}} \right) = \alpha\, \dot{q}_{\rm int}\,, \tag{77}$$

where

$$\Phi_{\rm sc} \equiv \frac{hc}{2e} \tag{78}$$

is the superconducting flux quantum and $\Phi_{\rm x}$ is the external magnetic flux. Then

$$\alpha^2 C' \ddot{\varphi} + E_{\rm J} \sin\varphi + \frac{\alpha^2}{L_{\rm sq}} \left(\varphi - 2\pi \frac{\Phi_{\rm x}}{\Phi_{\rm sc}} \right) = \alpha \, \frac{C_{\rm int}}{\mathcal{C}} \, \partial_t \bar{\rho}(x_0) \, , \qquad C' \equiv C + C_{\rm int} \, . \tag{79}$$

The equation of motion for the charge density can be obtained by modifying the continuity equation (66) to account for the current drain to the squid. It will be necessary to account for the finite width of the wire connecting the resonator to the coupling capacitor. Then

$$\partial_t \rho + \partial_x I = -\Delta(x - x_0) \, \dot{q}_{\rm int} \, . \tag{80}$$

The sign on the right-hand-side of (80) assures that the resonator sees the current $\dot{q}_{\rm int}$ flowing downward through the coupling capacitor as a current sink. Combining (80) with (64) and (65) leads to

$$\big[\mathcal{C}'(x) \, \partial_t^2 - \frac{1}{\mathcal{L}} \, \partial_x^2\big] \rho = \alpha \, C_{\rm int} \, \mathcal{C} \, \Delta(x - x_0) \, \dddot{\varphi} \, , \qquad \mathcal{C}'(x) \equiv \mathcal{C} + \Delta(x - x_0) \, C_{\rm int} \, . \tag{81}$$

To obtain (81) I have used the fact that (for the modes of interest) ρ is slowly varying on the scale b, so that $\Delta(x - x_0) \, \bar{\rho}(x_0) \approx \Delta(x - x_0) \, \rho(x)$.

As with our earlier investigation of a squid capacitively coupled to a parallel LC oscillator, there is no time-local Lagrangian that gives the equations of motion (79) and (81). To proceed, make a transformation from ρ to a dimensionless node-flux *field*

$$\phi(x,t) \equiv \frac{1}{\alpha \mathcal{C}} \int_{-\infty}^{t} dt' \, \rho(x, t') \tag{82}$$

and integrate the equation resulting from (81) over time. This leads to the set of coupled equations

$$\alpha^2 C' \ddot{\varphi} + E_{\rm J} \sin\varphi + \frac{\alpha^2}{L_{\rm sq}} \left(\varphi - 2\pi \frac{\Phi_{\rm x}}{\Phi_{\rm sc}} \right) = \alpha^2 \, C_{\rm int} \, \partial_t^2 \bar{\phi}(x_0) \tag{83}$$

and

$$\alpha^2 \big(\mathcal{C}' \, \partial_t^2 - \tfrac{1}{\mathcal{L}} \, \partial_x^2 \big) \phi = \alpha^2 \, C_{\rm int} \, \Delta(x - x_0) \, \ddot{\varphi} \, , \tag{84}$$

dropping an arbitrary constant of integration. A Lagrangian for the coupled system is

$$
\begin{aligned}
L = &\frac{\alpha^2 C'}{2}\dot{\varphi}^2 + E_{\rm J}\cos\varphi - \frac{\alpha^2}{2L_{\rm sq}}\left(\varphi - 2\pi\frac{\Phi_{\rm x}}{\Phi_{\rm sc}}\right)^2 \\
&+ \int_0^d dx \left[\frac{\alpha^2 \mathcal{C}'}{2}(\partial_t\phi)^2 - \frac{\alpha^2}{2\mathcal{L}}(\partial_x\phi)^2\right] \\
&- \int_0^d dx\, \alpha^2 C_{\rm int}\,\dot{\varphi}\,\Delta(x - x_0)\,\partial_t\phi(x)\,.
\end{aligned} \tag{85}
$$

The canonical momenta are

$$p = \alpha^2 C'\dot{\varphi} - \alpha^2 C_{\rm int}\,\partial_t\bar{\phi}(x_0) \tag{86}$$

and

$$\Pi(x) = \alpha^2\mathcal{C}'(x)\,\partial_t\phi(x) - \alpha^2 C_{\rm int}\,\dot{\varphi}\,\Delta(x - x_0)\,. \tag{87}$$

Note that the resonator momentum density $\Pi(x)$ is a field; it depends on x. The velocities in terms of these momenta are

$$\dot{\varphi} = \frac{\mathcal{C}'(x_0)\,p + C_{\rm int}\,\bar{\Pi}(x_0)}{\alpha^2[C'\mathcal{C}'(x_0) - C_{\rm int}^2\Delta(0)]} \tag{88}$$

and

$$\partial_t\phi(x) = \frac{\Pi(x)}{\alpha^2\mathcal{C}'(x)} + \frac{C_{\rm int}[\mathcal{C}'(x_0)\,p + C_{\rm int}\,\bar{\Pi}(x_0)]\Delta(x - x_0)}{\alpha^2\mathcal{C}'(x)[C'\mathcal{C}'(x_0) - C_{\rm int}^2\Delta(0)]}\,. \tag{89}$$

The Hamiltonian is

$$H = p\dot{\varphi} + \int dx\,\Pi\,\partial_t\phi - L = H_\varphi + H_\phi + \delta H\,, \tag{90}$$

where

$$H_\varphi \equiv \frac{\mathcal{C}'(x_0)\,p^2}{2\alpha^2[C'\mathcal{C}'(x_0) - C_{\rm int}^2\,\Delta(0)]} + U(\varphi)\,, \tag{91}$$

with

$$U(\varphi) \equiv -E_{\rm J}\cos\varphi + \frac{\alpha^2}{2L_{\rm sq}}\left(\varphi - 2\pi\frac{\Phi_{\rm x}}{\Phi_{\rm sc}}\right)^2, \tag{92}$$

$$H_\phi \equiv \int_0^d dx\left[\frac{\Pi^2}{2\alpha^2\mathcal{C}'} + \frac{\alpha^2(\partial_x\phi)^2}{2\mathcal{L}}\right] + \frac{C_{\rm int}^2[\bar{\Pi}(x_0)]^2}{2\alpha^2\mathcal{C}'(x_0)[C'\mathcal{C}'(x_0) - C_{\rm int}^2\,\Delta(0)]}\,, \tag{93}$$

and

$$\delta H = \frac{C_{\rm int}}{\alpha^2[C'\mathcal{C}'(x_0) - C_{\rm int}^2\,\Delta(0)]}\,p\,\bar{\Pi}(x_0)\,. \tag{94}$$

Quantization leads to the conditions

$$[\varphi, p] = i\hbar \quad \text{and} \quad [\phi(x), \Pi(x')] = i\hbar\delta(x - x') \,. \tag{95}$$

Next we make two approximations concerning the value of the coupling capacitance $C_{\rm int}$, namely

$$C_{\rm int} \ll C \quad \text{and} \quad C_{\rm int} \ll C^*, \tag{96}$$

where C is the JJ capacitance and C^* is defined in (60). With these assumptions the system Hamiltonian simplifies to

$$H = E_{\rm c}\mathcal{N}^2 + U(\varphi) + \int dx\, \mathcal{H}_{\rm res} + \frac{C_{\rm int}}{\alpha^2 C\mathcal{C}}\, p\, \Pi(x_0) \,. \tag{97}$$

Here $\mathcal{N} \equiv p/\hbar$ and $E_{\rm c} \equiv 2e^2/C$. The Hamiltonian density

$$\mathcal{H}_{\rm res} \equiv \frac{\Pi^2}{2\alpha^2\mathcal{C}} + \frac{\alpha^2(\partial_x\phi)^2}{2\mathcal{L}} \tag{98}$$

in (97) now describes an isolated resonator. The averaging over $\Pi(x_0)$ in the interaction term has been dropped, as it is assumed that we will use (97) only for resonator modes with wavelengths much larger than b.

The equation of motion resulting from (98) is the operator wave equation $(\partial_t^2 - v^2\partial_x^2)\phi = 0$, with velocity given in (59). According to (82), the boundary conditions on ϕ are that $\partial_x\phi = 0$ at the resonator ends. Therefore, the charge density eigenfunctions defined in (69) can be used here as a basis in which to expand the node-flux field ϕ and its conjugate momentum, as

$$\phi(x) = \sum_{n=1}^{\infty} \sqrt{\frac{\hbar}{2\alpha^2\mathcal{C}\omega_n}}\, f_n(x)\, \big[a_n + a_n^\dagger\big] \tag{99}$$

and

$$\Pi(x) = -i\sum_{n=1}^{\infty} \sqrt{\frac{\alpha^2\mathcal{C}\hbar\omega_n}{2}}\, f_n(x)\, \big[a_n - a_n^\dagger\big] \,. \tag{100}$$

Here a_n and $a_n^\dagger$ are bosonic creation and annihilation operators, and $f_n(x)$ and ω_n are the resonator eigenmodes and frequencies given in (69) and (71). These expansions neglect additive "zero-mode" contributions that are necessary for (99) and (100) to satisfy the second commutation relation in (95), because the eigenfunctions (69) do not themselves form a complete basis, but the zero-mode contributions have no effect here. Using (99) and (100), along with the orthonormality conditions (70) and the additional identity

$$\int_0^d dx\; f_n'(x)\, f_{n'}'(x) = \delta_{nn'}\, \frac{\pi^2 n^2}{d^2} \,, \tag{101}$$

leads to the expected result

$$H_{\rm res} \equiv \int dx \, \mathcal{H}_{\rm res} = \sum_{n=1}^{\infty} \hbar\omega_n \, a_n^\dagger a_n \,. \tag{102}$$

By retaining only the $n = 1$ fundamental-mode term in (100), projecting the squid momentum into the qubit subspace according to

$$p \rightarrow \frac{\hbar \, \varphi_{01} \epsilon}{2E_{\rm c}} \sigma^y \,, \tag{103}$$

and projecting the resonator fundamental mode into the ground and one-photon subspace according to

$$a_1 - a_1^\dagger = i\sigma^y \,, \tag{104}$$

the interaction term in (97) can now be written as (56) with the coupling constant given in (57).

The $x_0 \rightarrow 0$ limit of (57) has to be taken carefully because of our smearing of the qubit-resonator contact point. The derivation above assumes that $x_0 > b$, so the $x_0 \rightarrow 0$ limit should really be implemented by setting $x_0 \rightarrow b$. However, because $b \ll d$ we can ignore this technicality and let $x_0 = 0$ in (57).

4.3. *LC network resonator model*

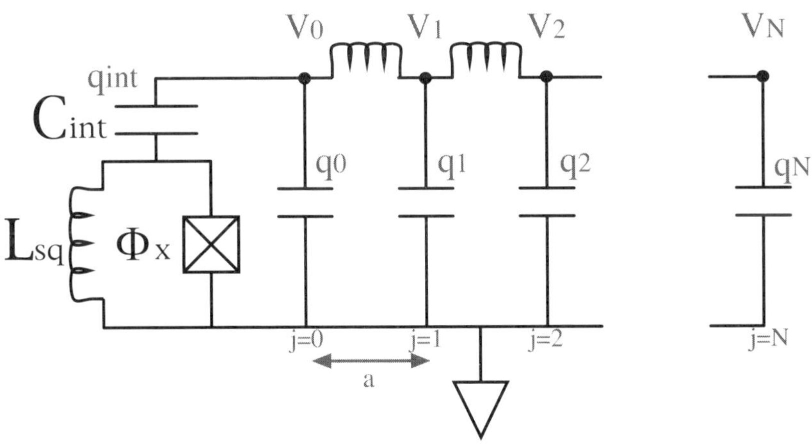

Fig. 10. LC network model of the qubit-resonator system. The ladder has N inductors l_0 and $N + 1$ capacitors c_0.

A discrete network model of the qubit-resonator system is illustrated in Fig. 10. The resonator is modeled as an LC ladder with N inductors l_0 and $N+1$ capacitors c_0. The size of each cell is

$$a \equiv \frac{d}{N}, \tag{105}$$

and

$$l_0 = \mathcal{L}a \quad \text{and} \quad c_0 = \mathcal{C}a, \tag{106}$$

with $\mathcal{L}$ and $\mathcal{C}$ the inductance and capacitance per unit length of the physical waveguide. In the continuum limit used below, we let $N \rightarrow \infty$ with d held fixed. This system has N independent resonator degrees of freedom, which I take to be the charges $\{q_1, q_2, \ldots, q_N\}$, and one squid degree of freedom φ. The charge $q_{\rm int}$ on the resonator side of the coupling capacitor is fixed by charge neutrality to be

$$q_{\rm int} = -(q_0 + q_1 + \cdots + q_N), \tag{107}$$

and by using the relation $q_{\rm int} = C_{\rm int}(\frac{q_0}{c_0} - \alpha\dot{\varphi})$, q_0 can be written in terms of the other degrees of freedom as

$$q_0 = \frac{\alpha C_{\rm int}\dot{\varphi} - (q_1 + q_2 + \cdots + q_N)}{1 + \frac{C_{\rm int}}{c_0}}. \tag{108}$$

The equation of motion for φ is

$$\alpha^2 C\ddot{\varphi} + E_{\rm J}\sin\varphi + \frac{\alpha^2}{L_{\rm sq}}\left(\varphi - 2\pi\frac{\Phi_{\rm x}}{\Phi_{\rm sc}}\right) = \alpha\,\dot{q}_{\rm int}, \tag{109}$$

or

$$\alpha^2 C'\ddot{\varphi} + E_{\rm J}\sin\varphi + \frac{\alpha^2}{L_{\rm sq}}\left(\varphi - 2\pi\frac{\Phi_{\rm x}}{\Phi_{\rm sc}}\right) = -\frac{\alpha C_{\rm int}}{c_0 + C_{\rm int}}\left(\dot{q}_1 + \dot{q}_2 + \cdots + \dot{q}_N\right), \tag{110}$$

with

$$C' \equiv C + \frac{C_{\rm int}}{1 + \frac{C_{\rm int}}{c_0}}. \tag{111}$$

The equations of motion for the LC ladder are

$$\begin{aligned}
l_0\,(\ddot{q}_1+\cdots+\ddot{q}_N)+\frac{q_1}{c_0}+\frac{q_1+\cdots+q_N}{c_0+C_{\text{int}}} &= \frac{\alpha C_{\text{int}}}{c_0+C_{\text{int}}}\,\dot{\varphi}\,,\\
l_0\,(\ddot{q}_2+\cdots+\ddot{q}_N)+\frac{q_2-q_1}{c_0} &= 0\,,\\
l_0\,(\ddot{q}_3+\cdots+\ddot{q}_N)+\frac{q_3-q_2}{c_0} &= 0\,,\\
&\vdots\\
l_0\,(\ddot{q}_{N-1}+\ddot{q}_N)+\frac{q_{N-1}-q_{N-2}}{c_0} &= 0\,,\\
l_0\,\ddot{q}_N+\frac{q_N-q_{N-1}}{c_0} &= 0\,.
\end{aligned} \tag{112}$$

Next transform to N "polarization" variables u_j, defined by

$$u_j \equiv -\sum_{n=j}^{N} q_n, \qquad j=1,2,\ldots,N\,. \tag{113}$$

The inverse relation is

$$q_j = \begin{cases} u_{j+1}-u_j & j<N \\ -u_N & j=N \end{cases}. \tag{114}$$

In particular,

$$\begin{aligned}
u_1 &= -(q_1+q_2+\cdots+q_N)\\
u_2 &= -(q_2+q_3+\cdots+q_N)\\
&\vdots\\
u_{N-1} &= -(q_{N-1}+q_N)\\
u_N &= -q_N
\end{aligned} \tag{115}$$

and

$$\begin{aligned}
q_1 &= u_2-u_1\\
q_2 &= u_3-u_2\\
&\vdots\\
q_{N-1} &= u_{N-1}-u_{N-1}\\
q_N &= -u_N\,.
\end{aligned} \tag{116}$$

In the continuum limit, the charge density is

$$\rho(x) = \partial_x u(x)\,, \tag{117}$$

which is why the u_j can be regarded as discrete polarization variables.

In terms of these polarization variables the equations of motion (110) and (112) are

$$\alpha^2 C' \ddot{\varphi} + E_{\rm J} \sin\varphi + \frac{\alpha^2}{L_{\rm sq}}\left(\varphi - 2\pi \frac{\Phi_{\rm x}}{\Phi_{\rm sc}}\right) = \frac{\alpha\, C_{\rm int}}{c_0 + C_{\rm int}}\, \dot{u}_1\,, \tag{118}$$

and

$$\begin{aligned} l_0\, \ddot{u}_1 - \frac{u_2 - u_1}{c_0} + \frac{u_1}{c_0 + C_{\rm int}} &= -\frac{\alpha\, C_{\rm int}}{c_0 + C_{\rm int}}\, \dot{\varphi}\,, \\ l_0\, \ddot{u}_2 - \frac{u_3 - 2u_2 + u_1}{c_0} &= 0\,, \\ l_0\, \ddot{u}_3 - \frac{u_4 - 2u_3 + u_2}{c_0} &= 0\,, \\ &\ \vdots \\ l_0\, \ddot{u}_{N-1} - \frac{u_N - 2u_{N-1} + u_{N-2}}{c_0} &= 0\,, \\ l_0\, \ddot{u}_N - \frac{-2u_N + u_{N-1}}{c_0} &= 0\,. \end{aligned} \tag{119}$$

A Lagrangian for this coupled system of equations is

$$\begin{aligned} L = {} & \frac{\alpha^2 C'}{2}\, \dot{\varphi}^2 - U(\varphi) + \sum_{j=1}^{N} \left(\frac{l_0}{2} \dot{u}_j^2 - \frac{u_j^2}{c_0} \right) + \frac{u_1 u_2 + u_2 u_3 + \cdots + u_{N-1} u_N}{c_0} \\ & - \frac{u_1^2}{2} \left(\frac{1}{c_0 + C_{\rm int}} - \frac{1}{c_0} \right) - \frac{\alpha C_{\rm int}}{c_0 + C_{\rm int}}\, \dot{\varphi}\, u_1\,, \end{aligned} \tag{120}$$

with $U(\varphi)$ defined as in (92). The canonical momenta are

$$p \equiv \frac{\partial L}{\partial \dot{\varphi}} = \alpha^2 C' \dot{\varphi} - \frac{\alpha\, C_{\rm int}}{c_0 + C_{\rm int}}\, u_1 \tag{121}$$

and

$$p_j \equiv \frac{\partial L}{\partial \dot{u}_j} = l_0 \dot{u}_j, \qquad j = 1, 2, \ldots, N\,. \tag{122}$$

Finally, the Hamiltonian is

$$H = H_\varphi + H_u + \delta H\,, \tag{123}$$

where

$$H_\varphi \equiv \frac{p^2}{2\alpha^2 C'} + U(\varphi), \tag{124}$$

$$H_u \equiv \sum_{j=1}^{N} \left(\frac{p_j^2}{2l_0} + \frac{u_j^2}{c_0} \right) - \frac{u_1 u_2 + u_2 u_3 + \cdots + u_{N-1} u_N}{c_0} + \frac{u_1^2}{C_1}, \tag{125}$$

with

$$C_1 \equiv \frac{2 c_0 C' (c_0 + C_{\rm int})^2}{C_{\rm int} [c_0 C_{\rm int} - (c_0 + C_{\rm int}) C']} \tag{126}$$

and

$$\delta H \equiv \frac{C_{\rm int}}{\alpha C' (c_0 + C_{\rm int})} p u_1 . \tag{127}$$

C' is defined in (111).

In addition to the expected squid-resonator interaction term δH, the resonator is itself modified by its coupling to the squid. That coupling results in an additional charging energy u_1^2/C_1 at the position where the squid is attached, modifying the resonator modes. There is also an additional capacitive loading of the squid, as described by the renormalized capacitance C'.

Now we assume

$$C_{\rm int} \ll C \quad \text{and} \quad C_{\rm int} \ll c_0 . \tag{128}$$

The second condition in (128) is perhaps counterintuitive, because in the continuum limit one would expect c_0 to vanish. However, there is a restriction on how small a can be for the network of Fig. 10 to describe a system with an extended squid-resonator contact region. Because the squid in Fig. 10 is electrically contacted to only a *single* cell of the network, we must require

$$a > b, \tag{129}$$

which implies [see (60) and (106)] the relation $C^* < c_0$. The requirement that $C_{\rm int} \ll C^*$ therefore leads to

$$C_{\rm int} \ll C^* < c_0 , \tag{130}$$

and hence to the second weak-coupling condition of (128).

With these assumptions the system Hamiltonian simplifies to

$$H = E_c N^2 + U(\varphi) + \sum_{j=1}^{N} \left(\frac{p_j^2}{2l_0} + \frac{u_j^2}{c_0} \right)$$

$$-\frac{u_1u_2+u_2u_3+\cdots+u_{N-1}u_N}{c_0}+\frac{C_{\rm int}}{\alpha Cc_0}pu_1\,. \tag{131}$$

In this weak coupling limit, the squid couples to the modes of the isolated resonator.

The Lagrangian for an isolated resonator ladder in the polarization representation (120) is

$$L_{\rm res}=\frac{l_0}{2}\sum_i \dot{u}_i^2-\frac{1}{2c_0}\sum_{ij}u_iK_{ij}u_j\,, \tag{132}$$

where

$$K\equiv\begin{pmatrix} 2 & -1 & 0 & 0 & \cdots & 0 & 0 & 0\\ -1 & 2 & -1 & 0 & \cdots & 0 & 0 & 0\\ 0 & -1 & 2 & -1 & \cdots & 0 & 0 & 0\\ & & & & \vdots & & & \\ 0 & 0 & 0 & 0 & \cdots & -1 & 2 & -1\\ 0 & 0 & 0 & 0 & \cdots & 0 & -1 & 2\end{pmatrix} \tag{133}$$

is an $N\times N$ matrix that can be recognized as a finite-difference representation of the operator $-\partial_x^2$, truncated in a manner consistent with the continuum boundary conditions (when acting on an eigenfunction). We wish to transform to a set of uncoupled generalized coordinates ξ_n. Let the $f_i^{(n)}$ [not to be confused with (69)] be the eigenvectors of K,

$$Kf^{(n)}=\lambda^{(n)}f^{(n)}\,, \tag{134}$$

with n labeling the eigenvectors, the fundamental mode being denoted by $n=1$. Because K is real and symmetric, its eigenvectors can be chosen to satisfy

$$\sum_i f_i^{(n)}f_i^{(n')}=\delta_{nn'}\quad\text{and}\quad\sum_n f_i^{(n)}f_j^{(n)}=\delta_{ij}\,. \tag{135}$$

Now we expand the polarization vector in this basis,

$$u_i=\sum_n \xi_n\,f_i^{(n)}\,, \tag{136}$$

and obtain

$$L_{\rm res}=\sum_n\left(\frac{l_0}{2}\dot{\xi}_n^2-\frac{\omega_n^2}{2c_0}\xi_n^2\right), \tag{137}$$

which describes independent harmonic oscillators; these are the eigenmodes of the resonator. The resonator frequencies are related to the eigenvalues λ

according to

$$\omega = \sqrt{\frac{\lambda}{l_0 c_0}} . \tag{138}$$

The fundamental mode solution of (134) can be found exactly. It is

$$f_i = A \sin\left(\frac{i\pi}{N+1}\right), \qquad i = 1, 2, \ldots, N . \tag{139}$$

Here

$$A \equiv \left[\sum_{i=1}^{N} \sin^2\left(\tfrac{i\pi}{N+1}\right)\right]^{-1/2} \tag{140}$$

is a normalization constant. The lowest eigenvalue is $\lambda = 2[1-\cos(\frac{\pi}{N+1})]$, and the fundamental-mode frequency is

$$\omega = \frac{2}{\sqrt{l_0 c_0}} \sin\left[\frac{\pi}{2(N+1)}\right] . \tag{141}$$

In the continuum (large N) limit,

$$\omega = \frac{\pi v}{d} , \tag{142}$$

in agreement with (58), and

$$A = \sqrt{\frac{2}{N}} . \tag{143}$$

Keeping only the fundamental mode and quantizing leads to

$$H_{\mathrm{res}} = \frac{p_\xi^2}{2 l_0} + \frac{l_0 \omega^2}{2} \xi^2 , \tag{144}$$

where p_ξ is the momentum conjugate to ξ. Expanding in bosonic creation and annihilation operators then leads to

$$\xi = \sqrt{\frac{\hbar}{2 l_0 \omega}} \left[a + a^\dagger\right] , \tag{145}$$

and

$$H_{\mathrm{res}} = \hbar\omega \, a^\dagger a . \tag{146}$$

In this representation the interaction is

$$\delta H = \frac{C_{\mathrm{int}} f_1}{\alpha C c_0} p \, \xi, \qquad f_1 = A \sin\left(\frac{\pi}{N+1}\right) . \tag{147}$$

After projecting the squid momentum into the qubit subspace according to (32), and projecting the resonator fundamental mode into the ground and one-photon subspace according to

$$\xi \to \sqrt{\frac{\hbar}{2l_0\omega}}\,\sigma^x\,, \tag{148}$$

we obtain the interaction

$$\delta H = g\,\sigma^y_\varphi \sigma^x_u\,, \tag{149}$$

with a coupling constant g that can be shown to be identical to (57) in the continuum limit.

4.4. *Relation between node-flux and polarization representations*

The result (149) appears to differ from (56), but these are written in different bases. In (56), the resonator degree of freedom has been expanded in a basis of eigenstates of node-flux, whereas in (149) a basis of polarization eigenstates is used. Because the transformation between node-flux and polarization is nonlocal in time, the connection between these bases is nontrivial.

Before proceeding, it is interesting to note that (149) and (56) are unitarily equivalent: They have the same spectrum and are therefore related by a unitary transformation. From this point of view it is natural to suspect that they are matrix representations of the same operator written in different bases.

To understand the relation between these representations, return to the description of the continuous isolated resonator in the node-flux representation. Keeping only the $n = 1$ fundamental mode terms in (99) and (100), they may be written as

$$\phi(x) = X\, f_1(x), \qquad X \equiv \sqrt{\frac{\hbar}{2\alpha^2 C\omega_{\rm res}}}\,\big[a_1 + a_1^\dagger\big]\,, \tag{150}$$

and

$$\Pi(x) = P\, f_1(x), \qquad P \equiv -i\sqrt{\frac{\alpha^2 C\hbar\omega_{\rm res}}{2}}\,\big[a_1 - a_1^\dagger\big]\,, \tag{151}$$

where

$$f_1(x) = \sqrt{\frac{2}{d}}\,\cos\!\left(\frac{\pi x}{d}\right). \tag{152}$$

X can be viewed as an operator describing the node-flux amplitude of the resonator fundamental mode, and P is its conjugate momentum. Inserting these projected quantities into the Hamiltonian density (98) and integrating, leads to a one-dimensional harmonic oscillator Hamiltonian for the fundamental-mode amplitude,

$$H_{\rm res} = \frac{P^2}{2\alpha^2 C} + \frac{\alpha^2 C \omega_{\rm res}^2}{2} X^2 \,. \tag{153}$$

In the node-flux representation, the fundamental-mode eigenstates are eigenfunctions of (153),

$$\psi_m(X) = (2^m m! \, \sqrt{\pi}\, \ell)^{-1/2} \, e^{-X^2/2\ell^2} \, \mathsf{H}_m\big(\tfrac{X}{\ell}\big) \,, \qquad \ell \equiv \sqrt{\frac{\hbar}{\alpha^2 C \, \omega_{\rm res}}}, \qquad m = 0, 1, 2, 3, \ldots \,, \tag{154}$$

where the H_m are Hermite polynomials.

For example, consider the matrix representation of the projected charge density operator $\rho(x)$ in this basis. According to (82),

$$\rho(x) = \alpha C \partial_t \phi(x) = \frac{\Pi(x)}{\alpha} = \frac{f_1(x)}{\alpha} P \,, \tag{155}$$

so the matrix elements in the basis (154) are

$$\begin{aligned} \langle m|\rho(x)|m'\rangle &= \frac{f_1(x)}{\alpha} \langle m|P|m'\rangle = -i\hbar \, \frac{f_1(x)}{\alpha} \times \int dX \, \psi_m \, \frac{\partial \psi_{m'}}{\partial X} \\ &= \sqrt{\frac{\hbar \omega_{\rm res} C}{d}} \cos\left(\frac{\pi x}{d}\right) \sigma^y_{mm'} \,, \end{aligned} \tag{156}$$

where we have further projected to the $m = 0, 1$ subspace.

Now let's compute the matrix elements of $\rho(x)$ in the polarization basis. In the continuum limit, (136) leads to

$$u(x) = \xi \sqrt{\frac{2a}{d}} \sin\left(\frac{\pi x}{d}\right), \tag{157}$$

where we have again projected into the fundamental mode. Similar to X, ξ can be viewed as an operator describing the polarization amplitude of the resonator fundamental mode. The resonator Hamiltonian in this representation is (144), and its eigenfunctions are

$$\psi_m(\xi) = (2^m m! \, \sqrt{\pi}\, \ell)^{-1/2} \, e^{-\xi^2/2\ell^2} \, \mathsf{H}_m\big(\tfrac{\xi}{\ell}\big), \qquad \ell \equiv \sqrt{\frac{\hbar}{l_0 \, \omega_{\rm res}}} \,. \tag{158}$$

Then according to (117), the matrix elements of the projected $\rho(x)$ in the basis (158) are

$$\langle m|\rho(x)|m'\rangle = \sqrt{\frac{2a}{d}} \frac{\pi}{d} \cos\left(\frac{\pi x}{d}\right) \times \int d\xi \, \psi_m \, \xi \, \psi_{m'}$$

$$= \sqrt{\frac{\hbar\omega_{\rm res}C}{d}} \cos\left(\frac{\pi x}{d}\right) \sigma^x_{mm'} \, . \tag{159}$$

Although I have used the same bracket notation in (156) and (159), it is to be understood that the $|m\rangle$ in these expressions refer to different basis functions.

Comparing (156) and (159), we conclude that the matrices σ^y_ϕ and σ^x_u appearing in (149) and (56) represent the same physical resonator operator written in different bases. Either basis can be used, although the choice of a measurement method may single out one as being more convenient.

Acknowledgments

This work was supported by the Disruptive Technology Office under grant W911NF-04-1-0204 and by the National Science Foundation under grant CMS-0404031. It is a pleasure to thank Andrew Cleland, Andrei Galiautdinov, John Martinis, Emily Pritchett, and Andrew Sornborger for useful discussions.

References

1. M.A. Neilsen and I.L. Chuang, Quantum Computation and Quantum Information (Cambridge University Press, Cambridge, England, 2000).
2. Y. Makhlin, G. Schön and A. Shnirman, Rev. Mod. Phys. **73**, 357–400 (2001).
3. J.Q. You and F. Nori, Superconducting circuits and quantum information, Physics Today, November 2005, p. 42.
4. M.R. Geller and A.N. Cleland, Phys. Rev. **A71**, 32311 (2005)
5. A. Blais, R.-S. Huang, A. Wallraff, S.M. Girvin and R.J. Schoelkopf, Phys. Rev. **A69**, 62320 (2004).

COHERENT CONTROL AND DECOHERENCE OF CHARGE STATES IN QUANTUM DOTS

P. MACHNIKOWSKI

Institute of Physics, Wrocław University of Technology,
50-370 Wrocław, Poland,
Pawel.Machnikowski@pwr.wroc.pl

This Chapter contains a review of the recent results, both experimental and theoretical, related to optical control of carriers confined in semiconductor quantum dots. The physics of Rabi oscillations of exciton and biexciton occupations, as well as time-domain interference experiments are discussed. Next, the impact of carrier–phonon interaction in a semiconductor structure is described and modern methods of theoretical description of the carrier–phonon kinetics and of the resulting dephasing are presented.

1. Introduction

The progress of semiconductor technology that took place in the 80s and 90s of the last century, in particular the rapid development of epitaxy and lithography techniques, has allowed physicists to manufacture and study structures in which carriers are confined to a small volume in space (tens of nanometers or even less).[1] In this way quantum dots (QDs), that is, artificial structures (boxes) containing a few particles with quantized energy levels, have been produced. Because of the similarity to natural atoms, such structures are also referred to as artificial atoms. However, the properties of QDs are more flexible in comparison with atoms: their shapes, size and various other features can be engineered at the stage of manufacturing by modifying the technological conditions and the number of confined electrons can be changed under laboratory conditions (e.g. by applying an external voltage) within a wide range of values.

Parallel to the laboratory investigations, a rapid process of miniaturization of commercial semiconductor structures (e.g., computer chips) took place. At this moment, the state-of-the-art commercially available microchips, manufactured using the immersion photolithography technique, are built of elements whose size can be reduced down to the 90 nm diffraction limit. The introduction of the 45 nm process is announced for 2007 or early 2008 and the implementation of the 16 nm technology is envisaged in the time frame 2013-14.[a] Thus, the characteristic size of elements in the microchips of our standard computer equipment has dropped below the size of the first QDs obtained in laboratories 20 years ago and rapidly approaches the size of the smallest structures described in today's research papers.

This progress of manufacturing technology is accompanied by a rapid development of optical spectroscopy. Currently, it is possible to study the optical properties of a single QD and to coherently control the evolution of a single carrier or a pair of carriers (electron-hole pair) in such a structure. Many experimental schemes of quantum optics have been implemented on QDs. Moreover, certain procedures relying on the specific structure of the energy levels of these artificial semiconductor structures have been demonstrated, which have no counterpart in atomic systems. These experimental achievements have motivated theoretical proposals for sophisticated quantum-optical schemes that may lead, for instance, to optical control of a single electron spin in a QD.

The goal of this chapter is twofold: First, to introduce the reader into the fascinating world of modern optical experiments on semiconductor quantum dots and to the astonishingly simple, yet nontrivial, theory underlying the phenomena observed in the labs on the fundamental, quantum-optical level. Second, to give a review of some more sophisticated theoretical methods that allow one to include the interaction with the lattice vibration modes (phonons) which are specific to semiconductor systems.

2. Essential properties of quantum dots

Quantum dots are semiconductor structures in which the carrier dynamics is restricted in all three dimensions to the length scales of several or a few tens of nanometers.[2,3] Various structures that have this property may be obtained by a variety of methods.

One of the most widely used techniques is the Stransky-Krastanov self-assembly. When a semiconductor compound is epitaxially (layer by layer) grown on a substrate with a different lattice constant (the InAs/GaAs pair is a typical example) each new layer must be squeezed to match the lattice constant of the substrate. At some point (at about 1.7 monolayers for InAs/GaAs) it is energetically favorable for the epitaxial layers to restruc-

[a]International Technology Roadmap for Semiconductors (www.itrs.net).

turize into a system of islands, which increases the free surface but relaxes strain.[4] The sample is then covered with the substrate material, leading to lens-shaped (or, sometimes, pyramidal) nanostructures as in Fig. 1a.[5] In general, the band edges of the nanostructure material are offset with respect to those of the substrate. Here we will only discuss structures as those in Fig. 1b-d, where the conduction band edge is shifted down and the valence band edge is shifted up, so that both electrons and holes are bound in the QD structure.

Another type of structures commonly used in the optical experiments are thickness fluctuations of a thin epitaxial layer of a semiconductor (so called quantum well) placed between thick structures of different semiconductor with a wider band gap. Here, again, the band edge offset leads to localization of carriers within the quantum well layer. If the epitaxial growth of the quantum well layer has been stopped after the formation of a new monolayer started, the quantum well has one-monolayer thickness fluctuations which weakly localize the carriers.

The QD nanostructures are typically 2–3 orders of magnitude larger than atoms. However, the effective mass of carriers in a semiconductor is often considerably lower than the free electron mass (e.g. $m^* = 0.07m_0$ in GaAs), and this degree of confinement is sufficient for quantization of carrier energies with electron inter-level spacing reaching 100 meV in self-assembled structures. This is definitely enough to resolve the states spectrally and to neglect thermal transitions to the excited states even at moderate temperatures. Therefore, we will restrict the discussion to the ground state of each kind of carriers.

The exact properties of the quantum states in a QD, e.g., the geometry of wave functions or Coulomb interaction between the confined carriers, may be found, e.g., by tight-binding or pseudopotential calculations. For lens-shaped QDs, a simple 2-dimensional harmonic model[2] has been shown to be a very good approximation.[6] The states of an interacting few-particle system may then be found by numerical configuration–interaction techniques.[7,8] However, these details are irrelevant for the discussion on the general quantum-optical level. Whenever a specific model is necessary (Secs. 6 and 7) we will use simple Gaussian wave functions. Also in these cases, the results do not depend essentially on this choice.

Even with the restriction to the lowest orbital states, the structure of energy levels and allowed optical transitions in a QD becomes quite complicated if the angular momenta of carriers are taken into account. The valence band in III-V semiconductors is composed of p-type atomic orbitals

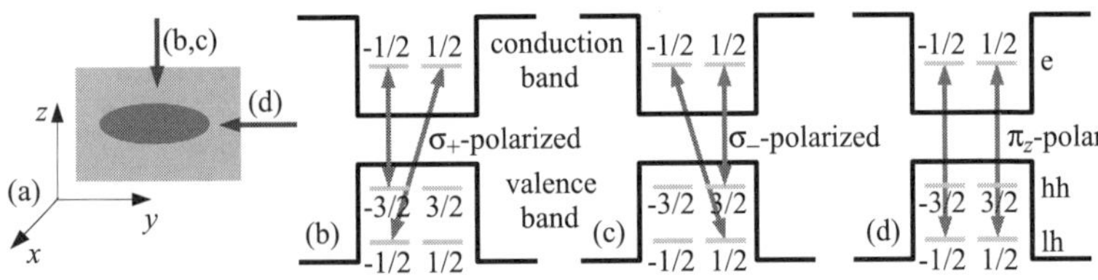

Fig. 1. (a) Schematic plot of a QD. The xy plane is referred to as the structure plane, while z is called growth direction and corresponds to the (approximate) symmetry axis of the structure. The arrows show the incidence of light corresponding to the three diagrams b-d. (b-d) The diagrams showing the transitions allowed by selection rules in a III-V QD, induced by light circularly polarized in the structure plane (b: right-polarized, c: left-polarized) and by light linearly polarized along z (d).

(orbital angular momentum 1), yielding six quantum states (taking spin into account) for each quasi-momentum $\boldsymbol{k}$. Due to considerable spin-orbit coupling the orbital angular momentum and spin are not separate good quantum numbers and the valence band states of a bulk crystal must be classified by the total angular momentum and its projection on a selected axis. Thus, the valence band is composed of three sub-bands corresponding to two different representations of the total angular momentum J. Out of these, the two states with $j = 1/2$ form a subband which is split-off by the spin–orbit interaction. The other four states with $j = 3/2$ are degenerate in a bulk crystal at $\boldsymbol{k} = 0$ but this degeneracy is lifted by size quantization and strain in a QD structure, with the heavy hole (hh) subband (angular momentum projection on the symmetry axis ($m = \pm 3/2$) lying above the light hole (lh) subband ($m = \pm 1/2$) in all known structures. The fundamental optical excitation of a QD consists in transferring optically an electron from the highest confined state in the valence band (thus leaving a hole) to the lowest confined state in the conduction band. The interacting electron–hole pair created in this way is referred to as exciton (lh-exciton or hh-exciton, depending on the kind of a hole involved). Angular momentum selection rules restrict the transitions allowed for a given propagation direction and polarization of the light beam, as depicted in Fig. 1. For instance, according to the selection rules represented in Fig. 1b, a right circularly polarized (σ_+-polarized) laser beam can only create an exciton with total momentum +1, referred to as "σ_+ exciton", in accordance with the angular momentum conservation (removing an electron with the angular momentum m is equivalent to the creation of a hole with the angular momentum $-m$). Similarly, a σ_--polarized (left circularly polarized) beam creates only a "σ_- exciton" with the angular momentum -1. Although there are still

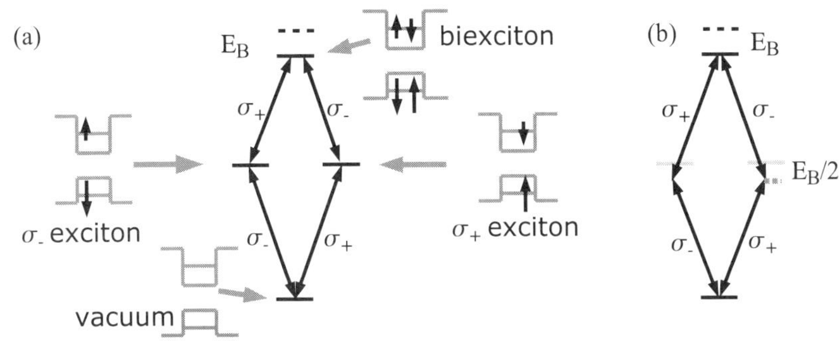

Fig. 2. (a) The four states of the heavy-hole biexciton system and the optical transitions between them. The band diagrams show the particles forming each state. The arrows in the valence band represent holes. (b) Schematic representation of the two-photon resonance between the ground state and the biexciton state, achieved with a properly tuned linearly polarized laser beam. Note that the single exciton transitions are forbidden.

two transitions allowed for a given circular polarization they can easily be distinguished since the lh states are well separated energetically from the lowest hh states.

In appropriately doped structures, QDs in the ground state of the system may be occupied by electrons. An optical excitation in this case corresponds to a transition between a single electron state and a negative trion state, i.e., the state of two electrons and one hole confined in a QD. From the Pauli exclusion principle it is clear that this transition is possible only if the state which is to be occupied by the photo-created electron is free. Hence, in the situation of Fig. 1b, a heavy hole trion may be created if the dot is initially occupied by a "spin up" ($m = +1/2$) electron but not if the electron in the dot is in the "spin down" ($m = -1/2$) state. This suppression of the optical transition depending on the spin of the electron in the QD is referred to as Pauli blocking and has been indeed observed experimentally.[9]

After exciting an electron from the heavy hole band with a σ_+-polarized laser beam ($m = -3/2 \to -1/2$), as in Fig. 1b, it is still possible to transfer also the $m = +3/2$ electron to the $m = +1/2$ conduction band state using σ_--polarized light (Fig. 1c). Thus, if the optical processes are spectrally restricted to heavy holes (as is the case in most experiments) there are four optically active states linked by allowed optical transitions as shown in Fig. 2a. The highest state, with two electrons and two holes present in the QD, is treated as composed of two electron-hole pairs and is called a biexciton. The Coulomb (dipole-dipole) interaction between the two excitons shifts the energy of the biexciton state[10] by the binding energy E_B, so that the exciton–biexciton transitions are non-degenerate with the ground state–exciton transitions. In this way, all four transitions in the diagram can be distinguished either spectrally or by polarizations.

3. Coherent control: experimental state of the art

The recent progress in ultrafast spectroscopy of semiconductor systems[11,12] made it possible to control and probe the quantum states of carriers confined in a QD on femtosecond time scales. In particular, high degree of control over carrier occupations in various kinds of QD structures has been demonstrated. One kind of an experiment consists in measuring the average occupation of the QD after a pulse of fixed length but variable amplitude. The QD occupation is defined as the probability of finding an exciton in the QD after the laser pulse, calculated as the fraction of cases in which an exciton was created over a large number of repetitions of the experiment. Within the linear absorption theory, the excitation (the QD occupation) grows proportionally to the pulse intensity. Obviously, this growth cannot be unlimited. When the occupation of the excited level is sufficiently large the spontaneous and induced emission processes suppress further increase of the occupation. Here, we are interested in the coherent limit, where the evolution is induced by a strong (essentially classical) laser field, inducing large occupation changes over time scales much shorter than the spontaneous emission time. Then, the system is driven from the ground to the excited state in a coherent way, and than back to the ground state via a coherently induced emission process. As a result, in an ideal case, the final occupation after a pulse should show sinusoidal oscillations between 0 and 1 as a function of the square root of pulse intensity (referred to as *pulse area* – see Sec.4.2). Such oscillations, known as pulse area dependent Rabi oscillations, have been indeed observed in a range of experiments on single QDs.[13–16] Similar effect can also be observed in ensembles of QDs.[17]

In Fig. 3a we show the results of such an experiment, performed with a single QD-based photodiode structure.[15,16,18] In this experiment the QD is placed in an electric field between a pair of electrodes (Fig. 3c). Each electron-hole pair generated by the optical excitation contributes to the photocurrent in the structure. If the laser pulse repetition rate is f (typically in kHz–MHz range) and the average QD occupation for a given pulse area is n, then the repeated pulsing results in a current $I = nef$. In this way, the Rabi oscillations of the average exciton number are reflected in the oscillating form of the photocurrent as a function of the pulse area, which provides a means of detection much more efficient than optical ones. These

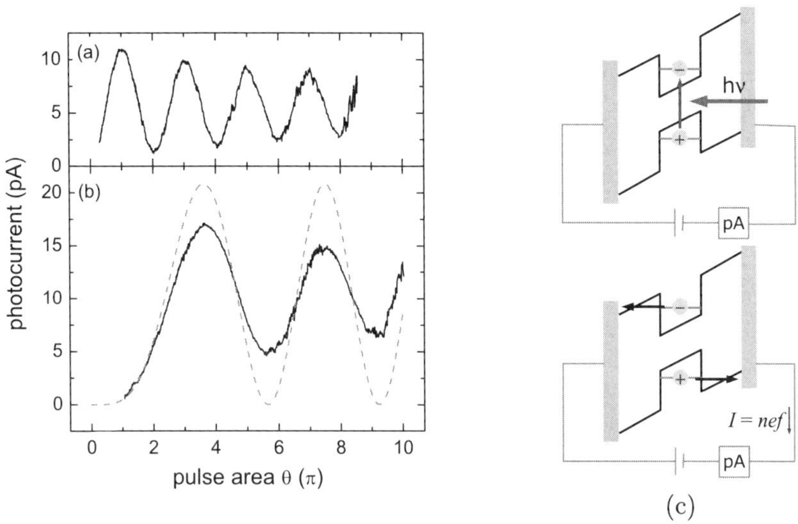

Fig. 3. (a) Pulse area dependent Rabi oscillations. (b) Two-photon Rabi oscillations. Solid line: experiment, dashed line: theory (without dephasing), assuming a sech pulse shape with a pulse length of 2.3 ps and biexciton binding energy $E_B = 2.7$ meV. Reprinted from Ref. [18], S. Stufler *et al.*, Phys. Rev. **B73**, 125304 (2006), ©American Physical Society 2006. (c) A schematic explanation of the functionality of the QD photodiode.

oscillations are clearly seen in Fig. 3a, although some damping, due to dephasing, is also visible.

Apart from the coherent control of exciton occupations, it has also been shown that phase control of carrier states in QDs is possible. A laser pulse detuned from the exciton resonance cannot induce real transitions and, therefore, does not change occupations. Nonetheless, as shown in an optical experiment with interface fluctuation QDs,[19] it can shift the energy levels via the AC (optical) Stark effect and affect the phases in a quantum superposition of empty dot and single-exciton states. Another way to control the phases is to drive the system with a slightly detuned laser pulse, which leads to a combination of occupation and phase evolution[20] (see Sec. 4.2). In this case, phase effects can be observed in the form of Ramsey interference fringes.[21]

It is possible to see in an experiment that the quantum state of an electron-hole pair in a QD maintains its phase coherence long after the laser pulse has been switched off. To this end, one splits the pulse into two parts. If the phase of the quantum state in the QD were random when the second pulse arrives the phase of the latter would be irrelevant. On the contrary, in the experiment one observes oscillations of the final QD occupation as a function of the relative phase shift between the two pulses[14,22,23] (the phase is shifted by tuning the delay between the pulses with a sub-femtosecond accuracy, that is, by a fraction of the optical oscillation period). Such an effect is referred to as time-domain interference, since it may be interpreted in terms of interference between the probability amplitudes for exciting the QD with the first or with the second pulse. A formal treatment of this class of experiments will be given in Sec. 4.3.

Apart from the fact that such interference experiments demonstrate coherent phase-sensitive quantum control with an amazingly precise timing, they are also of interest from a more general point of view. Obviously, in spite of the wave-like behavior manifested by the interference effect, a single measurement of the QD occupation always yields either 0 or 1, demonstrating the particle-like nature of the exciton. Thus, the time-domain interference experiments on QDs demonstrate quantum complementarity between the particle-like nature of an exciton and its ability to show quantum interference.[24]

Controlling a single quantum degree of freedom (an exciton) is just the first step towards large-scale nano-optoelectronic and quantum computing applications. The next step towards more complex implementations is to include coupling between two or more individual quantum subsystems. The simplest experimental realization of such a composite system is a biexciton. As explained in Sec. 2, due to the Coulomb interaction between the two excitons, the excitation energy of an exciton in the presence of the other one is different than in its absence. Thus, these two transitions can be addressed individually. Together with the polarization dependence of the allowed transitions (selection rules, see Sec. 2), this allows one to excite, say, the σ_+ exciton if and only if the the σ_- exciton is present. Such a conditional control procedure was indeed performed in an experiment,[25] where Rabi oscillations on the exciton–biexciton transition were demonstrated. If the two excitons are viewed as quantum bits (with 0 and 1 corresponding to their presence or absence in the QD) then such a conditional excitation constitutes an implementation of the controlled-NOT gate which is fundamental for quantum commuting.[26] Such a controlled-NOT gate was indeed implemented experimentally in a QD system.[25] In a similar way it is possible to coherently manipulate a biexciton system in two coupled QDs (with one exciton localized in each dot).[27]

It is interesting to see what happens if the frequency ω of a linearly polarized laser beam is chosen such that the energy of two photons matches the biexciton energy, $2\omega = 2E - E_{\rm B}$, while the single-exciton transitions are detuned by $E_{\rm B}/2$ (Fig. 2b). A linearly polarized pulse is a superposition of two circularly polarized components, so that both exciton transitions are allowed by the selection rules but neither of them satisfies the energy conservation. Perturbation theory would predict that the occupation of the biexciton state should grow proportionally to the square of the pulse intensity (that is, to the 4th power of the pulse area), as a result of a two-photon absorption process. Experimental results[18] presented in Fig. 3b indeed show such a behavior for low pulse intensities. However, beyond this perturbative regime, a pattern of two-photon pulse area dependent Rabi oscillations between the ground and biexciton states develops, which become almost periodic for large pulse intensities. The theory of such coherent phenomena in the biexciton system will be presented in Sec. 5.

4. Quantum dot as a two-level system

Let us now proceed to a theoretical description of the quantum evolution of carriers confined in a QD and subject to a laser field. We will start with the simplest situation when the QD is driven by a circularly polarized laser beam (say, σ_+) tuned to the fundamental heavy hole transition, as in Fig. 1a (vertical arrow). Then, only two states are involved in the evolution: the ground state (empty dot), denoted by $|0\rangle$, and the σ_+ exciton state, denoted $|1\rangle$ (see also Fig. 2a). Therefore, one effectively deals with a very simple two-level system.

4.1. *General considerations*

The discussion of such systems is made particularly transparent by introducing the concept of the Bloch sphere (actually, a ball). The state (pure or mixed) of such a system is represented by a density matrix: a 2×2, hermitian, and positive definite operator with unit trace. Any such operator can be written in the form

$$\rho = \frac{1}{2}(\mathbb{I} + \boldsymbol{n} \cdot \boldsymbol{\sigma}), \tag{1}$$

where $\boldsymbol{n}$ is a real three-dimensional vector with $n = |\boldsymbol{n}| \leq 1$, $\mathbb{I}$ is the unit operator, and $\boldsymbol{\sigma}$ is the vector of Pauli matrices in the basis $|0\rangle, |1\rangle$. Thus, the state of a two-level system is represented in a unique way by a unit ball in a three-dimensional real space (Fig. 4). Since $\operatorname{Tr} \sigma_i = 0$, $\operatorname{Tr} \mathbb{I} = 2$, and $(\boldsymbol{n} \cdot \boldsymbol{\sigma})^2 = n^2 \mathbb{I}$, one finds $\operatorname{Tr} \rho^2 = n^2$, so that pure states (represented by

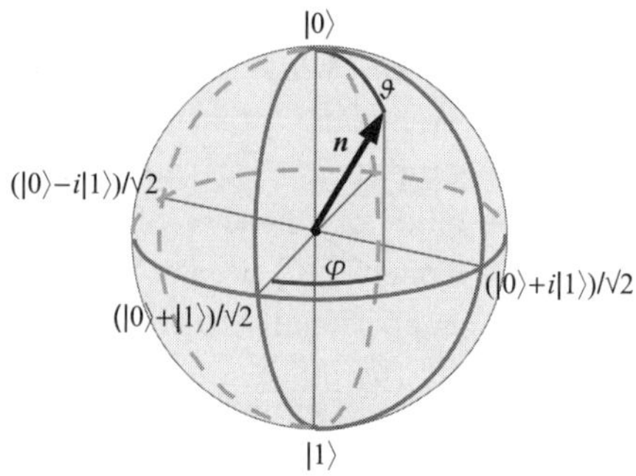

Fig. 4. The Bloch sphere.

projectors) correspond to unit vectors $\boldsymbol{n}$ and are mapped to the surface of the ball. It is easy to see that the point with spherical coordinates (ϑ, φ) represents the state vector $\cos(\vartheta/2)|0\rangle + e^{i\varphi}\sin(\vartheta/2)|1\rangle$.

The electric field of the laser beam is conveniently written as $\mathcal{E}(t)\cos(\omega t - \phi)$, where ω is the frequency of the light, ϕ is the phase of the pulse and $\mathcal{E}(t)$ is the envelope of the pulse, varying slowly compared to the oscillations of the optical field. The state of the system $|\Psi\rangle$ evolves according to the Schrödinger equation

$$i\hbar\frac{d}{dt}|\Psi\rangle = H|\Psi\rangle, \tag{2}$$

with the Hamiltonian

$$H = E|1\rangle\langle 1| + \mathcal{E}(t - t_{\mathrm{a}})\cos[\omega(t - t_{\mathrm{a}}) - \phi]\left(\mu|0\rangle\langle 1| + \mu^{*}|1\rangle\langle 0|\right), \tag{3}$$

where E is the energy of the interband transition in the QD, μ is the off-diagonal (interband) matrix element of the electric dipole moment (between the two relevant states), and we allow the pulse to arrive at an arbitrary time t_{a}. By a proper choice of the phase of the state $|1\rangle$ it is possible to have μ real, which will be assumed in the following. We will use the shorthand notation $f(t) = \mu\mathcal{E}(t)$.

In the absence of the driving, the phase of the state $|1\rangle$ rotates with the frequency $E/\hbar$ which is of the order of fs^{-1}. One gets rid of this trivial fast dynamics by describing the system in the frame rotating with the same frequency. This is similar to a transition to the interaction picture except that, for practical reasons, it is convenient to perform the transformation with the laser frequency ω instead of the system frequency $E/\hbar$ (these two frequencies are close to each other). Thus, we perform the transformation

$$|\tilde{\Psi}\rangle = U_{\mathrm{r}}|\Psi\rangle, \quad U_{\mathrm{r}} = \exp(i\omega t|1\rangle\langle 1|).$$

Using Eq. (2) one easily finds the evolution equation for the redefined states in the form $i\hbar|\dot{\tilde{\Psi}}\rangle = \tilde{H}|\tilde{\Psi}\rangle$, with the Hamiltonian

$$\begin{aligned}\tilde{H} &= -\hbar\omega|1\rangle\langle 1| + U_{\mathrm{r}} H U_{\mathrm{r}}^{\dagger} \\ &= -\Delta|1\rangle\langle 1| + \frac{1}{2} f(t - t_{\mathrm{a}}) \left[\left(e^{-i\omega t_0 - i\phi} + e^{-i\omega(2t - t_0) + i\phi} \right) |0\rangle\langle 1| \right. \\ &\quad \left. + \left(e^{i\omega t_0 + i\phi} + e^{i\omega(2t - t_0) - i\phi} \right) |1\rangle\langle 0| \right],\end{aligned} \tag{4}$$

where $\Delta = \hbar\omega - E$ is the detuning of the laser beam from the transition energy.

The next essential step is to note that the natural frequency scale of the system evolution is set by the detuning Δ and by the pulse amplitude $f(t)$ and is many orders of magnitude smaller than the optical frequency ω. Therefore, the quickly oscillating terms ($\sim e^{2i\omega t}$) are strongly off-resonant and will have very little impact on the system evolution. Therefore, they can be neglected, which leads to the rotating wave approximation (RWA).[28] As a result, one obtains the following RWA Hamiltonian

$$\begin{aligned}H_{\mathrm{RWA}} &= -\Delta|1\rangle\langle 1| + \frac{1}{2} f(t - t_{\mathrm{a}}) \left(e^{-i(\omega t_0 + \phi)} |0\rangle\langle 1| + e^{i(\omega t_0 + \phi)} |1\rangle\langle 0| \right), \\ &= -\Delta|1\rangle\langle 1| + \frac{1}{2} f(t - t_{\mathrm{a}}) \hat{\boldsymbol{u}} \cdot \boldsymbol{\sigma},\end{aligned} \tag{5}$$

where $\hat{\boldsymbol{u}} = [\cos(\omega t_{\mathrm{a}} + \phi), \sin(\omega t_{\mathrm{a}} + \phi), 0]$.

4.2. *Pulse area dependent Rabi oscillations*

Let us start with applying the formalism to the Rabi oscillations described in Sec. 3. First, assume that the pulse is resonant with the optical transition in the QD, that is, $\Delta = 0$. In this case, the evolution operator generated by the RWA Hamiltonian (5) can be found analytically. Indeed, in the resonant case the Hamiltonians at different times commute with one another, so that the evolution operator may be written as

$$U(t) = \exp\left[\frac{1}{2}\Phi(t)\hat{\boldsymbol{u}} \cdot \sigma\right] = \cos\frac{\Phi(t)}{2}\mathbb{I} - i\sin\frac{\Phi(t)}{2}\hat{\boldsymbol{u}} \cdot \boldsymbol{\sigma}, \tag{6}$$

where $\Phi(t) = \int_{t_0}^{t} d\tau f(\tau)$, t_0 is the initial time of the evolution, and the last identity is easily proven by expanding the exponent in a series and collecting the odd- and even-order terms using the identities $(\hat{\boldsymbol{u}} \cdot \boldsymbol{\sigma})^{2m} = \mathbb{I}$ and $(\hat{\boldsymbol{u}} \cdot \boldsymbol{\sigma})^{2m+1} = \hat{\boldsymbol{u}} \cdot \boldsymbol{\sigma}$ for any natural number m. We will always assume that the initial time is before any pulses were switched on, so that one can set $t_0 \to -\infty$. The value of $\alpha \equiv \Phi(\infty)$ is called the pulse area and

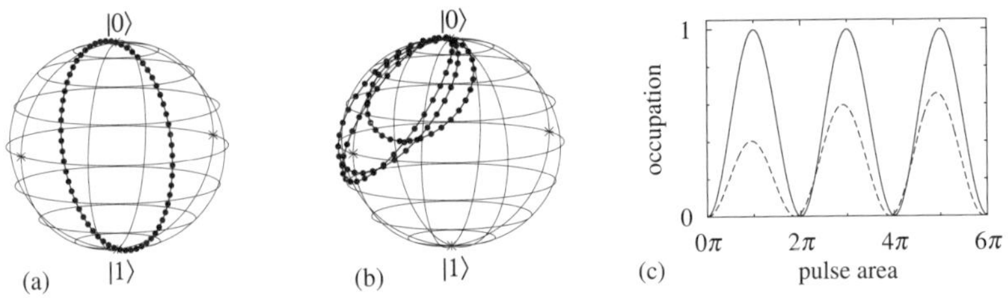

Fig. 5. (a) Pulse-area dependent Rabi oscillations at the resonance. The points represent final system states for different pulse areas $\alpha \in (0, 2\pi)$ with a step of $\Delta\alpha = 0.1$. (b) Pulse-area dependent Rabi oscillations off resonance. The points represent final system states for different pulse areas $\alpha \in (0, 6\pi)$ with a step of $\Delta\alpha = 0.2$. (c) The occupation of the state $|1\rangle$ as a function of the pulse area for resonant (solid) and non-resonant (dashed) driving.

determines the unitary transformation of the system state performed by the complete pulse. One speaks of π-pulses, $\pi/2$ pulses etc., referring to the value of the pulse area.

Starting from the ground system state $|0\rangle$ one obtains, after switching the pulse off,

$$|\tilde{\Psi}\rangle = U(\infty)|0\rangle = \cos\frac{\alpha}{2}|0\rangle - ie^{i\phi}\sin\frac{\alpha}{2}|1\rangle. \tag{7}$$

The occupation of the state $|1\rangle$ is, therefore,

$$|\langle 1|\Psi\rangle|^2 = |\langle 1|\tilde{\Psi}\rangle|^2 = \sin^2\frac{\alpha}{2}$$

and indeed oscillates between 0 and 1 as a function of the pulse area α. The final system states for a set of values of α in this resonant case are shown in Fig. 5a. Let us note here that the final state is determined by a simple function of just one quantity, the pulse area, and is independent of any details of the pulse shape. This fact is known as the area theorem.

Off resonance (for $\Delta \neq 0$), the evolution can be found in a closed analytical form only for rectangular pulse envelopes $f(t)$.[29] Instead of performing this simple exercise, let us look at the evolution for a family of Gaussian pulses

$$f(t) = \frac{\hbar\alpha}{\sqrt{2\pi}\tau_{\mathrm{p}}} e^{-\frac{1}{2}\left(\frac{t}{\tau_{\mathrm{p}}}\right)^2} \tag{8}$$

obtained by (also very simple) numerical integration of the Schrödinger equation (2), shown in Fig. 5b (see also Ref. [20]). In the detuned case, the system does not reach the $|1\rangle$ state. Instead, the state is rotated around a tilted axis and the final state does not show periodicity as a function of the pulse area. The difference between the resonant and non-resonant case is also visible in the dependence of the final occupation of the state $|1\rangle$ (Fig. 5c).

4.3. *Time-domain interference*

Another experiment that can easily be explained based on the two level model is that of time domain interference. As discussed in Sec. 3, such experiments are performed with two laser pulses selectively tuned to the exact resonance with one of the two fundamental optical transitions. The pulses are generated by splitting a single laser pulse and delaying one part with respect to the other. In this way, the two pulses are phase-locked, i.e., their relative phase is definite and determined by the delay time τ. The first pulse arrives at $t_{\mathrm{a}} = 0$ and prepares the initial superposition state. In a usual two-slit (space-domain) experiment, this would correspond to splitting the particle path. The second pulse arrives at $t_{\mathrm{a}} = \tau$. This pulse plays the role of "beam merger" providing, at the same time, a phase shift between the "paths".

The Hamiltonian describing the system driven by the two pulses is obtained by an obvious generalization of Eq. (5).

$$H = \frac{1}{2} f_1(t)(|0\rangle\langle 1| + |1\rangle\langle 0|) + \frac{1}{2} f_2(t-\tau)(e^{-i\omega\tau}|0\rangle\langle 1| e^{i\omega\tau}|1\rangle\langle 0|), \qquad (9)$$

where $f_i(t)$ are the envelopes of the two pulses and we have set $\phi = 0$. The pulses do not overlap in time so that the evolution can be split into two independent stages.

In an experiment, the system is initially in the state $|0\rangle$. The first pulse is a $\pi/2$ pulse that performs the transformation $U_1 = (\mathbb{I} - i\sigma_x)/\sqrt{2}$. This pulse leaves the system in the equal superposition state

$$|\psi\rangle = \frac{|0\rangle - i|1\rangle}{\sqrt{2}}. \qquad (10)$$

The second pulse is again a $\pi/2$ pulse,

$$U_2 = \frac{1}{\sqrt{2}} (\mathbb{I} - i\hat{\boldsymbol{n}} \cdot \boldsymbol{\sigma}), \qquad (11)$$

where $\hat{\boldsymbol{n}} = [\cos\omega\tau, \sin\omega\tau, 0]$, as follows from the general discussion in Sec. 4.1. After this pulse, the average number of excitons in the dot is

$$N(\phi) = |\langle 1|U_2|\psi\rangle|^2 = \frac{1}{2}(1 - \cos\omega\tau),$$

and changes periodically between $N_{\min} = 0$ and $N_{\max} = 1$ as a function of the delay time τ, thus producing an interference pattern.

The quality of the interference pattern is quantified in terms of visibility of interference fringes

$$\mathcal{V} = \frac{N_{\max} - N_{\min}}{N_{\max} + N_{\min}}.$$

The amplitude (visibility) of the fringes in an ideal experiment is $\mathcal{V} = 1$ for $\pi/2$ pulses, i.e., for an equal superposition state in between the pulses. Otherwise, some *a priori* information on the superposition state can be inferred and the visibility is reduced. Reduction of visibility occurs also if some information on the system state between the pulses (analogous to *which way* information) has been extracted either intentionally, in a controlled experiment (see Sec. 5.2), or as a result of uncontrolled dephasing.[24]

5. Beyond the two levels: optically driven evolution in a biexciton system

The quantum control of a two-level system formed by the ground state and a single exciton exploits only small part of the possibilities offered by a QD. Much more interesting physics, as well as potential applications, becomes available if all the states are involved in the dynamics. Below we study two quantum optical schemes involving the confined biexciton system. First, the theory of the experimentally observed two-photon Rabi oscillations is given.[18] Then, a theoretical proposal for an experiment highlighting the role of quantum complementarity in time-domain interference is described.[24]

5.1. *Two-photon Rabi oscillations*

If we apply circularly polarized excitation to a QD in the ground state we can only create single excitons, as follows from the selection rules discussed in Sec. 2 and from the Pauli exclusion principle. By using linear polarization, which is a superposition of two equal σ_+ and σ_- polarized components, one enables also biexciton generation. In this case, the ground state is coupled to the biexciton state via both exciton states. Here we will consider a situation in which the laser frequency is tuned to the two-photon resonance with the biexciton, while the single-exciton states are detuned by $\Delta = E_{\mathrm{B}}/2$, as shown in Fig. 2b.

In the rotating wave approximation, the Hamiltonian of the biexcitonic system described above can be written in the form

$$H = \frac{E_B}{2}\left(|+\rangle\langle+| + |-\rangle\langle-|\right) + \frac{f(t)}{2\sqrt{2}}\left[(|\mathrm{g}\rangle + |\mathrm{XX}\rangle)(\langle+| + \langle-|) + \mathrm{H.c.}\right], \quad (12)$$

where $|+\rangle$,$|-\rangle$ and $|\mathrm{XX}\rangle$ are the σ_+, σ_- and biexciton states, respectively. Thus, the laser pulse couples only one (bright) combination of the ground and biexciton states, $|\mathrm{B}\rangle = (|\mathrm{g}\rangle + |\mathrm{XX}\rangle)/\sqrt{2}$ to one (π_x–polarized) single exciton state $|\mathrm{X}\rangle = (|+\rangle + |-\rangle)/\sqrt{2}$.

The system evolution is easily described in the new basis $|Y\rangle = (|+\rangle - |-\rangle)/\sqrt{2}$,$|\mathrm{D}\rangle = (|\mathrm{g}\rangle - |\mathrm{XX}\rangle)/\sqrt{2}$, $|\mathrm{X}\rangle$,$|\mathrm{B}\rangle$, where the first two states are the (decoupled) π_y-polarized exciton state and the dark combination of the ground and biexciton states. In this basis, the Hamiltonian (12) can be written as

$$H = \begin{pmatrix} E_B/2 & 0 & 0 & 0 \\ 0 & 0 & 0 & 0 \\ 0 & 0 & E_B/2 & f(t)/\sqrt{2} \\ 0 & 0 & f(t)/\sqrt{2} & 0 \end{pmatrix}, \quad (13)$$

which corresponds to the detuned rotation in the invariant subspace spanned by the x–polarized exciton state $|X\rangle$ and the state $|\mathrm{B}\rangle$, while the first two states are decoupled and undergo only a trivial evolution. The initial (ground) state is now written as $|\mathrm{g}\rangle = \frac{|\mathrm{D}\rangle + |\mathrm{B}\rangle}{\sqrt{2}}$.

Let us consider the instantaneous (for fixed t) eigenstates of the Hamiltonian (13) as a function of $f(t)$. For a large binding energy E_{B} the two branches belonging to the non-trivial two-dimensional invariant subspace are always widely separated. Thus, under the action of a sufficiently slowly varying laser pulse the system undergoes an adiabatic evolution with the pulse envelope $f(t)$ playing the role of a slowly varying parameter of the Hamiltonian. Therefore, one can assume[30] that at each time t the state corresponds to the adiabatic (instantaneous) eigenstate of the Hamiltonian (13), i.e.,

$$|\psi(t)\rangle = \frac{|\mathrm{D}\rangle + \left[c_-(t)|\mathrm{X}\rangle + c_+(t)|\mathrm{B}\rangle\right] e^{-i\Lambda(t)}}{\sqrt{2}},$$

where

$$c_\pm(t) = \frac{1}{\sqrt{2}}\left(1 \pm \frac{E_B}{\sqrt{(E_B)^2 + 8f^2(t)}}\right)^{1/2},$$

and

$$\Lambda(t) = \frac{1}{4\hbar}\int_{-\infty}^{t} d\tau \left(E_B - \sqrt{(E_B)^2 + 8f^2(\tau)}\right). \quad (14)$$

After the laser pulse has been switched off, the state becomes

$$|\psi\rangle = \frac{|\mathrm{D}\rangle + e^{-i\Lambda(\infty)}|\mathrm{B}\rangle}{\sqrt{2}} = e^{-i\Lambda(\infty)/2}\left(\cos\frac{\Lambda(\infty)}{2}|\mathrm{g}\rangle + \sin\frac{\Lambda(\infty)}{2}|\mathrm{XX}\rangle\right). \tag{15}$$

and coincides with the biexciton state for $\Lambda(\infty) = \pi$. Thus, the system effectively performs Rabi oscillations between the ground state and the biexciton state, with the occupation of the biexciton state given by

$$N_{\mathrm{XX}} = |\langle\mathrm{XX}|\psi\rangle|^2 = \sin^2\frac{\Lambda(\infty)}{2}. \tag{16}$$

For $\theta \ll \tau_0\Delta E_B/\hbar$ one has

$$\Lambda(\infty) \approx \frac{1}{\hbar\Delta E_B}\int_{-\infty}^{\infty} dt f^2(t) = \frac{4\hbar\mathrm{Arcosh}\sqrt{2}}{\pi^2\Delta E_B\tau_0}\theta^2$$

and the biexciton occupation N_{XX} grows as θ^4, i.e., proportional to the square of the pulse intensity, as expected for a two-photon process (see Fig. 3).

It should be noted that the pulse area appearing as the parameter of the standard Rabi oscillations is now replaced by the adiabatic dynamical phase Λ, which is a nontrivial functional of the pulse shape [Eq. (14)]. In this way, the simple universal dynamics of a resonantly driven two level system, described by the area theorem, is replaced by a more complicated pattern of oscillations, depending on the detuning parameter $\tau_{\mathrm{p}}E_{\mathrm{B}}$ but also on the exact pulse shape (e.g. Gaussian vs. sech[18]). The absence of any single exciton occupation in the final state is obviously due to the fact that the adiabatic limit corresponds to weak and long (i.e., spectrally narrow) pulses and the excitation of single-exciton states becomes forbidden by energy conservation. Nonetheless, during the evolution the single-exciton states are also occupied.

5.2. *Quantum complementarity in time-domain interference experiments*

This Section discusses the essential modification to time domain interference experiments (Sec. 4.3) that allows one to attain partial information on the state of the system and to observe the related visibility reduction of the interference pattern. First, however, a measure of the partial information is introduced and the complementarity principle is formulated in a quantitative form.[31,32]

The notion of "partial information" is understood as follows. The system (S) of interest is coupled to another *quantum probe* (QP) system and conditional dynamics of the latter is induced, leading to correlations between the states of the systems S and QP. Next, a measurement on QP is performed and its result is used to infer the state of S, i.e., to predict the outcome of a subsequent measurement on S. The probability of a correct prediction ranges from 1/2 (guessing at random in absence of any correlations) to 1 (knowing for sure, when the systems are maximally entangled).

Quantitatively, an intrinsic measure of information on the system S extracted by QP is provided by the *distinguishability of states*,[31,32]

$$\mathcal{D} = 2\left(p - \frac{1}{2}\right), \tag{17}$$

where p is the probability of a correct prediction for the state of S maximized over all possible measurements on QP. In this way, guessing at random and knowing for sure correspond to $\mathcal{D} = 0$ and $\mathcal{D} = 1$, respectively. According to a general theory,[31,32] the complementarity relation between the knowledge of the system state and the visibility of the fringes may be written, using the distinguishability $\mathcal{D}$ as a measure of information, in the quantitative form,

$$\mathcal{D}^2 + \mathcal{V}^2 \leq 1. \tag{18}$$

The equality holds for systems in pure states.

In a QD, this formal scheme translates naturally into the conditional dynamics of a biexcitonic system (Fig. 2a), as described in Sec. 3. Thus, the exciton addressed in the interference experiment described in Sec. 4.3 is, say, the σ_+ exciton (system S). The other degree of freedom of the biexciton system (the σ_- exciton) will be used as the quantum probe.

We will use a tensor product notation with $|0\rangle$ and $|1\rangle$ denoting the absence and presence of the respective exciton, as previously, with the interfering system (S) always to the left. In the rotating basis with respect to both subsystems, the RWA Hamiltonian for the biexciton system is

$$H = H_1 \otimes \mathbb{I} + \Delta |1\rangle\langle 1| \otimes |1\rangle\langle 1| + \frac{1}{2} f_{\mathrm{QP}}(t - \tau_{\mathrm{QP}}) \mathbb{I} \otimes (|0\rangle\langle 1| + \mathrm{H.c.}) \tag{19}$$

where Δ is the bi-exciton energy shift and $E_{\mathrm{QP}}(t)$ is the envelope of the pulse coupled to the QP exciton. Here the first term denotes the Hamiltonian (9) and corresponds to the pulse sequence of the interference experiment described in Sec. 4.3, the second one accounts for the bi-excitonic energy shift and the third term describes the action of the pulse coupled to the second (QP) exciton and spectrally tuned to the exciton-biexciton

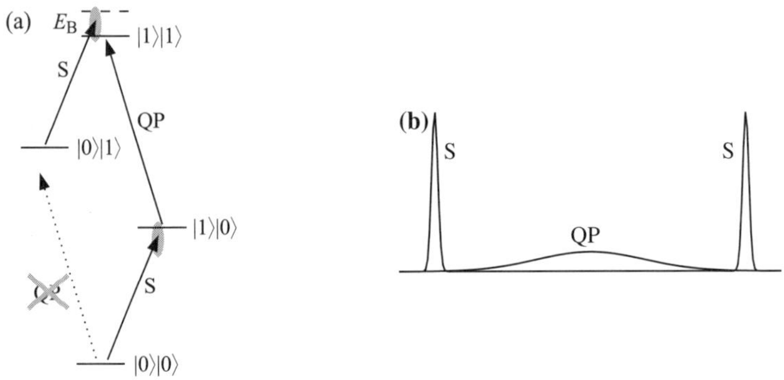

Fig. 6. (a) The diagram of energy levels and transitions in the system, assuming that the excitons are confined in neighboring dots with different transition energies. The pulses inducing transitions on the system S have broad spectrum so that both transitions are possible. The pulse acting on the QP system is spectrally selective and tuned to the biexcitonic transition, so that the single-exciton transition is energetically forbidden in this subsystem. (b) The sequence of pulses used in the experiment.

transition. This pulse arrives at $t = \tau_{\mathrm{QP}}$, between the other two pulses (that is, $0 < t_{\mathrm{QP}} < \tau$), and will induce the conditional dynamics. Its phase is irrelevant and will be assumed 0. The structure of system excitations and the sequence of pulses are shown in Fig. 6.

Assume that the exciton (system S) is in the equal superposition state (10). The probability of correctly guessing the result of a measurement in the $|0\rangle, |1\rangle$ basis without any additional information is obviously 1/2. Now, we can correlate this excitonic system with the other one (QP; initially in the state $|0\rangle$). To this end, one applies a selective (spectrally narrow) pulse with the area α (the arrow labelled 'QP' in Fig. 6a). This rotates the state of the second subsystem according to Eq. (6) (with $\hat{\boldsymbol{n}} = [1, 0, 0]$) if and only if the first system is in the state $|1\rangle$. The corresponding unitary transformation of the compound system is

$$U_{\mathrm{QP}} = |0\rangle\langle 0| \otimes \mathbb{I} + |1\rangle\langle 1| \otimes \left(\cos\frac{\alpha}{2}\mathbb{I} - i\sin\frac{\alpha}{2}\sigma_x \right),$$

and takes the state $|\psi\rangle$ into

$$|\psi'\rangle = \frac{1}{\sqrt{2}}|0\rangle \otimes |0\rangle - \frac{i}{\sqrt{2}}|1\rangle \otimes \left(\cos\frac{\alpha}{2}|0\rangle - i\sin\frac{\alpha}{2}|1\rangle \right).$$

For $\alpha = \pi$, this pulse performs a CNOT-like transformation[26] on the biexcitonic system. As a result, the total system is in the maximally entangled state $(|0\rangle|0\rangle - i|1\rangle|1\rangle)/2$ and a measurement on the QP system uniquely determines the state of the system S. Hence, due to quantum correlations between the systems, complete information on the state of S has been extracted to QP. On the other hand, if the biexciton is excited with a pulse with area $\alpha < \pi$ the correlation between the subsystems is weaker and a measurement on QP cannot fully determine the state of S, although the attained information may increase the probability for correctly predicting the result of a subsequent measurement on S. According to the discussion above, this means that partial information on the state of S is available.

In order to find the distinguishability measure in the biexciton scheme discussed here, we write the density matrix of the total system corresponding to the state $|\psi'\rangle$,

$$\varrho = \frac{1}{2}\sum_{nm} |n\rangle\langle m| \otimes \rho_{nm}, \tag{20}$$

with $\rho_{00} = |0\rangle\langle 0|$, $\rho_{11} = \frac{1}{2}(\mathbb{I} + \cos\alpha\, \sigma_z - \sin\alpha\, \sigma_y)$, $\rho_{01} = \rho_{10}^{\dagger} = i\cos\frac{\alpha}{2}|0\rangle\langle 0| - \sin\frac{\alpha}{2}|0\rangle\langle 1|$. Note that ρ_{00} and ρ_{11} (but not ρ_{01}) are density matrices.

According to the general theory,[31,32] the best chance for correctly guessing the state of S results from the measurement of the observable $\rho_{00} - \rho_{11}$ and the probability of the correct prediction is then $p = \frac{1}{2} + \frac{1}{4}\mathrm{Tr}|\rho_{00} - \rho_{11}|$ where $|\rho|$ is the modulus of the operator ρ. Using the explicit forms of the density matrices ρ_{00}, ρ_{11} and the definition (17) one finds for the distinguishability in our case

$$\mathcal{D} = \frac{1}{2}\mathrm{Tr}\,|\rho_{00} - \rho_{11}| = \left|\sin\frac{\alpha}{2}\right|. \tag{21}$$

Thus, the amount of information on the system S accessible via a measurement on QP increases from 0 (no QP pulse at all) to 1 (for a π pulse).

Next, we study the effect of extracting the which path information on the interference fringes. In the state (20), the reduced density matrix of the subsystem S is $\rho_{\mathrm{S}} = \mathrm{Tr}_{\mathrm{QP}}\varrho = (1/2)(\mathbb{I} - \cos(\alpha/2)\sigma_y)$. Upon applying the second pulse of the interference experiment scheme, namely the unconditional $\pi/2$ pulse [Eq. (11)], the average number of excitons in the dot is

$$N(\phi) = \langle 1|U_2\rho_{\mathrm{S}}U_2^{\dagger}|1\rangle = \frac{1}{2}\left(1 - \cos\frac{\alpha}{2}\cos\phi\right).$$

Now, the average occupation oscillates between the limiting values $(1 \pm |\cos\alpha|)/2$ and the visibility of the fringes is

$$\mathcal{V} = |\cos(\alpha/2)|. \tag{22}$$

Comparing Eq. (22) with Eq. (21) it is clear that the more certain one is whether the exciton *is there* or *not* ($\mathcal{D}$ increases), the less clear the interference fringes become ($\mathcal{V}$ decreases). Quantitatively, the relation $\mathcal{D}^2+\mathcal{V}^2=1$ holds which is consistent with the complementarity relation (18). It can be shown that, in the presence of coupling to the environment, where both subsystems are in mixed states due to dephasing, this relation will turn into inequality.[24] Let us notice that the impact of the which path information on interference fringes is the same no matter whether the QP subsystem is measured before or after generating and detecting the interference fringes and even whether it is measured at all.

The time-domain manifestation of quantum complementarity discussed here not only broadens the class of experiments in which fundamental aspects of the quantum world may be tested but also has the advantage of being independent of the position-momentum (Heisenberg's) uncertainty that has been historically tied to the space-domain discussions of complementarity.[33] In fact, it is independent of any uncertainty principles whatsoever. Indeed, the only two quantities which are measured in the time-domain interference experiment are the occupations of two different excitons. These quantities refer to *different subsystems* and, therefore, are obviously *commuting* and simultaneously measurable. Although the quantum probe exciton is created in a way that correlates it with the presence of the other exciton (S), this cannot be interpreted as a (projective) measurement on S, since the QP exciton is definitely a quantum (microscopic) system and not a classical, macroscopic measurement device. Thus, the experimental procedure described in this section demonstrates quantum complementarity in its pure form, involving only the notion of *information* on the system state and independent of any uncertainty relations between non-commuting observables.

Further discussion of the concept of complementarity in the context of optical experiments on QDs, including the analysis of its feasibility in terms of the currently available experimental techniques and of the parameters of the existing structures, as well as the analysis of the role of dephasing is given in Ref. [24].

6. Carrier-phonon interaction in quantum dots

This section presents the derivation of the coupling constants between bulk phonon modes and confined carriers in a semiconductor.[34,35] We will restrict ourselves to the deformation potential coupling, which is relevant for the discussion in the following sections. The treatment of other couplings (piezoelectric and polar) is reviewed in Ref. [36].

Any crystal deformation leads to shifts of the conduction (c) and valence (v) bands which are, to the leading order, proportional to the relative volume change. The corresponding contribution to the energy of electrons (e) and holes (h) in the long-wavelength limit is

$$H_{\mathrm{e/h}}^{(\mathrm{DP})} \equiv \pm\Delta E_{\mathrm{c/v}} = \mp\sigma_{\mathrm{e/h}}\frac{\delta V}{V},$$

where $\sigma_{\mathrm{e/h}}$ are the deformation potential constants for electrons and holes and V is the unit cell volume. Using the strain tensor $\hat{\sigma}$,

$$\sigma_{ij} = \frac{1}{2}\left(\frac{\partial u_i}{\partial r_j} + \frac{\partial u_j}{\partial r_i}\right),$$

one may write $H_{\mathrm{e/h}}^{(\mathrm{DP})} = \mp\sigma_{\mathrm{e/h}}\mathrm{Tr}\hat{\sigma} = \mp\sigma_{\mathrm{e/h}}\nabla\cdot\boldsymbol{u}(\boldsymbol{r})$, where $\boldsymbol{u}(\boldsymbol{r})$ is the local displacement field. The displacement is quantized in terms of phonons,

$$\boldsymbol{u}(\boldsymbol{r}) = i\frac{1}{\sqrt{N}}\sum_{\boldsymbol{k}}\sqrt{\frac{\hbar}{2\rho V\omega_{\boldsymbol{k}}}}\hat{\boldsymbol{e}}_{\boldsymbol{k}}\left(b_{\boldsymbol{k}} + b_{-\boldsymbol{k}}^{\dagger}\right)e^{i\boldsymbol{k}\cdot\boldsymbol{r}}, \tag{23}$$

where $\omega_{\boldsymbol{k}}$ is the frequency for the wave vector $\boldsymbol{k}$, $\hat{\boldsymbol{e}}_{\boldsymbol{k}} = -\hat{\boldsymbol{e}}_{-\boldsymbol{k}}$ is the corresponding real unit polarization vector, and ρ is the crystal density. Only the longitudinal branch contributes to $\nabla\cdot\boldsymbol{u}$ in (23) and the final interaction Hamiltonian in the coordinate representation for carriers is

$$H_{\mathrm{e/h}}^{(\mathrm{DP})} = \pm\sigma_{\mathrm{e/h}}\frac{1}{\sqrt{N}}\sum_{\boldsymbol{k}}\sqrt{\frac{\hbar k}{2\rho V\omega_{\boldsymbol{k}}}}\left(b_{\boldsymbol{k}} + b_{-\boldsymbol{k}}^{\dagger}\right)e^{i\boldsymbol{k}\cdot\boldsymbol{r}}. \tag{24}$$

In the second quantization representation with respect to the carrier states this reads

$$\begin{aligned} H_{\mathrm{e/h}}^{(\mathrm{DP})} &= \sum_{nn'}\langle n|V(\boldsymbol{r})|n'\rangle a_n^{\dagger}a_{n'} = \sum_{nn'}\int_{-\infty}^{\infty} d^3r\psi_n^*(\boldsymbol{r})V(\boldsymbol{r})\psi_{n'}(\boldsymbol{r})a_n^{\dagger}a_{n'} \\ &= \frac{1}{\sqrt{N}}\sum_{\boldsymbol{k}nn'} a_n^{\dagger}a_{n'}f_{\mathrm{e/h},nn'}^{(\mathrm{DP})}(\boldsymbol{k})\left(b_{\boldsymbol{k}} + b_{-\boldsymbol{k}}^{\dagger}\right), \end{aligned} \tag{25}$$

where $f_{\mathrm{e/h},nn'}^{(\mathrm{DP})}(\boldsymbol{k}) = \sigma_{\mathrm{e/h}}\sqrt{\frac{\hbar k}{2\rho Vc}}\mathcal{F}_{nn'}(\boldsymbol{k})$, with the formfactor

$$\mathcal{F}_{nn'}(\boldsymbol{k}) = \int_{-\infty}^{\infty} d^3\boldsymbol{r}\psi_n^*(\boldsymbol{r})e^{i\boldsymbol{k}\cdot\mathbf{r}}\psi_{n'}(\boldsymbol{r}) = \mathcal{F}_{n'n}^*(-\boldsymbol{k}). \tag{26}$$

While the common coefficient of the coupling Hamiltonian contains the fundamental and material-dependent constants and reflects general electrical and mechanical properties of the semiconductor system, the formfactor

(26) contains information about the geometry of the confinement and the resulting properties of wave functions. In this sense, it is the "engineerable" part of the carrier-phonon coupling.

From the orthogonality of single-particle states one immediately has $\mathcal{F}_{nn'}(0) = \delta_{nn'}$. If the wave functions are localized at a length l, then the extent of the formfactor is $\sim 1/l$. Thus, for carrier states localized in a QD over many lattice sites and smooth within this range, the functions $\mathcal{F}_{nn'}(\boldsymbol{k})$ will be localized in $\boldsymbol{k}$–space very close to the center of the Brillouin zone.

As an example, let us consider a Gaussian wave function,

$$\psi(\boldsymbol{r}) = \frac{1}{\pi^{3/4} l_z l_\perp} \exp\left[-\frac{1}{2}\left(\frac{r_\perp}{l_\perp}\right)^2 - \frac{1}{2}\left(\frac{z}{l_z}\right)^2\right], \tag{27}$$

where $r_\perp$ is the position component in the xy plane and $l_\perp, l_z$ are the localization widths in-plane and in the growth (z) direction. The corresponding formfactor is then easily found to be

$$\mathcal{F}(\boldsymbol{k}) = \exp\left[-\left(\frac{k_\perp l_\perp}{2}\right)^2 - \frac{1}{2}\left(\frac{k_z l_z}{2}\right)^2\right]. \tag{28}$$

The formulas derived above describe the interaction in the single-particle basis. However, most of the following deals with excitonic states, i.e. states of confined electron-hole pairs interacting by Coulomb potentials. Both carriers forming the exciton couple to phonons according to Eq. (25). For the further discussion in this chapter it is sufficient to discuss the lowest exciton state. It is reasonable to assume its wave function approximately in a product form,[37] i.e.,

$$|1\rangle = a^\dagger_{\mathrm{e},0} a^\dagger_{\mathrm{h},0}|0\rangle,$$

where $a^\dagger_{\mathrm{e/h},0}$ create an electron and a hole in the ground confined state in the QD. Then, one gets from Eq. (25) the following coupling constant between the confined exciton and phonons for the deformation potential interaction

$$g_{\boldsymbol{k}} = \sqrt{\frac{\hbar k}{2\rho V c_{\mathrm{l}}}} \left(\sigma_{\mathrm{e}} \mathcal{F}^{(\mathrm{e})}(\boldsymbol{k}) - \sigma_{\mathrm{h}} \mathcal{F}^{(\mathrm{h})}(\boldsymbol{k})\right). \tag{29}$$

Note that, due to different deformation potential constants $\sigma_{\mathrm{e/h}}$, this coupling does not vanish even if the electron and hole wave functions are the same, leading to identical single-particle formfactors.

Thus, the exciton (restricted to its ground state) driven by a laser field and interacting with LA phonons via the deformation potential coupling is described by the Hamiltonian

$$H = H_{\text{RWA}} + \sum_{\boldsymbol{k}} \hbar\omega_{\boldsymbol{k}} b_{\boldsymbol{k}}^\dagger b_{\boldsymbol{k}} + |1\rangle\langle 1| \sum_{\boldsymbol{k}} g_{\boldsymbol{k}} (b_{\boldsymbol{k}} + b_{-\boldsymbol{k}}^\dagger), \tag{30}$$

where the first term describes the carrier subsystem and is given by Eq. (5).

The Hamiltonian (30) is the basis for the microscopic modeling of dephasing effects in QDs. It turns out that this relatively simple model is very successful in reproducing the experimental data.[38,39] Hence, it may serve as a reliable starting point for describing the evolution of the combined system of confined carriers and lattice modes.

7. Theoretical methods for carrier–phonon kinetics

In this section we will study a few theoretical methods that have proven to be useful for the description of the carrier-phonon quantum kinetics in QDs. This discussion will be limited to the simplest case of a two-level system described by the Hamiltonian (30).

7.1. *Exact solution for ultrafast excitations*

Let us start with the case of an ultrafast excitation. A very short laser pulse prepares the system in a certain superposition (dependent on the pulse phase and intensity) of the ground state $|0\rangle$ (no exciton) and the single-exciton state $|1\rangle$. By *very short* we mean a pulse much shorter than the time scales of phonon dynamics, so that the preparation of the initial state may be considered instantaneous. This corresponds to the actual experimental situation with pulse durations of the order of 100 fs.[40] On the other hand, the pulse is long enough to assure a relatively narrow spectrum and to prevent the population of higher confined levels and excitation of optical phonons.[41] In this ultrafast limit the only role of the laser pulse is to prepare the initial state, while the subsequent evolution is generated by a time-independent Hamiltonian,

$$H = -\Delta|1\rangle\langle 1| + \sum_{\boldsymbol{k}} \hbar\omega_{\boldsymbol{k}} b_{\boldsymbol{k}}^\dagger b_{\boldsymbol{k}} + |1\rangle\langle 1| \sum_{\boldsymbol{k}} g_{\boldsymbol{k}} (b_{\boldsymbol{k}} + b_{-\boldsymbol{k}}^\dagger), \tag{31}$$

obtained by setting $H_{\text{RWA}} = -\Delta|1\rangle\langle 1|$ in Eq. (30).

In a superposition state created by the laser pulse [Eq. (7)] the interband component of the electric dipole moment has a non-vanishing average value oscillating at an optical frequency (hence referred to as *optical polarization*)[42] which leads to the emission of coherent electromagnetic radiation with an amplitude proportional to the oscillating dipole moment. In an unperturbed system (e.g., in an atom), the radiation would be emitted over time of the order of the lifetime of the superposition state, i.e., until the system relaxes to the ground state due to radiative energy loss.

In a semiconductor structure an additional effect, related to carrier-phonon coupling, appears on a time scale much shorter than the lifetime of the state. The last two terms in Eq. (31) describe a set of harmonic oscillators which, in the presence of the exciton, are displaced by an external force proportional to the coupling constant $g_{\boldsymbol{k}}$. This means that, due to the interactions between confined carriers and lattice ions, the ground state of the lattice in the presence of a charge distribution is different than in its absence. As a result, after the creation of a confined exciton the lattice relaxes to a new equilibrium, which is accompanied by the emission of phonon wave packets[37,43] that form a trace in the macroscopic crystal distinguishing the exciton state from an empty dot. This information broadcast via emitted phonons leads to a decay of the coherence of the superposition state[44] although the average occupations of the system states remain unaffected (hence the process is referred to as *pure dephasing*). Since coherent dipole radiation requires well-defined phase relations between the components of a quantum superposition, the amplitude of this radiation, measured in the experiment, gives access to the coherence properties of the quantum state of confined carriers itself. The dephasing of the quantum superposition is therefore directly translated into the decay of coherent optical radiation from the system.

As discussed in Sec. 6, the carrier-phonon interaction term in Eq. (31) is linear in phonon operators and describes a shift of the lattice equilibrium induced by the presence of a charge distribution in the dot. The stationary state of the system corresponds to the exciton and the surrounding coherent cloud of phonons representing the lattice distortion to the new equilibrium. The transformation that creates the coherent cloud is the shift $wb_{\boldsymbol{k}}w^{\dagger} = b_{\boldsymbol{k}} - f_{\boldsymbol{k}}/(\hbar\omega_{\boldsymbol{k}})$, generated by the Weyl operator[35,44]

$$w = \exp\left[\sum_{\boldsymbol{k}} \left(\frac{g_{\boldsymbol{k}}^*}{\hbar\omega_{\boldsymbol{k}}} b_{\boldsymbol{k}}^{\dagger} - \frac{g_{\boldsymbol{k}}}{\hbar\omega_{\boldsymbol{k}}} b_{\boldsymbol{k}} \right)\right]. \tag{32}$$

A straightforward calculation shows that the Hamiltonian (31) is diagonalized by the unitary transformation $W = |0\rangle\langle 0| \otimes \mathbb{I} + |1\rangle\langle 1| \otimes w$, where $\mathbb{I}$ is the identity operator and the tensor product refers to the carrier subsystem (first component) and its phonon environment (second component). As a result one gets

$$\widetilde{H} = WHW^{\dagger} = -\tilde{\Delta}|1\rangle\langle 1| + H_{\mathrm{ph}},$$

where $\tilde{\Delta} = \Delta + \sum_{\boldsymbol{k}} |g_{\boldsymbol{k}}|^2/(\hbar\omega_{\boldsymbol{k}})$ and $H_{\mathrm{ph}} = \sum_{\boldsymbol{k}} \hbar\omega_{\boldsymbol{k}} b_{\boldsymbol{k}}^{\dagger} b_{\boldsymbol{k}}$.

We assume that at the beginning ($t = 0$) the state of the whole system is $\rho_0 = (|\Psi\rangle\langle\Psi|) \otimes \rho_{\mathrm{ph}}$, where ρ_{ph} is the density matrix of the phonon subsystem (environment) at thermal equilibrium and $|\Psi\rangle$ is given by Eq. (7). For simplicity, we will assume an equal superposition state, setting $\alpha = \pi/2$ and $\varphi = -\pi/2$.

The evolution operator $U(t) = e^{-iHt/\hbar}$ may be written as

$$\begin{aligned} U(t) &= W^{\dagger}WU(t)W^{\dagger}W = W^{\dagger}\widetilde{U}(t)W = W^{\dagger}\widetilde{U}(t)W\widetilde{U}^{\dagger}(t)\widetilde{U}(t) \\ &= W^{\dagger}W(t)\widetilde{U}(t), \end{aligned}$$

where $\widetilde{U}(t) = e^{-i\widetilde{H}t/\hbar}$ and $W(t) = \widetilde{U}(t)W\widetilde{U}^{\dagger}(t)$. Since $\widetilde{U}(t)$ is diagonal the explicit form of $W(t)$ may easily be found and one gets

$$U(t) = \left[|0\rangle\langle 0| \otimes \mathbb{I} + |1\rangle\langle 1| \otimes w^{\dagger}w(t)\right] \widetilde{U}(t), \tag{33}$$

where $w(t) = e^{-iH_{\mathrm{ph}}t/\hbar} w e^{iH_{\mathrm{ph}}t/\hbar}$.

Using the evolution operator in the form (33) the system state at a time t may be written as

$$\rho(t) = \frac{1}{2}\begin{pmatrix} \rho_{\mathrm{E}} & e^{-i\tilde{\Delta}t/\hbar}\rho_{\mathrm{E}} w^{\dagger}(t) w \\ e^{i\tilde{\Delta}t/\hbar} w^{\dagger} w(t)\rho_{\mathrm{ph}} & w^{\dagger}w(t)\rho_{\mathrm{ph}} w^{\dagger}(t) w \end{pmatrix}, \tag{34}$$

where we used the tensor product notation in which an operator A is expanded as $A = \sum_{m,n} |m\rangle\langle n| \otimes A_{mn}$ with a set of operators A_{mn} acting on the second subsystem, and written in the matrix form with respect to the first subsystem. The density matrix for the carrier subsystem is obtained by tracing out the phonon degrees of freedom, i.e., $\rho_{\mathrm{S}} = \mathrm{Tr}_{\mathrm{E}}\rho$. Hence,

$$\rho_{\mathrm{S}}(t) = \frac{1}{2}\begin{pmatrix} 1 & e^{-i\tilde{\Delta}t/\hbar}\langle w^{\dagger}(t) w\rangle \\ e^{i\tilde{\Delta}t/\hbar}\langle w^{\dagger} w(t)\rangle & 1 \end{pmatrix}. \tag{35}$$

The average may be calculated using two rules for Weyl operators.[34,44] The multiplication of two operators of the form

$$w_i = \exp\left[\sum_{\boldsymbol{k}} \left(g_{\boldsymbol{k}}^{(i)*}\gamma_{\boldsymbol{k}}^{\dagger} - \gamma_{\boldsymbol{k}}^{(i)} b_{\boldsymbol{k}}\right)\right], \quad i = 1, 2, 3 \tag{36}$$

yields

$$w_1 w_2 = w_3 \exp\left[\frac{1}{2}\sum_{\boldsymbol{k}} \left(\gamma_{\boldsymbol{k}}^{(1)*}\gamma_{\boldsymbol{k}}^{(2)} - \gamma_{\boldsymbol{k}}^{(1)}\gamma_{\boldsymbol{k}}^{(2)*}\right)\right], \tag{37}$$

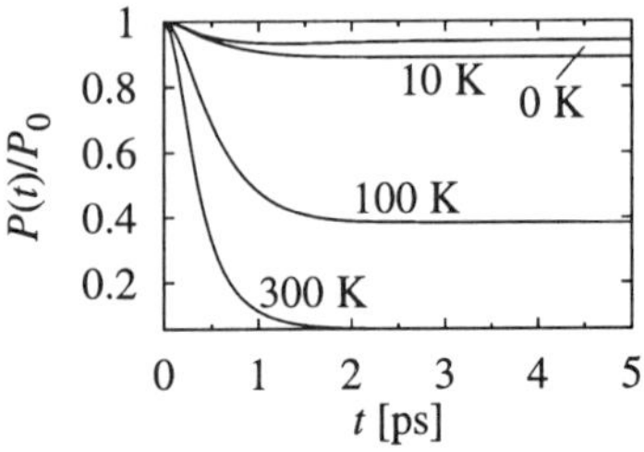

Fig. 7. Decay of the coherent radiation from a confined exciton at various temperatures, as shown. In our calculations we use typical parameters for a self-assembled InAs/GaAs structure: single particle wave functions $\psi(\boldsymbol{r})$ modelled by Gaussians with $l_{\perp} = 4$ nm and $l_z = 1$ nm [Eq.(27)], the deformation potential difference $\sigma_{\rm e} - \sigma_{\rm h} = 9.5$ eV, crystal density $\varrho = 5300$ kg/m^3, and the speed of sound $c = 5150$ m/s.

where w_3 is given by Eq. (36) with $\gamma_{\boldsymbol{k}}^{(3)} = \gamma_{\boldsymbol{k}}^{(1)} + \gamma_{\boldsymbol{k}}^{(2)}$. The rule for averaging of an operator given by Eq. (36) in the thermal equilibrium state is

$$\langle w_i \rangle = e^{-\frac{1}{2}\sum_{\boldsymbol{k}} |\gamma_{\boldsymbol{k}}^{(i)}|^2 (2n_{\boldsymbol{k}}+1)}, \tag{38}$$

where $n_{\boldsymbol{k}}$ are bosonic equilibrium occupation numbers. The final result is

$$\langle w^{\dagger}(t) w \rangle = \exp\left\{ -\sum_{\boldsymbol{k}} \left| \frac{g_{\boldsymbol{k}}}{\hbar\omega_{\boldsymbol{k}}} \right|^2 \left[i \sin\omega_{\boldsymbol{k}} t + (1 - \cos\omega_{\boldsymbol{k}} t)(2n_{\boldsymbol{k}} + 1) \right] \right\}.$$

The emitted coherent dipole radiation is proportional to the off-diagonal element of the density matrix $\rho_s(t)$ and its amplitude is

$$P(t) = P_0 |\langle w^{\dagger}(t) w \rangle|. \tag{39}$$

In Fig. 7 we show the normalized polarization amplitude $P(t)/P_0$ (originally derived in Ref. [45]). The interaction with the macroscopic crystal environment leads to a reduction of coherent radiation due to pure dephasing of the exciton state, reflected by the reduced value of the off-diagonal element of the density matrix $\rho_{\rm S}$. At $t = 0$ one has $\langle w^{\dagger}(t) w \rangle = 1$, while at large values of t, $\cos\omega_{\boldsymbol{k}} t$ oscillates very quickly as a function of $\boldsymbol{k}$ and averages to 0 (see also Ref. [45]). Thus, for long times, the polarization amplitude tends to a temperature-dependent finite value

$$P(t) \to P_0 \exp\left[-\sum_{\boldsymbol{k}} \left| \frac{g_{\boldsymbol{k}}}{\hbar\omega_{\boldsymbol{k}}} \right|^2 (2n_{\boldsymbol{k}} + 1) \right] < P_0.$$

This partial decay of coherence is a characteristic feature of short-time dephasing for carrier-phonon couplings encountered in real systems.[40]

7.2. Perturbation theory for driven systems

In this section we derive the equations for the reduced density matrix of the carrier subsystem to the leading order in the phonon coupling, assuming that the unperturbed evolution is known. This approach yields very simple and intuitive formulas that may easily be applied to a range of problems. Here, we will again restrict the formalism to a single exciton level. A more general discussion is given in Ref. [36].

As already mentioned, the evolution of the carrier subsystem is generated by the (time dependent) Hamiltonian H_{RWA} [Eq. (5)], describing the properties of the system itself as well as its coupling to driving fields. The evolution of the phonon subsystem (reservoir) is described by the Hamiltonian H_{ph} [the second term in Eq. (30)]. The evolution operator for the driven carrier subsystem and free phonon modes, without the carrier-phonon interaction, is

$$U_0(t) = U_{\mathrm{RWA}}(t) \otimes e^{-iH_{\mathrm{ph}}(t-s)},$$

where $U_{\mathrm{RWA}}(t)$ is the operator for the unperturbed evolution of the carrier subsystem and s is the initial time.

The carrier-phonon coupling may be written in the form $V = S \otimes R$, where S acts in the Hilbert space of the carrier subsystem while the time-independent R affect only the environment. For instance, in the special case of Eq. (30), $S = |1\rangle\langle 1|$ and

$$R = \frac{1}{\sqrt{N}} \sum_{\boldsymbol{k}} g_{\boldsymbol{k}} \left(b_{\boldsymbol{k}} + b_{-\boldsymbol{k}}^{\dagger} \right). \tag{40}$$

We will assume that at the initial time s the system is in the product state $\varrho(s) = |\psi_0\rangle\langle\psi_0| \otimes \rho_{\mathrm{ph}}$, where $|\psi_0\rangle$ is a certain state of the carrier subsystem and ρ_{ph} is the thermal equilibrium distribution of phonon modes. Physically, such an assumption is usually reasonable due to the existence of two distinct time scales: the long one for the carrier decoherence (e.g. 1 ns ground state exciton lifetime[40,46]) and the short one for the reservoir relaxation (1 ps dressing time[37,40,45]).

The starting point is the evolution equation for the density matrix of the total system in the interaction picture with respect to the externally driven evolution U_0, in the second order Born approximation with respect to the carrier-phonon interaction[47]

$$\tilde{\varrho}(t) = \tilde{\varrho}(s) + \frac{1}{i\hbar} \int_s^t d\tau [V(\tau), \varrho(s)] - \frac{1}{\hbar^2} \int_s^t d\tau \int_s^{\tau} d\tau' [V(\tau), [V(\tau'), \varrho(s)]], \tag{41}$$

where $\tilde{\varrho}(t) = U_0^\dagger(t)\varrho(t)U_0(t)$, $\quad V(t) = U_0^\dagger(t)VU_0(t)$ (it should be kept in mind that, in general, V may depend on time itself).

The reduced density matrix of the carrier subsystem at time t is $\rho(t) = U_{\mathrm{RWA}}(t)\tilde{\rho}(t)U_{\mathrm{RWA}}^\dagger(t)$, $\quad \tilde{\rho}(t) = \mathrm{Tr}_{\mathrm{R}}\tilde{\varrho}(t)$, where the trace is taken over the phonon degrees of freedom. The first (zeroth order) term in (41) gives

$$\rho^{(0)}(t) = U_{\mathrm{RWA}}(t)|\psi_0\rangle\langle\psi_0|U_{\mathrm{RWA}}^\dagger(t) = |\psi_0(t)\rangle\langle\psi_0(t)|. \tag{42}$$

The second term vanishes, since it contains the thermal average of an odd number of phonon operators. The third (second order) term describes the leading phonon correction to the dynamics of the carrier subsystem,

$$\tilde{\rho}^{(2)}(t) = -\frac{1}{\hbar^2}\int_s^t d\tau \int_s^\tau d\tau' \mathrm{Tr}_{\mathrm{R}}[V(\tau),[V(\tau'),\varrho(s)]]. \tag{43}$$

First of the four terms resulting from expanding the commutators in (43) is (I) $= -Q_t|\psi_0\rangle\langle\psi_0|$, where

$$Q_t = \frac{1}{\hbar^2}\int_s^t d\tau \int_s^\tau d\tau' S(\tau)S(\tau')\langle R(\tau-\tau')R\rangle. \tag{44}$$

The operators S and R are transformed into the interaction picture in the usual way $S(t) = U_0^\dagger(t)SU_0(t)$,$R(t) = U_0^\dagger(t)RU_0(t)$ and $\langle\hat{O}\rangle = \mathrm{Tr}_{\mathrm{R}}[\hat{O}\rho_{\mathrm{ph}}]$ denotes the thermal average (obviously $[U_0(t),\rho_{\mathrm{ph}}] = 0$).

The second term is

$$(\mathrm{II}) = \frac{1}{\hbar^2}\int_s^t d\tau \int_s^\tau d\tau' |\psi_0\rangle\langle\psi_0|S(\tau')S(\tau)\langle R(\tau'-\tau)R\rangle = -|\psi_0\rangle\langle\psi_0|Q_t^\dagger,$$

where we used the relation $\langle R(\tau'-\tau)R\rangle = \langle R(\tau-\tau')R\rangle^*$.

In a similar manner, the two other terms may be combined into

$$(\mathrm{III}) + (\mathrm{IV}) = \hat{\Phi}_t\left[|\psi_0\rangle\langle\psi_0|\right].$$

where

$$\hat{\Phi}_t\left[\rho\right] = \frac{1}{\hbar^2}\int_s^t d\tau \int_s^t d\tau' S(\tau')\rho S(\tau)\langle R(\tau-\tau')R\rangle. \tag{45}$$

In terms of the new Hermitian operators

$$A_t = Q_t + Q_t^\dagger, \quad h_t = \frac{1}{2i}(Q_t - Q_t^\dagger), \tag{46}$$

the density matrix at the final time t (42,43) may be written as

$$\rho(t) = U_{\mathrm{RWA}}(t)\left(|\psi_0\rangle\langle\psi_0| - i\left[h_t,|\psi_0\rangle\langle\psi_0|\right]\right. \tag{47}$$
$$\left. -\frac{1}{2}\left\{A_t,|\psi_0\rangle\langle\psi_0|\right\} + \hat{\Phi}_t[|\psi_0\rangle\langle\psi_0|]\right)U_{\mathrm{RWA}}^\dagger.$$

The first term is a Hamiltonian correction which does not lead to irreversible effects and, in principle, may be compensated by an appropriate modification of the control Hamiltonian $H_{\rm RWA}$. The other two terms describe processes of entangling the system with the reservoir, leading to the loss of coherence of the carrier state.

Let us introduce the spectral density of the reservoir,

$$R(\omega) = \frac{1}{2\pi\hbar^2} \int dt \langle R(t) R\rangle e^{i\omega t}. \tag{48}$$

For the operators given in Eq. (40) one has explicitly

$$R(\omega) = \frac{1}{\hbar^2} |n_{\rm B}(\omega)+1| \frac{1}{N} \sum_{\boldsymbol{k}} |g_{\boldsymbol{k}}|^2 \left[\delta(\omega-\omega_{\boldsymbol{k}}) + \delta(\omega+\omega_{\boldsymbol{k}})\right], \tag{49}$$

where $n_{\rm B}(w) = -n_{\rm B}(-\omega) - 1$ is the Bose distribution function. The spectral density $R(\omega)$ depends on the material parameters and system geometry and characterizes the properties of the lattice subsystem. For the deformation potential coupling to LA phonons, it has the long-wavelength behavior $R(\omega) \sim \omega^3[n_{\rm B}(\omega)+1]$.

With the help of (48) one may write

$$\hat{\Phi}_t\left[\rho\right] = \int d\omega R(\omega) Y(\omega) \rho Y^\dagger(\omega) \tag{50}$$

where the frequency-dependent operators have been introduced,

$$Y(\omega) = \int_s^t d\tau S(\tau) e^{i\omega\tau}. \tag{51}$$

Using (48) again one has

$$Q_t = \int d\omega \int_s^t d\tau \int_s^t d\tau' \theta(\tau-\tau') S(\tau) S(\tau') R(\omega) e^{-i\omega(\tau-\tau')}.$$

Next, representing the Heaviside function as

$$\theta(t) = -e^{i\omega t} \int \frac{d\omega'}{2\pi i} \frac{e^{-i\omega' t}}{\omega'-\omega+i0^+},$$

we write

$$\begin{aligned} Q_t &= -\int d\omega R(\omega) \int \frac{d\omega'}{2\pi i} \frac{Y^\dagger(\omega') Y(\omega')}{\omega'-\omega+i0^+} \\ &= -\int d\omega R(\omega) \int \frac{d\omega'}{2\pi i} Y^\dagger(\omega') Y(\omega') \left[-i\pi\delta(\omega'-\omega) + \mathcal{P}\frac{1}{\omega'-\omega}\right], \end{aligned}$$

where $\mathcal{P}$ denotes the principal value.

Hence, the two Hermitian operators defined in (46) take the form

$$A_t = \int d\omega R(\omega) Y^\dagger(\omega) Y(\omega) \tag{52}$$

and

$$h_t = \int d\omega R(\omega) \mathcal{P} \int \frac{d\omega'}{2\pi} \frac{Y^\dagger(\omega') Y(\omega')}{\omega' - \omega}. \tag{53}$$

In order to provide an example of an application of the perturbative method presented in this section let us consider a $\pi/2$ rotation of an excitonic two-level system in the presence of carrier-phonon coupling. As we have already pointed out, any fast change of the state of the carrier subsystem leads to spontaneous processes of lattice relaxation that affect the coherence of the carrier state. It is reasonable to expect that coherent control is recovered if the evolution of the carrier subsystem is slow (adiabatic) compared to the typical timescales of the lattice dynamics. Thus, the requirement to avoid traces of carrier dynamics in the outside world favors slow operation on the carrier subsystem, contrary to other decoherence processes (of Markovian character), like radiative decay of the exciton or thermally activated processes of phonon-assisted transitions to higher states. The latter have the character of an exponential decay and, for short times, contribute an error $\tilde{\delta} = \tau_{\mathrm{g}}/\tau_{\mathrm{d}}$, where τ_g is the gating time and τ_d is the decay time constant. Here we consider the interplay between these two contributions to the error for the solid-state qubit implementation using excitonic (charge) states in quantum dots (QDs),[48] with computational states defined by the absence ($|0\rangle$) or presence ($|1\rangle$) of one exciton in the ground state of the dot, operated by resonant coupling to laser light. We show that it leads to a trade-off situation with a specific gating time corresponding to the minimum decoherence for a given operation.[49]

To quantify the quality of the rotation, we use the *fidelity*[26]

$$F = \langle \psi_0 | U^\dagger_{\mathrm{RWA}}(\infty) \rho(\infty) U_{\mathrm{RWA}}(\infty) | \psi_0 \rangle^{1/2} \tag{54}$$

which is a measure of the overlap between the ideal (pure) final state without perturbation, $U_{\mathrm{RWA}}(\infty)|\psi_0\rangle$, and the actual final state of the system given by the density matrix $\rho(\infty)$. If the procedure is performed ideally, i.e. without discrepancies from the desired qubit operation, then $F = 1$. The fidelity loss $\delta = 1 - F^2$ is referred to as the *error* of the quantum gate. Using the definition (54) and the Master equation (47), the error may be written in a general case as

$$\delta = \langle \psi_0 | A_t | \psi_0 \rangle - \left\langle \psi_0 \middle| \hat{\Phi} \left[|\psi_0\rangle\langle\psi_0| \right] \middle| \psi_0 \right\rangle .$$

It should be noted that the unitary correction generated by h_t does not contribute to the error at this order. Using the definitions (45,52) this may be further transformed into

$$\delta = \int d\omega R(\omega) \left| \langle \psi_\perp | Y(\omega) | \psi_0 \rangle \right|^2 , \tag{55}$$

where $\psi_\perp$ is a state orthogonal to $|\psi_0\rangle$ in the two-dimensional space of interest.

Since the coherence of superpositions induced by short pulses is unstable due to phonon-induced pure dephasing (Sec. 7.1), it seems reasonable to perform operations on *dressed states*, i.e. on the correctly defined quasi-particles of the interacting carrier-phonon system.[50] This may be formally achieved by employing the solid-state-theory concept of adiabatic switching on/off the interaction[51] (as done in Ref. [52]) to transform the states of the noninteracting system into the states of the interacting one. Thus, we assume adiabatic switching on/off of the interaction with phonons by appending the appropriate exponent to the original interaction Hamiltonian [Eq. (30)],

$$H_{\text{int}} = e^{-\varepsilon|t|} \left[|1\rangle\langle 1| \sum_{\boldsymbol{k}} g_{\boldsymbol{k}} \left(b_{\boldsymbol{k}} + b^\dagger_{-\boldsymbol{k}} \right) \right] , \tag{56}$$

where $\varepsilon = 0^+$. The operator S now becomes

$$S(t) = U_{\text{RWA}}(t) e^{-\varepsilon|t|} |1\rangle\langle 1| U^\dagger_{\text{RWA}}(t),$$

where the free evolution operator is generated by an optical pulse at the resonance (Sec. 4.2). The general formula (55) may now be used with the bare initial state $|\psi_0\rangle$. The adiabatic procedure assures that it is transformed to the dressed state before comparing it to the density matrix ρ, so that the fidelity is defined with respect to stable, dressed states.

The operator $Y(\omega)$ can be written in the form

$$\begin{aligned} Y(\omega) = & \frac{1}{4i\omega} F(\omega) (|1\rangle\langle 0| - |0\rangle\langle 1| + |0\rangle\langle 0| - |1\rangle\langle 1|) \\ & + \frac{1}{4i\omega} F^*(-\omega) (|0\rangle\langle 1| - |1\rangle\langle 0| + |0\rangle\langle 0| - |1\rangle\langle 1|), \end{aligned}$$

where

$$F(\omega) = \int_{-\infty}^{\infty} d\tau e^{i\omega\tau} \frac{d}{d\tau} e^{i\Phi(\tau)}.$$

Since in quantum information processing applications the initial state of the quantum bit is in general not known, it is reasonable to consider the error averaged over all input states. Let us introduce the function

$$S(\omega) = \omega^2 |\langle \psi_\perp | Y(\omega) | \psi_0 \rangle|^2_{\mathrm{av}},$$

where the average is taken over the Bloch sphere. According to Eq. (55), the average error may now be written as

$$\delta = \int \frac{d\omega}{\omega^2} R(\omega) S(\omega). \tag{57}$$

The averaging is most conveniently performed by noting that

$$Y(\omega) = \frac{1}{2i\omega} F(\omega) |+\rangle\langle -| + \frac{1}{2i\omega} F^*(-\omega) |-\rangle\langle +|,$$

where $|\pm\rangle = (|0\rangle \pm |1\rangle)/\sqrt{2}$. Choosing $|\psi_0\rangle = \cos\frac{\theta}{2}|+\rangle + e^{i\varphi}\sin\frac{\theta}{2}|-\rangle$, $|\psi_0^\perp\rangle = \sin\frac{\theta}{2}|+\rangle - e^{i\varphi}\cos\frac{\theta}{2}|-\rangle$, one gets

$$S(\omega) = \frac{1}{4} \left| F(-\omega) e^{i\varphi} \cos^2\frac{\theta}{2} - F^*(\omega) e^{-i\varphi} \sin^2\frac{\theta}{2} \right|^2,$$

which, upon averaging over the angles θ, φ on the Bloch sphere, leads to

$$S(\omega) = \frac{1}{12} \left(|F(\omega)|^2 + |F(-\omega)|^2 \right).$$

Let us now consider a Gaussian pulse for performing the quantum gate, $f(t) = [\alpha/(\sqrt{2\pi}\tau_{\mathrm{p}})] e^{-\frac{1}{2}(t/\tau_{\mathrm{p}})^2}$, were τ_{p} is the gate duration, while α is the angle determining the rotation. The function $|F(\omega)|^2$ that carries all the needed information about spectral properties of the system's dynamics may be approximately written as

$$|F_\pm(\omega)|^2 \approx \alpha^2 e^{-\tau_{\mathrm{p}}^2 \left(\omega \pm \frac{\alpha}{\sqrt{2\pi}\tau_{\mathrm{p}}} \right)^2}. \tag{58}$$

As may be seen from (57) and (58), for a spectral density $R(\omega) \sim \omega^n$ the error scales with the gate duration as τ_{p}^{-n+1} and τ_{p}^{-n+2} at low and high temperatures, respectively. Therefore, for $n > 2$ (typical for various types of phonon reservoirs) the error grows for faster gates. Assuming the spectral density of the form $R(\omega) = R_{\mathrm{DP}}\omega^3$ for low frequencies (as for the deformation potential coupling at low temperatures), we obtain from (57) and (58)

$$\delta = \frac{1}{12} \alpha^2 R_{\mathrm{DP}} \tau_{\mathrm{p}}^{-2}, \quad \text{at } T = 0$$

This leading order formula holds for $\delta \ll 1$. Also, if we introduce the upper cut-off, the error will be finite even for an infinitely fast gate (see Fig. 8); this is the ultrafast limit discussed in Sec. 7.1.

Thus, the phonon-induced error indeed decreases for slow driving. This could result in obtaining arbitrarily low error by choosing a suitably low gate speed. However, if the system is also subject to other types of noise this becomes impossible.[49] Indeed, assuming an additional contribution growing with rate γ_{M}, the total error per gate is

$$\delta = \frac{\gamma_{\mathrm{nM}}}{\tau_{\mathrm{p}}^2} + \gamma_{\mathrm{M}}\tau_{\mathrm{p}}, \quad \gamma_{\mathrm{nM}} = \frac{1}{12}\alpha^2 R_{\mathrm{DP}}, \quad \gamma_{\mathrm{M}} = \frac{1}{\tau_{\mathrm{r}}}, \tag{59}$$

where τ_{r} is the characteristic time of Markovian decoherence (recombination time in the excitonic case). As a result, the overall error is unavoidable and optimization is needed. The formulas (59) lead to the optimal values of the form (for $T = 0$)

$$\delta_{\min} = \frac{3}{2}\left(\frac{2\alpha^2 R_{\mathrm{DP}}}{3\tau_{\mathrm{r}}^2}\right)^{1/3}, \quad \text{for } \tau_{\mathrm{p}} = \left(\frac{2}{3}\alpha^2 R_{\mathrm{DP}}\tau_{\mathrm{r}}\right)^{1/3}. \tag{60}$$

For the specific material parameters of GaAs, the optimal gate time and minimal decoherence resulting from Eqs. (60) are

$$\tau_{\mathrm{p}} = \alpha^{2/3} 1.47\ \mathrm{ps}, \quad \delta_{\min} = \alpha^{2/3} 0.0035.$$

The exact solution within the proposed model, taking into account the cut-off and anisotropy (flat shape) of the dot and allowing finite temperatures, is shown in Fig. 8. The size-dependent cut-off is reflected by a shift of the optimal parameters for the two dot sizes: larger dots allow faster gates and lead to lower error.

It should be noted that these optimal times are longer than the limits imposed by level separation.[48,53,54] Thus, the non-Markovian reservoir effects (dressing) seem to be the essential limitation to the gate speed. On the other hand, in the above discussion we used simple Gaussian pulses and a straightforward way of encoding the qubit. The error resulting from the phonon dynamics may be reduced by optimizing the shape of the control pulse[55,56] or by encoding the qubit into a state of an array of QDs.[57]

7.3. *Correlation expansion*

The correlation expansion technique is a standard method used for the description of quantum kinetics of interacting carriers and phonons in semiconductor systems of any dimensionality.[58] It has been successfully applied to carrier-phonon kinetics in quantum dots driven by an optical field, beyond the instantaneous excitation limit described in Sec. 7.1 and beyond the weak perturbation case which allows a perturbative treatment (Sec. 7.2).[59–61] Compared to higher-dimensional systems, in quantum

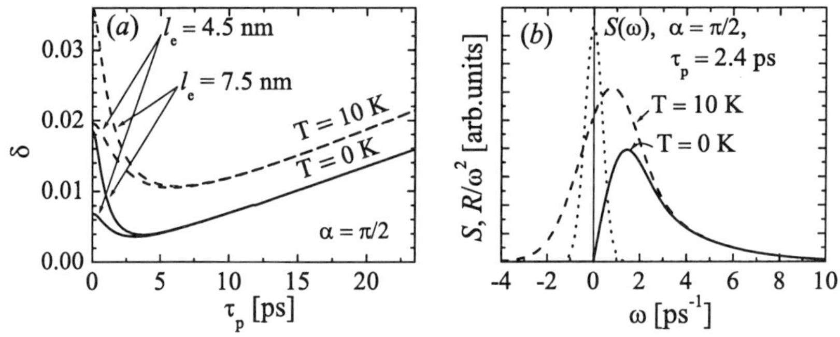

Fig. 8. (a) Combined Markovian and non-Markovian error for a $\alpha = \pi/2$ rotation on a qubit implemented as a confined exciton in a InAs/GaAs quantum dot, for $T = 0$ (solid lines) and $T = 10$ K (dashed lines), for two dot sizes (dot height is 20% of its diameter). The Markovian decoherence times are inferred from experimental data.[40] (b) Spectral density of the phonon reservoir $R(\omega)$ at these two temperatures and the gate profile $S(\omega)$ for $\alpha = \pi/2$.

dots coherent and non-equilibrium phonons play a larger role because of the localized polaron effect. Therefore, a reliable description of the carrier phonon-kinetics in these systems requires a high enough degree of the correlation-expansion technique.[61]

Various implementations of this technique differ in notation and in the choice of dynamical variables. Here, let us start from the three dynamical variables x, y, z describing the carrier state, $x = \langle \sigma_x(t) \rangle, \ldots$, where $\sigma_i(t) = e^{i\tilde{H}t/\hbar} \sigma_i e^{-i\tilde{H}t/\hbar}$ are Pauli operators, written in the $|0\rangle, |1\rangle$ basis, in the Heisenberg picture. These three variables are the coordinates of the evolving Bloch vector, uniquely determining the reduced density matrix of the carrier subsystem according to Eq. (1).

From the Heisenberg equations of motion one finds the dynamical equations for these three variables,

$$\dot{x} = i\langle [H, \sigma_x] \rangle = -\Delta y - 4y \sum_{\boldsymbol{k}} \operatorname{Re} B_{\boldsymbol{k}} - 4y \sum_{\boldsymbol{k}} \operatorname{Re} y_{\boldsymbol{k}}, \tag{61}$$

and analogous for y and z (from now on, the time dependence will not be written explicitly). Obviously, this set of equations is not closed, but involves new the phonon variables $B_{\boldsymbol{k}} = g_{\boldsymbol{k}} \langle b_{\boldsymbol{k}} \rangle$, as well as phonon-assisted variables of the form $y_{\boldsymbol{k}} = g_{\boldsymbol{k}} \langle\langle \sigma_y b_{\boldsymbol{k}} \rangle\rangle = \langle \sigma_y b_{\boldsymbol{k}} \rangle - \langle \sigma_y \rangle \langle b_{\boldsymbol{k}} \rangle$. The double angular brackets, $\langle\langle \ldots \rangle\rangle$, denote the correlated part of a product of operators, obtained by subtracting all possible factorizations of the product.

Next, one writes down the equations of motion for the new variables

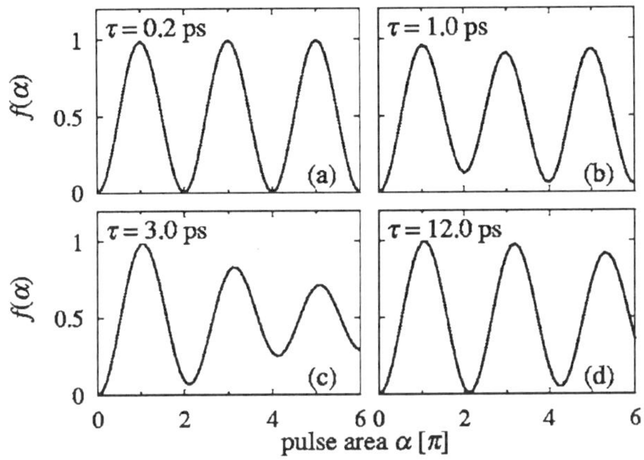

Fig. 9. Rabi oscillations of the exciton occupation $f = (1+z)/2$ as a function of the nominal pulse area α, calculated using the correlation expansion at $T = 0$. Reprinted from Ref. [61], A. Krugel *et al.*, Phys. Rev. B. **73**, 035302 (2006), ©American Physical Society 2006.

that appeared in the previous step, for instance,

$$\dot{y}_{\boldsymbol{k}} = i\langle [H, y_{\boldsymbol{k}}]\rangle = \Delta x_{\boldsymbol{k}} - 2V z_{\boldsymbol{k}} - i\omega_{\boldsymbol{k}} y_{\boldsymbol{k}} + |g_{\boldsymbol{k}}|^2 (iyz + x) \\ +2\sum_{q}(x_{q\boldsymbol{k}} + \tilde{x}_{q\boldsymbol{k}}) + 4x_{\boldsymbol{k}}\sum_{q}\operatorname{Re} B_{q} + 2x\sum_{q}(B_{q\boldsymbol{k}} + \tilde{B}_{q\boldsymbol{k}}), \tag{62}$$

where the new two-phonon and two-phonon-assisted variables are defined as $B_{q\boldsymbol{k}} = g_q g_{\boldsymbol{k}}\langle\langle b_q b_{\boldsymbol{k}}\rangle\rangle$, $\tilde{B}_{q\boldsymbol{k}} = g_q^* g_{\boldsymbol{k}}\langle\langle b_q^\dagger b_{\boldsymbol{k}}\rangle\rangle$, $x_{q\boldsymbol{k}} = g_q g_{\boldsymbol{k}}\langle\langle \sigma_x b_q b_{\boldsymbol{k}}\rangle\rangle$, $\tilde{x}_{q\boldsymbol{k}} = g_q^* g_{\boldsymbol{k}}\langle\langle \sigma_x b_q^\dagger b_{\boldsymbol{k}}\rangle\rangle$, etc. In the next step, one writes the equation of motion for these new variables, introducing three-phonon variables. It is clear that the resulting hierarchy of equations in infinite and has to be truncated at a certain level. Here we do this by setting all the correlated parts of the three-phonon and three-phonon assisted variables equal to zero. This amounts to neglecting the correlations involving three or more phonons or, physically, to neglecting three-phonon processes (that is, emission or absorption of three or more phonons within the memory time of the phonon reservoir, which is of order of 1 ps). The motivation for this procedure is that higher order correlations should play a decreasing role in the dynamics. From the equations of motion it is also clear that such higher order correlations develop in higher orders with respect to the coupling constants $g_{\boldsymbol{k}}$.

As an example of an application of this technique, Fig. 9 shows the results for the Rabi oscillations of a coherently driven exciton (see Sec. 4) interacting with phonons.[60] An interesting feature of this result is that the quality of oscillations decreases for moderate pulse durations but then increases again for longer pulses. This can be understood (using the perturbative approach of Sec. 7.2) in terms of a resonance between the oscillations of the charge distribution and the phonon modes.[62]

8. Conclusion

This chapter presented an overview of some recent results related to optical control and decoherence of carriers in semiconductor quantum dots. The problems discussed here are of interest not only from the scientific point of view but are also important for possible applications of nanostructure-based devices in the field of nano-electronics, optoelectronics and spintronics.[63,64]

Obviously, a chapter of limited length cannot give an exhaustive review of this broad and rapidly developing field. Consequently, the goal of the present review was rather to give an introduction to the field rather than a complete, encyclopedic account of all the achievements. In particular, the two level (or few-level) model on which the presented discussion was based, is obviously merely an approximation to the complex semiconductor system. In fact, any interaction of the confined carriers with the external driving fields not only leads to the desired quantum transitions but also can induce some unwanted ones. If both the desired and unwanted transitions have a discrete nature (e.g., the exciton vs. biexciton transition in a QD) the latter may be suppressed by a suitable choice of control pulses.[53,54] A more demanding problem of transitions involving the macroscopic phonon continuum is treated in Ref. [56].

Another class of experiments and theories that are not covered by this review concerns optical methods for coherent spin control in QDs. The capability of encoding and manipulating information at the single-spin level is of great importance for spintronic and quantum information processing applications. Recent experimental progress includes the generation[65] and optical control[66,67] of the spin coherence together with a possible readout of the state of a single confined spin in a QD system.[68] It was also demonstrated that spin states in QDs may be prepared with high fidelity (exceeding 99.8%) by optical coupling of electronic spin states. This was done by resonant excitation of the trion transition in the presence of small heavy-light hole mixing.[69] Corresponding to these achievements, a range of schemes for optical spin control has been proposed theoretically.[70–75] These schemes are based on quantum-optical procedures, exploiting the structure of selection rules in a QD discussed in this chapter. In particular, the spin rotation is performed by coupling af a single electron state to a trion state. Since the spin control is achieved by means of spin-dependent charge evolution these procedures result in phonon-induced dephasing that may be described using the methods described here.[76] An introductory review of optical spin control in QDs is given elsewhere.[77]

References

1. M.A. Reed, *Quantum dots*, Scientific American, **3**, 40, (1993).
2. L. Jacak, P. Hawrylak and A. Wojs, *Quantum Dots.* (Springer Verlag, Berlin, 1998).
3. D. Bimberg, M. Grundmann and N. Ledentsov, *Quantum Dot Heterostructures.* (Wiley, Chichester, 1999).
4. P.M. Petroff and S.P. Denbaars, Superlattices and Microstructures. **15**, 15, (1994).
5. P.W. Fry, I.E. Itskevich, D.J. Mowbray, M.S. Skolnick, J.J. Finley, J.A. Baker, E.P. O'Reilly, L.R. Wilson, I.A. Larkin, P.A. Maksym, M. Hopkinson, M. Al-Khafaji, J.P.R. David, A.G. Cullis, G. Hill and J.C. Clark, Phys. Rev. Lett. **84**, 733, (2000).
6. A. Wójs, P. Hawrylak, S. Fafard and L. Jacak, Phys. Rev. **B54**, 5604, (1996).
7. A. Wójs and P. Hawrylak, Phys. Rev. **B51**, 10880, (1995).
8. P. Hawrylak, A. Wójs and J.A. Brum, Phys. Rev. **B54**, 11397, (1996).
9. G. Chen, N. H. Bonadeo, D.G. Steel, D. Gammon, D.S. Katzer, D. Park and L.J. Sham, Science. **289**, 1906, (2000).
10. A. Hartmann, Y. Ducommun, E. Kapon, U. Hohenester and E. Molinari, Phys. Rev. Lett. **84**, 5648, (2000).
11. V.M. Axt and T. Kuhn, Rep. Prog. Phys. **67**, 433, (2004).
12. H.J. Krenner, S. Stufler, M. Sabathil, E.C. Clark, P. Ester, M. Bichler, G. Abstreiter, J.J. Finley and A. Zrenner, New J. Phys. **7**, 184, (2005).
13. T.H. Stievater, X. Li, D.G. Steel, D. Gammon, D.S. Katzer, D. Park, C. Piermarocchi and L.J. Sham, Phys. Rev. Lett. **87**, 133603, (2001).
14. H. Htoon, T. Takagahara, D. Kulik, O. Baklenov, A.L. Holmes Jr. and C.K. Shih, Phys. Rev. Lett. **88**, 087401, (2002).
15. A. Zrenner, E. Beham, S. Stufler, F. Findeis, M. Bichler and G. Abstreiter, Nature. **418**, 612, (2002).
16. S. Stufler, P. Ester, A. Zrenner and M. Bichler, Phys. Rev. **B72**,R121301, (2005).
17. P. Borri, W. Langbein, S. Schneider, U. Woggon, R.L. Sellin, D. Ouyang and D. Bimberg, Phys. Rev. **B66**, R081306, (2002).
18. S. Stufler, P. Machnikowski, P. Ester, M. Bichler, V.M. Axt, T. Kuhn and A. Zrenner, Phys. Rev. **B73**, 125304, (2006).
19. T. Unold, K. Mueller, C. Lienau, T. Elsaesser and A.D. Wieck, Phys. Rev. Lett. **92**, 157401, (2004).
20. S.M. de Vasconcellos, S. Stufler, S.-A. Wegner, P. Ester, A. Zrenner and M. Bichler, Phys. Rev. **B74**, 081304, (2006).
21. S. Stufler, P. Ester, A. Zrenner and M. Bichler, Phys. Rev. Lett. **96**, 037402, (2006).

22. N.N. Bonadeo, J. Erland, D. Gammon, D.S. Katzer, D. Park and D.G. Steel, Science. **282**, 1473, (1998).
23. H. Kamada, H. Gotoh, J. Temmyo, T. Takagahara and H. Ando, Phys. Rev. Lett. **87**, 246401, (2001).
24. P. Machnikowski, Phys. Rev. **B72**, 205332, (2006).
25. X. Li, Y. Wu, D. Steel, D. Gammon, T. Stievater, D. Katzer, D. Park, C. Piermarocchi and L. Sham, Science. **301**, 809, (2003).
26. M.A. Nielsen and I.L. Chuang, *Quantum Computation and Quantum Information*, (Cambridge University Press, Cambridge, 2000).
27. T. Unold, K. Mueller, C. Lienau, T. Elsaesser and A.D. Wieck, Phys. Rev. Lett. **94**, 137404, (2005).
28. M.O. Scully and M.S. Zubairy, *Quantum Optics*, (Cambridge University Press, Cambridge, 1997).
29. L. Allen and J.H. Eberly, *Optical resonance and two-level atoms*, (Dover, New York, 1987).
30. A. Messiah, *Quantum Mechanics*, (North-Holland, Amsterdam, 1966).
31. G. Jaeger, A. Shimony and L. Vaidman, Phys. Rev. **A51**, 54, (1995).
32. B.-G. Englert, Phys. Rev. Lett. **77**, 2154, (1996).
33. W.K. Wootters and W.H. Zurek, Phys. Rev. **D19**, 473, (1979).
34. G.D. Mahan, *Many-Particle Physics*, (Kluwer, New York, 2000).
35. H. Haken, *Quantum Field Theory of Solids. An Introduction*, (North-Holland, Amsterdam, 1976).
36. A. Grodecka, L. Jacak, P. Machnikowski and K. Roszak. *Phonon impact on the coherent control of quantum states in semiconductor quantum dots*; In P.A. Ling (Ed.), *Quantum Dots: Research Developments*, p. 47. Nova Science, NY, (2005); Preprint cond-mat/0404364.
37. L. Jacak, P. Machnikowski, J. Krasnyj and P. Zoller, Eur. Phys. J. **D22**, 319, (2003).
38. A. Vagov, V.M. Axt and T. Kuhn, Phys. Rev. **B67**, 115338, (2003).
39. A. Vagov, V.M. Axt, T. Kuhn, W. Langbein, P. Borri and U. Woggon, Phys. Rev. **B70**, R201305, (2004).
40. P. Borri, W. Langbein, S. Schneider, U. Woggon, R.L. Sellin, D. Ouyang and D. Bimberg, Phys. Rev. Lett. **87**, 157401, (2001).
41. P. Machnikowski and L. Jacak, Semicond. Sci. Technol. **19**, S299, (2004).
42. W. Schäfer and M. Wegener, *Semiconductor Optics and Transport Phenomena*, (Springer, Berlin, 2002).
43. A. Vagov, V.M. Axt and T. Kuhn, Phys. Rev. **B66**, 165312, (2002).
44. K. Roszak and P. Machnikowski, Phys. Lett. **A351**, 251, (2006).
45. B. Krummheuer, V.M. Axt and T. Kuhn, Phys. Rev. **B65**, 195313, (2002).
46. M. Bayer and A. Forchel, Phys. Rev. **B65**, R041308, (2002).
47. C. Cohen-Tannoudji, J. Dupont-Roc and G. Grynberg, *Atom-Photon Interactions* (Wiley-Interscience, New York, 1998).
48. E. Biolatti, R.C. Iotti, P. Zanardi and F. Rossi, Phys. Rev. Lett. **85**, 5647, (2000).
49. R. Alicki, M. Horodecki, P. Horodecki, R. Horodecki, L. Jacak and P. Machnikowski, Phys. Rev. **A70**, R010501, (2004).

50. P. Machnikowski, V.M. Axt and T. Kuhn, Phys. Rev. **B75**, 052330, (2007).
51. D. Pines and P. Nozières, *The Theory of Quantum Liquids.* (Addison-Wesley, Redwood, 1989).
52. R. Alicki, M. Horodecki, P. Horodecki and R. Horodecki, Phys. Rev. **A65**, 062101, (2002).
53. Pochung Chen, C. Piermarocchi and L.J. Sham, Phys. Rev. Lett. **87**, 067401, (2001).
54. C. Piermarocchi, Pochung Chen, Y.S. Dale and L.J. Sham, Phys. Rev. **B65**, 075307, (2002).
55. U. Hohenester and G. Stadler, Phys. Rev. Lett. **92**, 196801, (2004).
56. V.M. Axt, P. Machnikowski and T. Kuhn, Phys. Rev. **B71**, 155305, (2005).
57. A. Grodecka and P. Machnikowski, Phys. Rev. **B73**, 125306, (2006).
58. F. Rossi and T. Kuhn, Rev. Mod. Phys. **74**, 895, (2002).
59. J. Förstner, C. Weber, J. Danckwerts and A. Knorr, Phys. Rev. Lett. **91**, 127401, (2003).
60. A. Krügel, V.M. Axt, T. Kuhn, P. Machnikowski and A. Vagov, Appl. Phys. **B81**, 897, (2005).
61. A. Krügel, V.M. Axt and T. Kuhn, Phys. Rev. **B73**, 035302, (2006).
62. P. Machnikowski and L. Jacak, Phys. Rev. **B69**, 193302, (2004).
63. M. Skolnick and D. Mowbray, Annu. Rev. Mater. Res. **34**, 181, (2004).
64. P. Bhattacharya, S. Ghosh and A. Stiff-Roberts, Annu. Rev. Mater. Res. **34**, 1, (2004).
65. M.V.G. Dutt, Jun-Cheng, B. Li, X. Xu, X. Li, P.R. Berman, D.G. Steel, A.S. Bracker, D. Gammon, S.E. Economou, Ren-Bao-Liu and L.J. Sham, Phys. Rev. Lett. **94**, 227403, (2005).
66. A. Greilich, R. Oulton, E.A. Zhukov, I.A. Yugova, D.R. Yakovlev, M. Bayer, A. Shabaev, A.L. Efros, V. Merkulov, V. Stavarache, D. Reuter and A. Wieck, Phys. Rev. Lett. **96**, 227401, (2006).
67. M.V.G. Dutt, J. Cheng, Y. Wu, X. Xu, D.G. Steel, A.S. Bracker, D. Gammon, S.E. Economou, R.-B. Liu and L.J. Sham, Phys. Rev. **B74**, 125306, (2006).
68. M. Atatüre, J. Dreiser, A. Badolato and A. Imamoglu, Nature Physics. **3**, 101, (2007).
69. M. Atature, J. Dreiser, A. Badolato, A. Hogele, K. Karrai and A. Imamoğlu, Science. **312**, 551, (2006).
70. E. Pazy, E. Biolatti, T. Calarco, I. D'Amico, P. Zanardi, F. Rossi and P. Zoller, Europhys. Lett. **62**, 175, (2003).
71. T. Calarco, A. Datta, P. Fedichev, E. Pazy and P. Zoller, Phys. Rev. **A68**, 12310, (2003).
72. F. Troiani, E. Molinari and U. Hohenester, Phys. Rev. Lett. **90**, 206802, (2003).
73. Pochung Chen, C. Piermarocchi, L. J. Sham, D. Gammon and D. G. Steel, Phys. Rev. **B69**, 075320, (2004).
74. A. Nazir, B.W. Lovett, S.D. Barrett, T.P. Spiller and G.A.D. Briggs, Phys. Rev. Lett. **93**, 150502, (2004).
75. B.W. Lovett, A. Nazir, E. Pazy, S.D. Barrett, T.P. Spiller and G.A.D. Briggs, Phys. Rev. **B72**, 115324, (2005).

76. K. Roszak, A. Grodecka, P. Machnikowski and T. Kuhn, Phys. Rev. **B71**, 195333, (2005).
77. P. Machnikowski, A. Grodecka, C. Weber and A. Knorr, *Optical control and decoherence of spin qubits in quantum dots*, Preprint arrXiv:0706.0276.

BASICS OF SPINTRONICS: FROM METALLIC TO ALL-SEMICONDUCTOR MAGNETIC TUNNEL JUNCTIONS

Jacek A. Majewski

Institute of Theoretical Physics, Physics Department, University of Warsaw
ul. Hoża 69, PL-00-681 Warszawa, POLAND
E-mail: jacek.majewski@fuw.edu.pl
www.fuw.edu.pl

This article provides an overview of the experimental and theoretical research concerning the tunneling magnetoresistance effect in metallic and novel all-semiconductor magnetic tunnel junctions. First, the experimental findings concerning various magnetic tunnel junctions are analyzed to extract physical mechanisms governing the tunneling magnetoresistance effect, paying attention to the material issues. Next, the simple qualitative models and quantitative theories of tunneling magnetoresistance are presented and illustrated by examples of predictions they can provide.

Keywords: spintronics, magnetic tunnel junctions, tunneling magnetoresistance, magnetization, magnetoelectronic devices

1. Introduction

Spintronics is a relatively new emerging multidisciplinary field of solid state physics.[1] The central theme is the active manipulation of spin degrees of freedom in solid-state, and/or molecular systems,[2] in addition to the degrees of freedom connected to the electron charge. This offers new opportunities for novel devices that could combine standard electronics with the spin dependent effects that arise from the interaction between spin of the carriers and the magnetic properties of the material. It is hoped that the advantages of these new devices would be among others nonvolatility, increased data processing speed, decreased electric power consumption, and increased integration densities. The vision of the spintronic systems and the main directions of the future development have been formulated in Ref. [1]. The actual stage of research has been already described in reviews,[3–6] monographs[7–14] and even a text book.[15]

1.1. *Spintronics and magnetoelectronics*

The spintronics deals essentially with a single spin of electrons[6] or with the average spin of an ensemble of particles, which can result in non-zero spin polarization of a system, therefore non-zero magnetization. In magnetic materials, the non-zero magnetization is an equilibrium property, that can be influenced by external factors such as temperature or external magnetic field. The subfield of spintronics dealing with devices that make use of ferromagnetic materials (e.g. read-write heads of computer hard drives) is very often called magnetoelectronics[4,5] and will be the topic we are dealing with in this review. However, in many situations considered in spintronics, the generation of spin polarization in a material means that the non-equilibrium spin polarization has been induced. This can be achieved either electrically or optically. In the first case, spin polarized carriers from a magnetic material are injected into non-magnetic one. One can also obtain oriented spins in a non-magnetic material by optical techniques utilizing the effect that the angular momenta of circularly polarized photons are transferred to the electrons in the process of photonic absorption. The life-time of the spin polarized electrons depends on the effectiveness of the relaxation processes which try to bring the spin polarized population of electrons back to equilibrium. Therefore, the efficient spin injection and possibility to tune spin lifetime of electron population are essential prerequisites of functional spintronic devices. Such a generic spintronic device is the spin field effect transistor (SFET) proposed some time ago by Datta and Das.[16] This device consists of two ferromagnetic electrodes and the semiconducting non magnetic channel. In the SFET the spin polarized electrons from a ferromagnet are injected into semiconducting channel. There the spin polarization of electrons in the channel could be steered by the gate voltage. The attempts to build the SFET have been a driving force for the development of spintronics.[4] This also initialized search for new materials for spintronic applications that would facilitate effective spin injection into nonmagnetic part of the device. In result, large class of materials have been proposed that could bring new functionalities to the traditional devices. Among many others, we would like to stress the role of ferromagnetic semiconductors[17–21] and to mention organic semiconductors and ferromagnets,[22–24] high-temperature superconductors[25] and carbon nanotubes.[26,27]

The beginning of spin-based electronics (or magnetoelectronics) is typically connected to the discovery of giant magnetoresistance (GMR) in 1988 by Albert Fert and Peter Grünberg, who have been awarded the Nobel Prize in 2007. The magnetoresistance is the basic effect of magnetoelectronics. It is generally defined as the ratio of the resistance of a material in the absence a magnetic field to its resistance in a magnetic field. The GMR effect is observed in man-made artificial thin-film materials composed of alternate ferromagnetic and nonmagnetic (usually metallic) layers. The resistance of the sample depends on the relative orientation of magnetizations in two adjacent ferromagnetic layers. Originally the orientation of the magnetizations may be antiparallel. When the external magnetic field of a certain strength is applied to the sample, it obviously forces the intrinsic magnetizations in each layer to be aligned in the direction of magnetic field. The resistance of the sample in the absence of the magnetic field differs from the resistance of the sample with external magnetic field applied, $R(H) \neq R(0)$. This effect is observed when the current is perpendicular (CPP configuration) or parallel (CIP) to interfaces between layers. The GMR effect is pronounced at room temperature and for relatively weak magnetic fields (typically of the strength of 100 to 1000 Oe). When the configuration of magnetizations changes from antiparallel to parallel the resistance changes by several percent. A large change in the electrical resistance in response to an applied magnetic field is of technological relevance for the development of magnetic sensors and memories. For example, the read heads based on GMR, commercialized by IBM in 1997, caused huge miniaturization of the hard disks. However, nowadays the GMR based hard-disk technology is making quickly place for the mass storage involving semiconductor flash memories (so-called Solid State Drives).[28] It looks that in the nearest future the hard disks based on GMR effect will disappear. However, GMR based magnetic field sensors will find broad application in automotive industry. The physical mechanisms of GMR effect and its bright applications have been reviewed in detail in several monographs (for example, see Refs. [7,8,29]). Here, we will not follow the GMR topic. Instead, we concentrate on the tunneling magnetoresistance (TMR) effect in magnetic tunnel junctions (MTJ). We provide an overview of the current research status of the TMR effect, which is closely related to the GMR and extremely important in the field of spintronics.

1.2. *Magnetic tunnel junctions*

The ever-increasing demand for larger hard disk drive storage capacities caused the intensive studies of magnetic tunnel junctions (MTJ) that exhibit a large TMR effect at room temperature.[30] In typical TMR junctions, two ferromagnetic metallic layers are separated by an insulator. They will be indicated as FM/I/FM (FM = ferromagnetic metal, I = insulator) throughout this review. Tunneling current depends strongly on the relative direction of magnetizations in the ferromagnetic electrodes.[30] The TMR effect has been utilized in mass storage devices, magnetic nonvolatile random-access-memories (MRAMs),[31] magnetic sensors and novel programmable logic devices.[32] Since the first report of relatively large TMR effect of 17% in CoFe/Al_2O_3/Fe MTJ,[33] the enormous progress has been achieved. Now MTJs with a crystalline MgO tunnel barrier exhibit TMR effect of more than few hundred per cent at room temperature[30] and their commercialization in various applications has just begun.

The diversity of the physical phenomena which govern functioning of these magnetoresistive devices makes MTJs also very attractive from the fundamental physics point of view. This stimulated tremendous activity in the experimental and theoretical investigations of the electronic, magnetic, and transport properties of MTJs. Here, we present a brief overview of the development of magnetic tunnel junctions, introducing the underlying physics and theoretical description of the TMR effect. It includes the simple models providing qualitative understanding of the TMR effect in various types of MTJs and quantitative, atomistic scale, theories based on tight-binding method and *ab initio* calculations. An emphasis is placed on the material aspects of MTJs.

The most technologically advanced MTJs are based on ferromagnetic metallic electrodes and insulating barriers in between. However, the MTJs with electrodes made of manganites show extremely large TMR effect and are definitely promising candidates for spintronic devices. Very interesting group of MTJs constitute structures based entirely on semiconductors. The discovery of the first ferromagnetic semiconductors (In,Mn)As[34] and (Ga,Mn)As[35] has arisen hopes for all semiconductor spintronic devices. This resulted in huge activity in this field leading to the high quality homogeneous ferromagnetic material (Ga,Mn)As with the highest observed Curie temperature T_c of 173 K. It is even claimed that (Ga,Mn)As is the best known ferromagnet. However, the semiconductor spintronics could be dramatically boosted by fabricating ferromagnetic semiconductor with T_c larger than 300 K. During intensive searches for such material a dozen of compounds has been found that exhibit histeresis and spontaneous magnetization above room temperature, however, the origin of ferromagnetism has been generally not understood.[36] Only recently, it becomes clear that the origin of ferromagnetism in many of such compounds may lie in the spinodal decomposition that leads to nanostructures with higher concentration of magnetic ions than in the rest of the compound.[37] Nevertheless, the all-semiconductor TMJs of the type FS/S/FS (where FS = ferromagnetic semiconductor and S = semiconductor) have been fabricated and exhibit very high TMR ratios.[38–41]

Another key factor for developing novel functional semiconductor spintronic devices is an efficient electrical injection of spin polarized carriers. Here again the III-V ferromagnetic p-type semiconductor (Ga,Mn)As with its high spin polarization[42] appears as a promising material. The electrical spin injection from p-(Ga,Mn)As into non-magnetic semiconductor has first been achieved by injection of spin polarized holes.[43] Later, injection of spin polarized electrons was demonstrated employing interband tunneling from the valence band of (Ga,Mn)As into the conduction band of an adjacent n-GaAs in a Zener-Esaki diode.[44,45] Recently, a very high spin polarization of the injected electron current (ca 80%) was obtained in such devices.[46] All these are very interesting and important issues of spintronics. However, in this review, we focus on experimental and theoretical studies of magnetic tunnel junctions with ferromagnetic semiconductors constituting the electrodes. Hereafter these MTJs will be indicated as FS/S/FS in analogy to the metallic junctions FM/I/FM. We will describe both types of MTJs and compare their physical properties.

This review is organized as follows. The Secs. 2 and 3 are devoted to the TMR effect in the FM/I/FM MTJs. To establish some basic language, we introduce first the definition of TMR effect. Then we summarize briefly pletora of experimental observations to provide a background for theories of the TMR effect. We discuss the influence of such factors as bias voltage, type of ferromagnetic electrodes, type of insulating barriers, and the structure and quality of FM/I interfaces on the TMR ratio. The survey of experimental data demonstrates the rich physics connected to the TMR effect. Further we discuss theories with various sophistication level of the TMR effect. The semiconductor MTJs of the type FS/S/FS are described in Sec. 4. We present there experimental findings and summarize the recent theoretical studies of these systems. Finally, we conclude the review in Sec.5.

2. TMR in FM/I/FM magnetic tunnel junctions

Tunneling magnetoresistance (TMR) in ferromagnet/insulator/ferromagnet (FM/I/FM) magnetic tunnel junctions (MTJ) was first observed experimentally by Julliere in 1975.[47] He measured the conductance $G(V)$ of the Fe/Ge/Co structure at low temperature $T \leq 4.2$ K for parallel and antiparallel orientations of the magnetizations in ferromagnetic layers. Relative change of the conductance (i.e. TMR) can be defined as

$$TMR = \frac{G_P(V) - G_{AP}(V)}{G_{AP}(V)} = \frac{R_{AP} - R_P}{R_P}, \tag{1}$$

where G_P (R_P) and G_{AP} (R_{AP}) are the conductances (resistances, $R = 1/G$) for parallel and antiparallel magnetization orientation, respectively. The principle of TMR experiments is illustrated in Figs. 1 and 2. A given voltage is applied to the ferromagnetic electrodes and resistance (conductance) of the tunneling current is measured. Application of strong enough magnetic field always aligns magnetizations in the two ferromagnetic electrodes. Therefore, the antiparallel orientation of magnetizations corresponds to zero magnetic field. TMR is a consequence of spin-dependent

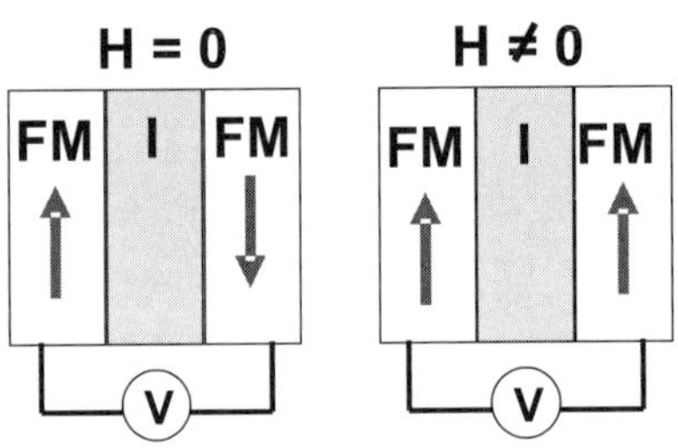

Fig. 1. The principle of a TMR measurement. The parallel orientation of magnetizations in electrodes is achieved by application of the external magnetic field to the junction

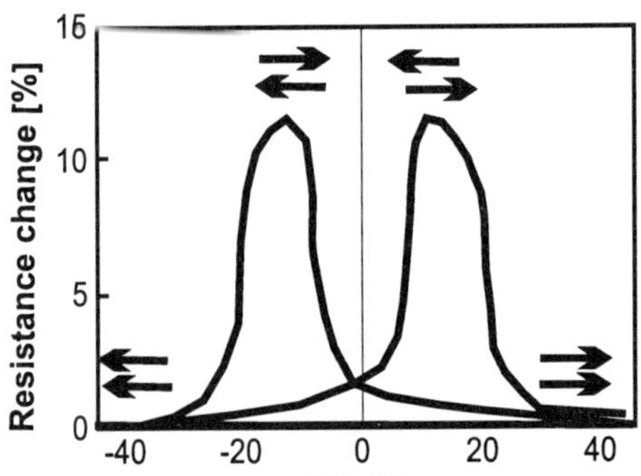

Fig. 2. The difference in the resistance versus magnetic field in a typical MTJ. The arrows indicate relative orientation of magnetizations. The figure after Ref. [33] for $CoFe/Al_2O_3/Co$ MTJ.

tunneling (SDT). The essence of SDT is an imbalance in the electric current carried by up- and down-spin electrons tunneling from a ferromagnet through a tunneling barrier.

The TMR definition as given in Eq. (1) is equivalent to the following definition TMR = $(I_P - I_{AP})/I_{AP}$, where I_P and I_{AP} are the tunneling currents for parallel and antiparallel magnetization orientation, respectively. It is worth to mention here that sometimes the TMR is defined as TMR $= (G_P - G_{AP})/(G_P + G_{AP})$ or TMR $= (G_P - G_{AP})/G_P$ (this actually corresponds to the quantity $\Delta G/G$ in Julliere's paper). However, the so-called optimistic definition of TMR given in Eq. (1) is commonly used by the majority of researchers in this field and will be employed throughout this review. Using this definition, the TMR values may lie in the range $(-\infty, +\infty)$. The experiments of Julliere[47] gave the maximal value of the ratio $\Delta G/G$ equal to 14% (coresponding to TMR as defined in Eq. (1) equal to 16%) for nearly zero bias V. The experiments demonstrated also fast decrease of the TMR with increasing external bias (see Fig. 3). Following the

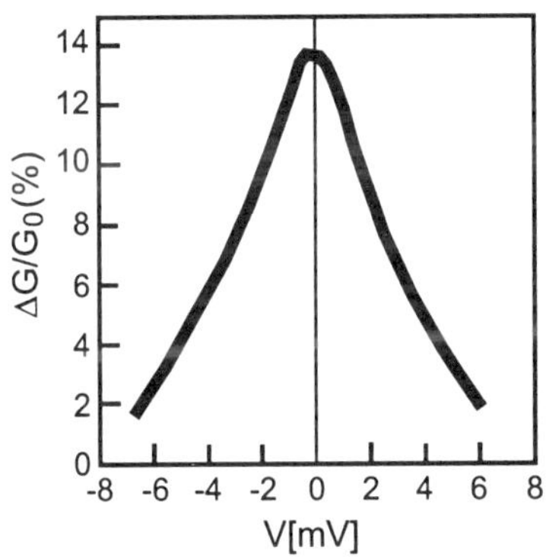

Fig. 3. The original demonstration of TMR effect in a Fe/Ge/Co junction at 4.2 K. The relative change of the conductance due to the applied magnetic field versus applied bias (after Ref. [47]).

earlier analysis of the tunneling current in the layer structure consisting of superconducting layer of Al, Al_2O_3 barrier, and ferromagnetic Ni from Ref. [48], Julliere related TMR for zero bias to the spin polarizations P_1 and P_2 of two ferromagnetic bulk metals constituting the electrodes,

$$P_i = \frac{N_{i\uparrow} - N_{i\downarrow}}{N_{i\uparrow} + N_{i\downarrow}}, \tag{2}$$

where $N_{i\uparrow}$ and $N_{i\downarrow}$ are density of states (DOS) for the spin-up and spin-down electrons in metal "i" at the Fermi energy E_F. If one assumes that the conductance for a particular magnetization orientation is proportional to the product of the densities of states of the two ferromagnetic metals, the conductances for the parallel and antiparallel alignment, G_P and G_AP, respectively, can be written as

$$G_\mathrm{P} \propto N_{1\uparrow}N_{2\uparrow} + N_{1\downarrow}N_{2\downarrow}, \tag{3}$$

$$G_\mathrm{AP} \propto N_{1\uparrow}N_{2\downarrow} + N_{1\downarrow}N_{2\uparrow} \, . \tag{4}$$

It leads to the famous expression for TMR in FM/I/FM tunnel junctions, the so-called Julliere's formula[47]

$$TMR = \frac{2P_1P_2}{1-P_1P_2} \, . \tag{5}$$

Using known values of spin polarization of Fe and Co bulk metals in Eq. (5), Julliere deduced the TMR value for Fe/Ge/Co MTJ of 26%, i.e., larger than the maximal experimentally measured value. Later on it turned out that the validity of this formula is very restricted. However, it is still used (especially for identical ferromagnetic electrodes in the junction) to express the experimentally measured values of TMR in MTJs through the effective spin polarization of the ferromagnetic material of the electrodes.

The Julliere's results inspired many researchers to study MTJs. Over the next two decades, several groups attempted to perform experiments on MTJs at low temperatures using different ferromagnetic electrodes and barrier layers. For example, in Ref. [49] the Ni/NiO/Co TMJ was studied, in Ref. [50] the fabricated NiFe/oxide/Co MTJs, with oxide layer constituting of Al_2O_3, Al-Al_2O_3, or MgO, whereas, in Ref. [51] Gd/GdO_x/Fe and Fe/GdO_x/Fe were measured. However, the measured values of TMR were usually of the order of few per cent, i.e., much smaller than the values reported by Julliere.[47] An interesting review of these earlier experiments may be found in recent reviews.[52,53]

Only in the middle of nineties, two groups have succeeded to demonstrate the TMR above 10% in FM/I/FM TMJs at room temperature. Miyazaki reported TMR of 18% (at 300 K) and 30% (at 4.2 K) in Fe/Al_2O_3/Fe TMJ,[54] whereas, Moodera found TMR in the range of 11.8 - 24 % (at 300 K - 4.2 K) in the MTJs CoFe/Al_2O_3/Co and CoFe/Al_2O_3/NiFe.[33] Since then the TMR effect has been attracting intensive attention because of its great potential in information technology. The enormous progress in the growth techniques, has allowed for fabrication of MTJs consisting of the whole pletora of materials. The progress that has been reached may be illustrated in Fig. 4. It is worth to note, that the huge increase of TMR is connected to the usage of MgO, instead of Al_2O_3, as the

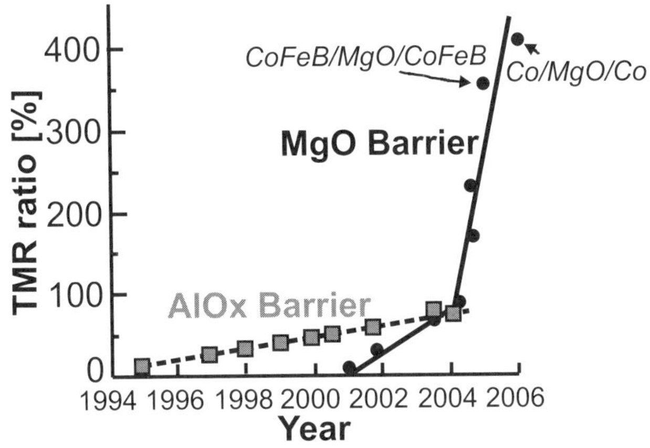

Fig. 4. Reported room-temperature TMR ratios over the past decade for amorphous AlOx-based MTJs (squares) and crystalline MgO-based MTJs (dots).

insulating barrier material in MTJs. Up to date, the highest TMR values at room temperature in TMJs with alumina tunnel barriers ranged up to 70% and have been obtained using CoFeB electrodes.[55] The mile stones in the development of TMJs that brought researchers to the fabrication of the effective nonvolatile magnetic random access memories (MRAMs) are listed in Ref. [53]. Here the scope of this review does not allow detailed report on all these attempts to fabricate FM/I/FM junctions with the highest possible TMR value and possibility of tuning. Excellent overview of the progress that has been achieved the reader can find in many recently published review articles.[52,53] In the experimental studies of FM/I/FM MTJs the researchers concentrated mostly on the dependence of the TMR effect on, (i) the magnetic field needed to switch relative magnetizations in FM electrodes, (ii) external bias, (iii) temperature, (iv) ferromagnetic material used in the electrodes, (v) the barrier material (vi) crystallographic growth direction of the TMJs, (vii) roughness of the interfaces between ferromagnet and the insulator (role of disorder and impurity tunneling). All these issues have been addressed in the Ref. [52]. Here we would like to say few words about the dependence of TMR on the factors listed above.

2.1. *Dependence of TMR on external bias*

In the most of MTJs the magnitude of the TMR decreases strongly with increasing bias voltage, similarly to that observed originally by Julliere (see Fig. 3). The figure of merit is the voltage at which the TMR is reduced by a factor of two, the so-called 'half-voltage'. In the first experiments of Julliere[47] only 3mV was needed to reduce the TMR by half. Later several groups were able to increase the 'half-voltage' to 200 mV[33] or even up to more than 500 mV.[56,57] These increase of the 'half-voltage' was mostly ascribed to the structural quality of the structures, particularly to the quality of interfaces. The physical origin of the TMR decrease with applied bias remains controversial up to now. Whereas, some studies suggested that the bias dependence is due to magnon excitations on the interfaces,[58] others just excluded such mechanism.[59] On the other hand new suggestions appear that impurities at interfaces are responsible for the TMR drop with bias.[60–62] Interestingly, the impurities at the interfaces can also lead to the increase of TMR in comparison to TMJs with pure interfaces.[63]

The electronic structure of the ferromagnets is another factor that can influence the voltage dependence of the conductance and further TMR. If external bias is applied to a MTJ, the tunneling current has the contribution from electrons which tunnel from the occupied states below the Fermi energy of one electrode to the empty states above the Fermi energy of the other electrode. The electronic structure of the ferromagnets depends on energy range, therefore, the conductance should be energy dependent resulting in the variation of TMR with the applied voltage. Recently, this has been demonstrated in Co/Al_2O_3/Co MTJs with fcc and hcp Co electrodes,[64] where it was possible to establish relationship between TMR observed in MTJs with electrodes in two different crystalline phases and calculated band structure. The relation between observed TMR and electronic structure of the FM electrodes has been also found out in other systems.[65,66]

2.2. *Dependence of TMR on ferromagnet*

The measurements of TMR in MTJs of the type FM/I/FM with (alumina as barrier and alloys of ferromagnetic metals) showed that the measured TMR correlated rather well with the value obtained from the phenomenological Julliere's formula [Eq. (5)].[52] According to this formula, one should expect the strongest TMR in MTJs with electrodes exhibiting the largest spin polarization. This explains a great deal of the recent interest in so-called 'half-metallic' ferromagnets, i.e., materials for which only one spin band is occupied at the Fermi level, resulting in perfect 100% spin polarization.[67] Many compounds have been predicted to be half-metalls,[52] but only some of them, such as manganites $La_{0.67}Sr_{0.33}MnO_3$ (LSMO),[68] NiMnSb,[69] and CrO_2[70,71] exhibited this feature in experiments. LSMO and LCMO (e.g., $La_{0.7}Ca_{0.3}MnO_3$) manganites have been successfully used as electrodes in

TMJs with barriers consisting of $SrTiO_3$, $PrBaCu_{2.8}Ga_{0.2}O_7$, CeO_2, and $NdGaO_3$. In such TMJs the measured TMR was above 400%.[72–74] Recently, the TMR of order of 1800% has been observed in the LSMO/$SrTiO_3$/LSMO TMJ.[76] An extensive review of the magnetotransport phenomenon in magnetic oxides with its intriguing physics can be found in Refs. [75] and [76].

The dependence of TMR on the ferromagnetic material of electrodes, rises a question whether the growth direction of the junction does play a role. It has been shown previously that the spin polarization of electrons emitted from a ferromagnetic metal depends on the crystallographic direction.[77] A strong dependence of TMR on the crystal orientation of ferromagnetic electrodes was confirmed experimentally in Ref. [56]. There, the authors studied the TMR of Fe/Al_2O_3/$Fe_{0.5}Co_{0.5}$ tunnel junctions with Fe electrode being a single crystal with (001), (110), and (211) planes. They obtained TMR of these junctions (at 2 K) being equal to 13%, 32%, and 42% for the Fe(001), Fe(110), and Fe(211) orientations, respectively. The crystal orientation dependence of TMR can be attributed to the crystal anisotropy of the spin polarization of Fe and the "momentum filtering effect" of the tunnel barrier.[56] Typical temperature and bias dependences of TMR are pronounced as the Al_2O_3 barrier becomes thicker, indicating that spin scattering in the barrier layer is one of the factors that modify the spin polarization of tunneling electrons at finite temperatures and bias voltages. In last decade, also ferromagnetic semiconductors have been used as electrodes in TMJ junctions exhibiting large values of TMR.[38] Unfortunately, Curie temperature of ferromagnetic semiconductors are still lower than 300 K, reaching in (Ga,Mn)As 173 K. The detailed discussion of these interesting systems will be postponed to the Sec. 4.

2.3. *Dependence of TMR on the barrier*

In the earlier stages of TMJs studies, mostly Al_2O_3 barriers have been used. Significant efforts have been invested to characterize, understand, and improve properties of alumina barriers. However, the highest observed TMR at room temperature does not exceed 70%.[55] This low magnetoresistance seriously limits the feasibility of spintronics devices, therefore the better materials for MTJs have been needed. Recently the TMJs with MgO barriers (and the same electrodes) dominated the field, and now exhibit TMR that exceeds 180%[78] or even 220%[79] at room temperature and 300% at low temperatures. Simultaneously, the influence of the MgO layer width d_{MgO} on the TMR has been studied.[78] The dependence of the TMR ratio on d_{MgO} (see Fig. 5) sheds light on the physical mechanism of the TMR effect. As can

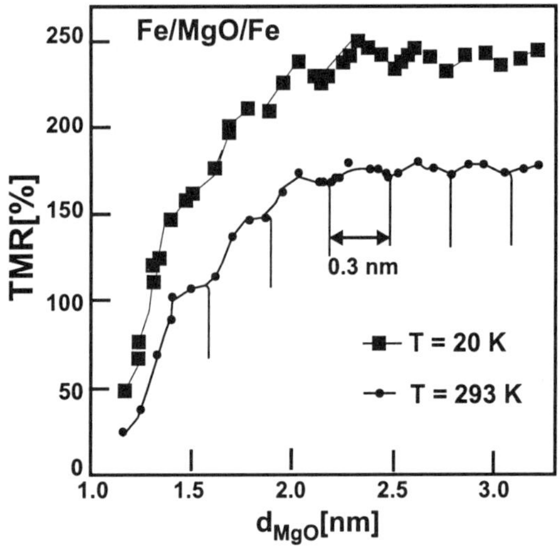

Fig. 5. Tunnel magnetoresistance of Fe(001)/MgO(001)/Fe(001) junctions. TMR at T = 293 K and 20 K (measured at a bias voltage of 10 mV) as a function of the barrier material width d_{MgO} (after Ref. [78]).

be seen in Fig. 5, the TMR increases with the increasing width of MgO barrier. Additionally the TMR exhibits clear oscillations as a function of d_{MgO}. The period of the oscillations (0.30 nm) was observed to be independent of temperature and bias voltage, but the amplitude of the oscillation decreased with increasing bias voltage. The increase of TMR with the increased d_{MgO} observed in experiment[78] agrees well with the theoretical calculations.[80,81] This can be understood as follows. When the tunnel barrier is thick, the tunneling current is dominated by electrons with momentum vectors normal to the barrier, because tunneling probability decreases rapidly when the momentum vectors deviate from the barrier-normal direction. The Fe-Δ_1 band has its highest spin polarization in the barrier normal ([001]) direction, resulting in a huge MR ratio. When the tunnel barrier is thin, electrons with momentum vectors deviating from the barrier-normal direction have a finite tunneling probability that decreases both the spin polarization and the MR ratio.[78] The origin of TMR oscillations lies in the coherent nature of the tunneling in these Fe/MgO/Fe tunnel junctions. The coherent tunneling is sensitive to the band structure details and theoretical explanations has been given in Ref. [80]. This strongly indicates that coherency of wave functions is conserved across the tunnel barrier. The coherent TMR effect is a key to making spintronic devices with novel quantum-mechanical functions, and to developing gigabit-scale MRAMs.

In the MTJs with MgO barriers, one can increase TMR ratio even further by substituting Fe electrodes with bcc Co (001) ones.[82] Such fully epitaxial Co(001)/MgO(001)/Co(001) junctions with metastable bcc Co (001) have been fabricated with molecular beam epitaxy along [001] cubic crystallographic direction.[82] The TMR for these MTJs reaches up to 410% at room temperature. In addition these MTJs exhibit smaller temperature dependence of TMR than Fe/MgO/Fe junctions. On the other hand, TMR in junctions with Co electrodes is more sensitive to the applied bias voltage than in the junctions with Fe electrodes.[82] One can increase TMR even further using Co and Fe alloy. Recently, Huyakawa *et al.* reported TMR ratio as high as 472% at room temperature and even 804% at 5 K n CoFeB/MgO/CoFeB MTJs.[83] It is the highest value of TMR in the tunnel junctions with MgO barrier achieved up to now.

In recent years, other, alternative to alumina and MgO barriers have been also employed. Very often MTJ with these new barriers show qualitatively new magnetoresistance behavior. In the FM/I/FM MTJs discussed up to now, the TMR as defined in Eq. (1) is usually positive so-called *normal* TMR. It means that the conductance (resistance) is larger (smaller) for parallel orientation of magnetizations in the electrodes. As expected, it has been found that the TMR is *positive* in Co/Al_2O_3/LSMO MTJs.[66] However, if the barrier material is changed from alumina to $SrTiO_3$ the Co/$SrTiO_3$/LSMO tunnel junction exhibits *negative* (so-called *inverse*) TMR in some range of external bias.[66] In this concrete junction, the *inverse* TMR has been observed for bias $V < 0.8$V, whereas for $V \geq 0.8$V the *normal* TMR was measured.[66] It is obvious that this effect can not be explained by Julliere's formula [Eq. (5)], which deals only with the spin polarizations of the bulk ferromagnets. Therefore, the authors introduced the concept of interface polarization, which should be negative for Co/$SrTiO_3$ interface and positive for the Co/Al_2O_3 one. Because of the half-metallic second electrode, the polarizations of the Al_2O_3/LSMO and $SrTiO_3$/LSMO interfaces should be in both cases positive. In order to show this more conclusively, they investigated Co/Al_2O_3/$SrTiO_3$/LSMO MTJs, with the expectation that a normal positive TMR would result for all biases. Experiments indeed confirmed their hypothesis. The measured TMR has been positive with bias dependence essentially identical to standard Co/Al_2O_3/Co junctions. These findings have been also confirmed in other experiments.[84,85] Further, experiments utilizing Ta_2O_5/Al_2O_3 composite barriers[65] have shown that, as the bias voltage is increased, the TMR decreases rapidly and changes its sign from positive to negative at a finite voltage. When the bias voltage is further increased, the TMR becomes concave-up with a tail bending upwards with the increase of the applied voltage (see Fig. 2 in Ref. [65]). The latter reminds us that the tail may be a physical symbol that the TMR would be oscillatory, if the bias voltage went out of the measurable range of the applied voltage.[86] In addition, a junction with an Al_2O_3 barrier shows monotonic decrease of the TMR with the increase of bias voltage but remains positive before the breakdown of the junction.[65] It implies that the measurable range of bias voltage for the Al_2O_3-barrier MTJ is in such a low-bias region that the sign change and oscillation could not be observed. It is obviously not the case in the Ta_2O_5-barrier junction, where the sign of the spin polarization at Ta_2O_5 interfaces varies with the applied bias voltage in similar manner to the Co/$SrTiO_3$ interface.

The dependence of TMR on the interface between insulator and ferromagnet gives possibility to engineer properties of MTJs. On the other hand, to be able to predict these properties one should know the microscopic details of the interfaces. Here, the reliable *ab initio* calculations of tunneling junctions could be of great help to understand the physics of the interfaces and facilitate design of interfaces leading to the required properties of the MTJs. Let us mention in this place two typical examples of such calculations, (i) the dependence of the interface transparency on the crystal orientation,[87] and (ii) spin-polarized tunneling in the Co/$SrTiO_3$/Co MTJs with the interfaces discussed above.[88]

At the end of this subsection, we would like to mention that FM/MgO/FM MTJs can exhibit negative TMR in certain bias range.[89,90] In high quality epitaxial Fe/MgO/Fe MTJs with extremely smooth bottom Fe-MgO interface, the tunnel magnetoresistance is found to be positive below 0.2 V and negative above. This feature is ascribed to an interfacial resonance state located in the minority band of Fe(001). When coupled to a metallic bulk state, this spin-polarized interfacial state enhances the band matching at the interface and, therefore, increases strongly the conductivity in the antiparallel magnetization configuration. Recently, the possibility to invert the sign of TMR with applied external bias has been also reported in hybrid MTJs using semiconductor as a barrier, namely Fe/GaAs/Fe.[91]

For TMJs to be applied as MRAMs, not only the high TMR is necessary but also, a reduced value of the resistance-area (RA) product. Typical values which are required for MRAMS and field sensors are 100 $\Omega\mu m^2$ and less than $0.5\Omega\mu m^2$, respectively.[52] This has motivated a search for other suitable barriers. Indeed, using for the barriers such materials as AlN, AlO_xN_y,

ZrO_2, and $ZrAlO_x$, it was possible to fabricate MTJs with similar values of TMR but considerably lower the resistance-area product.[52]

2.4. *Influence of interfaces on TMR*

The experimental observations presented above strongly suggest that the tunneling current in magnetic junctions is very often strongly influenced by the microscopic structural details of the interface region and, as a consequence, by the interface induced changes in the electronic structure. One of the ways to explore this issue is to fabricate MTJ with ultrathin layers (so-called 'dusting' layers) inserted at the electrode–barrier interfaces. In this context, the spin polarization of Al/Al_2O_3/Au/Fe junctions as a function of Au interlayer thickness has been studied.[92] Later, the tunneling structure Co/Cr/Al_2O_3/Co with ultrathin layer of Cr have been investigated.[93] In structures with 'dusting' layer, one observes roughly exponential decay of TMR. However, if one builds in only few monolayer of ferromagnetic Co between the 'dusting' layer and the Al_2O_3 barrier, i.e., one growths the structure Co/Cr/Co/Al_2O_3/Co the TMR the TMR is nearly completely restored with 3-5 monolayers of Co between Cr and Al_2O_3. In other words, the Co/Cr/Co/Al_2O_3/Co MTJ behaves as the Co/Al_2O_3/Co one. This shows that just the few monolayers of the electrode adjacent to the insulator dominate MTJ properties. In MTJs described above, there was no experimental evidence of the formation of the quantum-well states within the Cr layer. To see quantum effects requires definitely structures of very high quality. However, the Co(001)/Cu(001)/Al_2O_3/$Ni_{80}Fe_{20}$ junctions with bottom epitaxial Co/Cu electrodes really exhibit the true quantum-well oscillations of

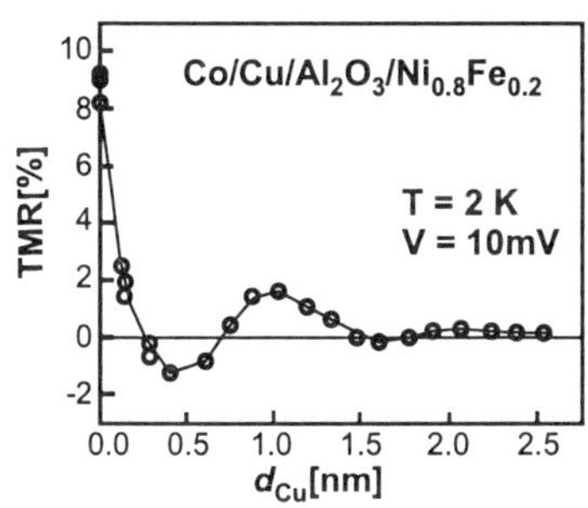

Fig. 6. TMR at 2 K and a bias of 10 mV as a function of Cu interlayer thickness for Co(001)/Cu(001)/Al_2O_3/$Ni_{0.8}Fe_{0.2}$ junctions. The period of the oscillation observed, 11.4 Å, is in agreement with the Fermi surface of Cu (after Ref. [94]).

the TMR (see Fig. 6).[94] A clear (damped) oscillations of the TMR with a period of 11.4Å are evident. Further, independent measurements on similarly grown Co/Cu/Co trilayers gave oscillations of the interlayer exchange coupling with a similar period of 11Å. This clear correlation suggests that the oscillations do indeed arise from spin-dependent reflection at the Co/Cu interface due to the formation of spin-polarized quantum-well states within the Cu interlayer.

2.5. *Various TMR structures*

The magnetoresistance effects have been also investigated in a Sr_2FeMoO_6 (SFMO)-based tunnel junction.[95] In a $Sr_2FeMoO_6/SrTiO_3$ /Co junction, the clear *positive* TMR of 50% has been observed. Since the polarization of the $SrTiO_3$ /Co interface is known to be negative (see Sec. 2.3), it implies that the polarization of the second interface $Sr_2FeMoO_6/SrTiO_3$ is also negative. Using commonly accepted value of the spin polarization for the $SrTiO_3$/Co interface equal to 20%, the authors ascribe *negative* spin polarization of 85% to half-metallic Sr_2FeMoO_6.

Huge positive TMR of 570% has been recently measured at low temperature in a MTJ consisting of a stacking structure of Co_2MnSi/Al–O/Co_2MnSi, which has been fabricated using magnetron sputtering system.[96] This is the highest value of TMR in a structure with amorphous Al-O tunneling barrier reported up to now. The observed dependence of tunneling conductance on bias voltage clearly reveals the half-metallic energy gap of Co_2MnSi. Interestingly, the TMR in those structures exhibits much stronger temperature dependence than in the standard $Co_{0.75}Fe_{0.25}$/Al-O/$Co_{0.75}Fe_{0.25}$ MTJs. For Co_2MnSi/Al–O/Co_2MnSi, the TMR that is equal to 570% at 2 K dramatically drops to 67% at room temperature, whereas the TMR for conventional MTJ remains nearly constant in this temperature range (being of order of 60%). The strong temperature dependence of the TMR ratio observed in the MTJ based on Co_2MnSi/Al–O/Co_2MnSi is an important problem that needs to be solved before these structures can find application in devices. Such strong temperature dependences of the TMR ratio are generally attributed to spin-flip tunneling caused by magnetic impurities at the FM/barrier interface or pinholes in the barrier layer.[97] Since the creation of magnetic impurities is a critical problem in these MTJs, it may be possible to suppress the temperature dependence of the TMR ratio by improving the quality of the Co_2MnSi/Al-O interface.

3. Theory of TMR in FM/I/FM tunnel junctions

The reachness of intriguing physical phenomena observed in the MTJs consisting of two ferromagnetic electrodes separated by a insulating barrier, i.e., of the typeFM/I/FM, stimulated, of course, intensive theoretical studies. On the other hand, deep understanding of all these phenomena is necessary to design novel MTJ-based devices. As one can expect from the survey of experimental data, a realistic description of spin dependent transport in MTJ requires taking into account accurate atomic, electronic, and magnetic structure of junctions. In general, the quantitative description is rather complicated because transport properties depend sensitively on interfacial roughness, impurity at the interface and in the barrier, and other types of disorder. Some of these factors can be taken into account in modern *ab initio* calculations. However, the physical insight can be also reached by studying simplified models of "ideal" (or perfect) MTJs.

To calculate tunneling current in FM/I/FM tunneling junctions and further TMR, the coherent transport approach has been employed in many theoretical studies. It means that any type of electron scattering which can affect tunneling conductance has been usually neglected. Thereby, purely ballistic transport over the whole MTJs has been assumed.[98] These theories have been reviewed in a series of papers.[52,99,100] A large part of theoretical activities concentrated on so-called free electron model, i.e., a model where the metallic electrodes are modelled by spin polarized free electron gas. In the many cases, however, it was necessary to take into account the realistic band structure of ferromagnetic metals. This can be done in relatively simple manner just employing tight-binding theory.[100] The largest insight into the role of atomistic details of tunneling junctions can be obtained from *ab initio* methods. On the other hand, the parameters obtained from the *ab initio* calculations can be used in qualitative models.[99]

In this section, we summarize the main theoretical achievements for the TMR in FM/I/FM TMJs. This allows later on an explicit comparison with the theoretical description of structures where the ferromagnetic electrodes consist of ferromagnetic semiconductors. Let us start the discussion with the simple free electron models of TMR in FM/I/FM MTJs.

3.1. *Free electron models*

Let us consider a FM/IB/FM tunnel junction structure, in which two identical ferromagnetic electrodes (FM) are separated by middle non metallic (insulating) barrier layer of thickness d (see a scheme of such a structure in

Fig. 7). All these layers are assumed to be homogeneous in the x-y plane

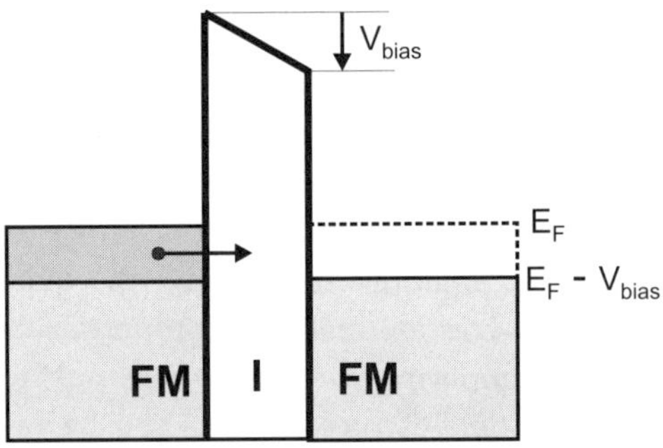

Fig. 7. Simple model of a magnetic tunnel junction at finite bias. Sloping barrier results from shifting bands on right side down by the bias voltage.

and to be stacked along the z direction. Consider two FM layers to have different magnetization directions, with θ being the angle between them. In the simplest case these magnetizations are parallel (i.e., $\theta = 0$) or antiparallel (i.e., $\theta = \pi$) to each other. In this model, it is assumed that the spin is a good quantum number in the whole structure, i.e., one neglects the spin-orbit coupling in these considerations.

If the width of the barrier is smaller than the coherence length and mean free path, which is usually the case, it is natural to analyze charge and spin currents through an insulator barrier separating two FM electrodes employing quantum-mechanical coherent approach. In the case of the coherent transport considered here, not only the total electron energy of the tunneling electron (E) is conserved but also the parallel to the interfaces component of the electron wave vector ($\vec{k}_{\|}$). This implies that the interfaces between ferromagnetic and insulating layers are ideally smooth and there is no diffusive scattering at interfaces. It is a common practise as well, to assume that the band structure of the ferromagnetic metals is described by two spin-polarized free electron s-type bands. This assumption can be very reasonable, as in the case of Fe where bands responsible for tunneling are nearly parabolic, albeit with effective masses that differ from the electron mass (see Fig. 8). Although details of the barrier structure are not embodied in this simple model, it is expected that essential features of tunneling currents across the barriers are captured. In the spin-polarized free-electron

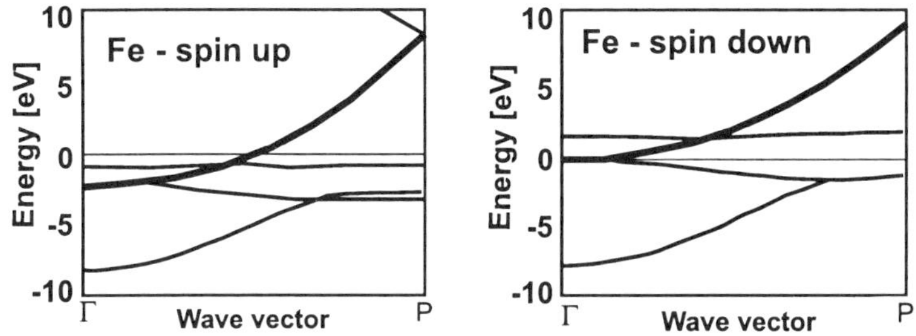

Fig. 8. Spin up and down band structures for Fe along the [100] or Γ-P direction. The bands responsible for the tunneling are represented by the thick line. The Fermi energy is at 0 eV.

approximation, the electron Hamiltonian is given by

$$\hat{H} = -\frac{\hbar^2}{2m(z)}\left(\frac{\partial^2}{\partial x^2} + \frac{\partial^2}{\partial y^2} + \frac{\partial^2}{\partial z^2}\right) + U(z) - \vec{h}(z)\cdot\vec{\sigma}. \tag{6}$$

Here the first term on right-hand side is the kinetic energy, the second one is the potential energy, and the third one is the internal exchange energy with $\vec{h}(z)$ the molecular field (or magnetization) and $\vec{\sigma}$ the conventional Pauli spin operator. Since the structure is homogeneous in the x-y plane, the wave function $\psi(x,y,z)$ separates, being equal to $\psi(x,y,z) = \exp(\vec{k}_{\|}\cdot\vec{r}_{\|})\Psi(z)$, with $\vec{r}_{\|} = (x,y)$. This leads to the Hamiltonian

$$\hat{H} = -\frac{\hbar^2}{2m(z)}\left(\vec{k}_{\|}^2 + \frac{\partial^2}{\partial z^2}\right) + U(z) - \vec{h}(z)\cdot\vec{\sigma}. \tag{7}$$

To the wave vector $\vec{k}_{\|}$, one can also ascribe the transverse mode energy $E_{\|}$, where $E_{\|} = \frac{\hbar^2}{2m}\vec{k}_{\|}^2$. Then the total energy can be written as $E = E_{\|} + E_z$. However, in many papers, only states with $\vec{k}_{\|}$=0 are studied. This corresponds to a strictly one dimensional problem. In the model, it is usually assumed that $\vec{h}(z) = \vec{h}_{\mathrm{R}}$ for $z < 0$ and $\vec{h}(z) = 0$ for $0 < z < d$, $\vec{h}(z) = \vec{h}_{\mathrm{L}}$ for $z > d$, with $\vec{h}_{\mathrm{R}}$ and $\vec{h}_{\mathrm{L}}$ being equal in the magnitude but generally having different direction. The spin splitting energy of "up" and "down" parabolic bands, which is caused by the internal exchange energy, is indicated by 2Δ. The potential energy $U(z)$ may contain the external bias V applied to the structure. This potential profile is typically assumed to be constant in the ferromagnetic electrodes, and, depending on the model, has relatively simple form in the barrier. Either it is taken to be a constant U_{B} (when the zero bias limit is considered),[101] or may have a trapezoidal shape $U(z)$ $= U_{\mathrm{B}}$ - $\frac{eVz}{d}$ for $0 < z < d$. However, the δ-like barrier potential, or two δ-like potentials for the insulator barriers have been also considered.[102] The effective electronic mass can differ in the ferromagnetic electrodes and in the barrier, $m(z) = m_{\mathrm{FM}}$ for $z < 0$ and $z > d$, and $m(z) = m_{\mathrm{B}}$ for $0 < z < d$. The difference in effective masses can have large consequences for the bias dependence of the TMR signal and the reversal of the TMR sign.[102]

If there is a non-zero angle between magnetizations $\vec{h}_{\mathrm{L}}$ and $\vec{h}_{\mathrm{R}}$, the spin quantization axis is assumed to be changed at plane $z = d$. A spin "up" (i.e., with the spin parallel to $\vec{h}_{\mathrm{L}}$) electron incident from the left FM will produce four sets of the waves: spin-up and spin-down electrons reflected from the barrier into the left electrode and spin-up and spin-down electrons transmitted into the right electrode. The spin-up incident plane wave can generate a spin-down reflecting wave only owing to the change in spin quantization axis. In the electrode regions, the wave function has the form

$$\Psi_{\mathrm{L}}=\begin{pmatrix}1\\0\end{pmatrix}\exp(ik_{\mathrm{Lz}\uparrow}z)+r_{\uparrow\uparrow}\begin{pmatrix}1\\0\end{pmatrix}\exp(-ik_{\mathrm{Lz}\uparrow}z)+r_{\uparrow\downarrow}\begin{pmatrix}0\\1\end{pmatrix}\exp(-ik_{\mathrm{Lz}\downarrow}z), \quad (8)$$

$$\Psi_{\mathrm{R}}=t_{\uparrow\uparrow}\begin{pmatrix}1\\0\end{pmatrix}\exp(ik_{\mathrm{Rz}\uparrow}z)+t_{\uparrow\downarrow}\begin{pmatrix}0\\1\end{pmatrix}\exp(ik_{\mathrm{Rz}\downarrow}z) \quad (9)$$

The wave vectors $k_{\mathrm{Lz}\uparrow}$, $k_{\mathrm{Lz}\downarrow}$, $k_{\mathrm{Rz}\uparrow}$, and $k_{\mathrm{Rz}\downarrow}$ describe the plane waves with "up" and "down" spin direction in the left and right electrode, respectively. They can be easily determined as in the textbook scattering problem.[86,102] The wave function in the barrier Ψ_{B} depends on the barrier potential. For simple $U(z)$ mentioned above, the wave function Ψ_{B} can be given analytically, either as a linear combination of plane waves[103] or as the linear combination of the Airy functions.[86,102] The reflection coefficients $r_{\uparrow\uparrow}$ and $r_{\uparrow\downarrow}$ and also transmission coefficients $t_{\uparrow\uparrow}$ and $t_{\uparrow\downarrow}$ can be established by the following boundary conditions

$$\Psi_{\mathrm{L}}|_{z=0} = \Psi_{\mathrm{B}}|_{z=0}, \quad (10)$$

$$\Psi_{\mathrm{B}}|_{z=d} = \hat{T}\Psi_{\mathrm{R}}\cdot|_{z=d}, \quad (11)$$

$$\frac{1}{m_{\mathrm{FM}}}\frac{\partial\Psi_{\mathrm{L}}}{\partial z}|_{z=0} = \frac{1}{m_{\mathrm{B}}}\frac{\partial\Psi_{\mathrm{B}}}{\partial z}|_{z=0}, \quad (12)$$

$$\frac{1}{m_{\mathrm{B}}}\frac{\partial\Psi_{\mathrm{B}}}{\partial z}|_{z=d} = \frac{1}{m_{\mathrm{FM}}}\hat{T}\frac{\partial\Psi_{\mathrm{B}}}{\partial z}|_{z=d}, \quad (13)$$

where $\hat{T}$ is the spinor transformation matrix

$$\hat{T} = \begin{pmatrix}\cos(\frac{\theta}{2}) & \sin(\frac{\theta}{2})\\ -\sin(\frac{\theta}{2}) & \cos(\frac{\theta}{2})\end{pmatrix}. \quad (14)$$

For a spin-down incident plane wave from the left FM, similar expressions for the outgoing waves of Eq. (8) are readily obtained and the transmission and reflection coefficients $t_{\downarrow\downarrow}$, $t_{\downarrow\uparrow}$, $r_{\downarrow\downarrow}$, $r_{\downarrow\uparrow}$ extracted. The knowledge of the all transmission coefficients allows for calculation of the charge current crossing the barrier. According to the Landauer-Büttiker theory the total current is given by

$$J = J_{\uparrow\uparrow} + J_{\downarrow\downarrow} + J_{\uparrow\downarrow} + J_{\downarrow\uparrow}, \tag{15}$$

where $J_{\mathrm{ss}'}$ (with $s, s' \in \{\uparrow, \downarrow\}$) can be evaluated by[104]

$$J_{\mathrm{ss}'} = \frac{em}{4\pi^2\hbar^3} \int dE_{\parallel} \int dE_z T_{\mathrm{ss}'} [f_{\mathrm{s}}(E_{\parallel} + E_z) - f_{\mathrm{s}'}(E_{\parallel} + E_z + eV)]. \tag{16}$$

Here $T_{\mathrm{ss}'}$ are related to the transmission coefficients $t_{\mathrm{ss}'}$ by the ratio of velocities v_{zs} in the left and right electrodes. Having in mind that $v_{zs} = \frac{1}{\hbar}\frac{\partial E_s}{\partial k_{\mathrm{zs}}}$ and that parabolic dispersion relations have been assumed, $T_{\mathrm{ss}'} = t_{\mathrm{ss}'}\frac{k_{\mathrm{zs}'}}{k_{\mathrm{zs}'}}$. The function f_{s} is the Fermi-Dirac distribution functions. In the equation given above the difference of the Fermi-Dirac distribution functions for the left and right electrodes is taken. The values of the electrochemical potentials in the electrodes differ by the applied external bias $\mu_{\mathrm{R}} = \mu_{\mathrm{L}} - eV$. Owing to the spin splitting of the bands, the Fermi energies for spin-up and spin-down electrons are different and, as a consequence, the corresponding Fermi-Dirac functions, which has been indicated by the proper spin index. The Fermi energies for spin-up and spin-down electrons are equal to $E_{\mathrm{F}\uparrow} = \mu + \Delta$, and $E_{\mathrm{F}\downarrow} = \mu - \Delta$, respectively. Note that in the Eq. (16) for the partial current $J_{\mathrm{ss}'}$, the integration over $E_{\parallel}$ and E_{z} can be substituted by the integration over $k_{\parallel}$ and the total energy E.[105]

To calculate the TMR, one has to calculate the total current for the simpler case than discussed above, namely for the situation when the magnetizations of the left $\vec{h}_{\mathrm{L}}$ and the right electrode $\vec{h}_{\mathrm{R}}$ are collinear, being parallel or antiparallel to each other. Indicating the suitable currents by $J_{\mathrm{P}}(V)$ and $J_{\mathrm{AP}}(V)$, respectively, the TMR can be defined as

$$TMR = \frac{J_{\mathrm{P}}(V) - J_{\mathrm{AP}}(V)}{J_{\mathrm{P}}(V)}. \tag{17}$$

The transmission coefficients $t_{\mathrm{ss}'}$, and therefore also corresponding currents $J_{\mathrm{ss}'}$, are dependent on the angle θ between the magnetizations in the left and right electrodes. For specific values of θ, some current components $J_{\mathrm{ss}'}$ vanish. It is easy to show that only $J_{\uparrow\uparrow}$ and $J_{\downarrow\downarrow}$ contribute to the J_{P} and $J_{\uparrow\downarrow}$ and $J_{\downarrow\uparrow}$ to the J_{AP}.

3.2. *Physical insight gained by free electron models*

The two-band model of free electrons in ferromagnetic metals explains a number of experimental results obtained in studies of the TMR in FM/I/FM tunnel junctions. For example, Slonczewski in his pioneering paper[101] has found that the conductance is (in the limit of thick barrier) a linear function of the cosine of angle θ between the magnetic moments of the films

$$G(\theta) = G_0(1 + P^2 \cos(\theta)), \tag{18}$$

where P is the effective polarization of tunneling electrons given by

$$P = \frac{k_\uparrow - k_\downarrow}{k_\uparrow + k_\downarrow} \times \frac{\kappa^2 - k_\uparrow k_\downarrow}{\kappa^2 + k_\uparrow k_\downarrow}, \tag{19}$$

with κ being the constant of decay of the wave function into the barrier which is determined by the potential barrier height U, $\kappa = \sqrt{(2m/\hbar^2)(U - E_F)}$. This result explains very well the observed dependence of the conductance on the angle between magnetizations in two electrodes as depicted in Fig. 9 for the CoFe/Al_2O_3/Co MTJ.[106] The simple

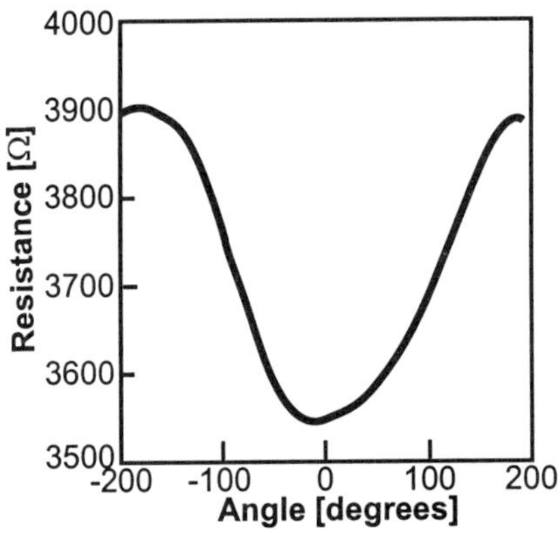

Fig. 9. The angular dependence of the magnetoresistance of a CoFe/Al_2O_3/Co junction measured in a magnetic field lower than the coercive field of one electrode but higher than the coercive field of the other electrode. Therefore, the magnetizations are never parallel or antiparallel to each other (after Ref. [106]).

model provides valuable predictions for a number of effects, including the dependence of the TMR on the electron effective mass of the insulator and an increasing TMR with increasing bias voltage. Among other things, it has been demonstrated that in order to decrease the decay rate of the TMR with increasing bias voltage, it is necessary to use insulators with small values of the effective electron mass.[86,102] Specifically, this model allows us to describe the dependence of the TMR sign reversal effect on both the bias voltage and the type of insulator.[86] The results for various parameters of the junctions are depicted in Fig. 10. It seems that TMR oscillations have

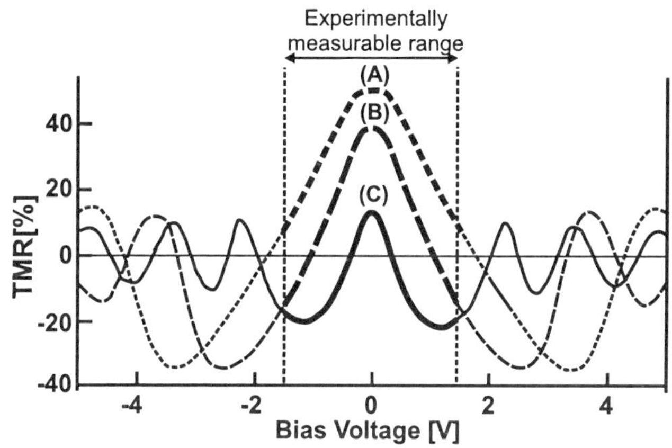

Fig. 10. The curves of TMR versus bias voltage for three different kinds of junction (after Ref. [86]).

rather universal character. However, in an experimentally accessible window of bias voltages, in certain junctions TMR can be apparently positive, or change the sign only once.

3.3. *Green's function tight-binding models*

In spite of the fact that simple two bands models of TMR in FM/I/FM MTJs give insight into physical mechanisms that govern TMR effect, the details of the electronic band structure of the involved materials also matter. One of the possible approaches is to employ tight-binding method for the description of the multiband electronic structure, which usually gives realistic band structure albeit the computational burden is much smaller than in the case of *ab initio* calculations. In the Ref. [100] such calculations for Fe/MgO/Fe junction have been reported. The conductances of the junction in its ferromagnetic and antiferromagnetic configurations have been determined from the real-space Kubo-Landauer formula[107] in terms of one-electron Green's functions at the Fermi surface ($E = E_F$).[108,109] The mixed representation has been used to represent the scattering states. The

Bloch-like representation has been used in the direction parallel to the layers and atomiclike in the perpendicular direction. The parameters of the tight-binding Hamiltonian have been fitted to an *ab initio* band structure of iron and MgO.[100] The couplings up to third nearest neighbors have been included in the TB-Hamiltonian. For the purpose of simplification, it has been assumed that the whole Fe/MgO/Fe (001) structure with bcc iron is pseudomorphic, i.e., the small lattice mismatch between Fe and MgO has been neglected.[100]

The calculations described above,[100] give optimistic TMR ratio that is of the order of 1000% for an MgO barrier of 20 atomic planes. This theoretical TMR ratio is an order of magnitude larger than observed experimentally. The calculation show[100] that spin polarization of the tunneling current is positive for all MgO thicknesses. It has been also found that spin-dependent tunneling in an Fe/MgO/Fe junction is not entirely determined by states at the Γ point ($k_{\|} = 0$) even for MgO thicknesses as large as 20 atomic planes. All these results are explained qualitatively in terms of the Fe majority- and minority-spin surface spectral densities and the complex MgO Fermi surface through the analysis of the parts of the lateral Brillouin Zone that give the largest contributions to the conductance in the each spin channel. From the performed TMR calculations,[100] it is completely clear that the large value of TMR obtained for the Fe/MgO/Fe junction is due to a very low spectral density of minority-spin electrons at the Γ point. Such a large TMR ratio could be only observed in structures with perfect Fe/MgO interfaces so that the theoretical minority-spin spectral density is well reproduced. The behavior of TMR at small thicknesses of MgO is determined by large peaks of the spectral density (hot spots) located outside the Γ point. These are very sensitive not only to interfacial roughness but also to the symmetry of the junction. Therefore, the physical insight can be only reached by performing reliable first-principles calculation, where the microscopic details of the metal-insulator interface are taken into account. We discuss *ab initio* calculations of the FM/I/FM MTJs in the following subsection.

3.4. *First-principles calculations*

First-principles methods employed for the studies of the TMR effect in MTJs are mostly based on density functional theory within the local spin density approximation (LSDA) for the electronic structure and the Landauer-Büttiker formula for the conductance. First, such approach can take into account the microscopic details of the FM/insulator interface. Second, it describes details of the resulting multiband electronic structure, the interfacial localized states, the variation of the potential across the barrier, and the evanescent states in the insulator. In many cases, it is the only computational scheme that allows for quantitative predictions.

One of the first *ab initio* calculations of the TMR effect were performed by MacLaren *et al.* for Fe/ZnSe/Fe(001) tunnel junction.[110] They used layer Korringa-Kohn-Rostoker (KKR) approach to calculate the transmission (and reflection) coefficient of the Bloch-waves, which were further used to calculate spin-dependent tunneling conductance in MTJs. These studies brought very interesting findings. It turned out that the spin asymmetry in the conductance increases dramatically with increasing barrier thickness. They showed that the difference in the decay rates for the majority-and minority-spin channels follows from the symmetry of the Bloch states at the Fermi energy, which have different spin injection (extraction) efficiencies and different decay rates when tunneling across the barrier. The ease of injection and extraction depends upon the character of the band in the electrode. For example, in the majority channel, the Δ_1 band, because it is compatible with the *s* character, couples efficiently with a decaying *sp* state in the ZnSe, and, thus, this band dominates the conductance. The much smaller tunneling conductance for the minority spins is a direct result of there being no Δ_1 band present at the Fermi energy.[110]

The calculations for Fe/ZnSe/Fe revealed also that that the spin asymmetry in the tunneling conductance should depend on the substrate crystal face. In the case of Fe, e.g., an examination of the band structure shows that for [100], [111], and [110] directions all have majority bands with s character present, and for all but the [100] direction, a band with this symmetry also crosses the Fermi energy for the minority channel. Thus, the [100] direction should exhibit the largest asymmetry in the tunnel conductance. Indeed, the dependence of the TMR on the crystal face of the epitaxial Fe electrodes in Fe/Al_2O_3/CoFe junctions was recently observed by Yuasa *et al.*.[56] However, in experiments the larger values of TMR for tunneling from Fe(110), rather than from Fe(100) electrodes, were found. Still, further analysis of bands contributing to the tunneling is needed to obtain consistency between the theory and experiment.

Later on Butler *et al.*[80] calculated TMR in Fe/MgO/Fe junctions using the layered KKR approach described in the previous paragraph. Their conclusions essentially support findings reported in Ref. [110]. In particular, they found that due to the absence of the minority Δ_1 band at the Fermi energy of Fe(100), the majority-spin conductance dominates tunneling which leads to a very high TMR for thick enough barriers.

The values of the TMR ratio in Fe/MgO/Fe MTJs predicted in theoretical calculations are much larger than the values predicted experimentally. This might be due to the formation of a partially oxidized FeO layer at the interface which was found in the experiments.[111] Motivated by this experimental work, first-principles calculations of the electronic structure and TMR in the Fe/FeO/MgO/Fe MTJs were performed and compared to that for Fe/MgO/Fe tunnel junctions.[112] Indeed, it was found that an atomic layer of iron-oxide at the interface between Fe substrate and the MgO layer greatly reduces the tunneling magnetoconductance. It happens mostly due to the bonding of Fe with O. This reduces the conductance when the magnetizations in the two electrodes are aligned but has little effect when the magnetizations are antiparallel. The TMR ratio decreases monotonically and exponentially with the increasing O concentration in the FeO layer. Further first-principle calculations of TMR in Fe/FeO/MgO/Fe tunneling junctions as functions of bias voltage were performed.[113] It turned out that the change in the electronic structure is minimal as a function of bias. The effective capacitance is consistent with the dielectric constant of MgO. The tunneling conductance is highly nonlinear. At low biases the TMR ratio is greatly reduced due to the large contribution to the tunneling current from interface resonance states. This contribution diminishes as the bias voltage increases, leading to an increase of the TMR ratio as a function of bias.

The influence of the interface structure on the bias dependence of TMR was also studied in Ref. [114]. Employing the formalism of Ref. [113], the authors performed *ab initio* calculations in the framework of the density functional theory for TMR in planar Fe/MgO/Fe junctions. The bias dependence of the tunneling conductance and the TMR was calculated in the limit of coherent tunneling. Positive and negative TMR ratios were obtained as a function of interface structure and even a sign reversal of TMR as a function of bias was found in agreement with experiments (see Sec. 2.3). The effect of interface structure on electronic structure and the conductance was studied by considering mixed Fe/O interfaces in the Fe/MgO/Fe tunnel junction. Guided by the experimental results,[78,111] three types of the junctions were discussed.[114] In the first type of the tunnel junction, two ideal Fe/MgO interfaces, i.e., without mixing, were chosen. In the second junction both interfaces consisted of an FeO layer and the system remains symmetric. The third crystal structure contained both the ideal and the FeO interface and was, of course, asymmetric. This way, structural differences between the left and right interface due to specific growth conditions[111] were modelled. The resulted bias dependence of the TMR are depicted in Fig. 11. In the case of the ideal interfaces TMR is positive for the whole con-

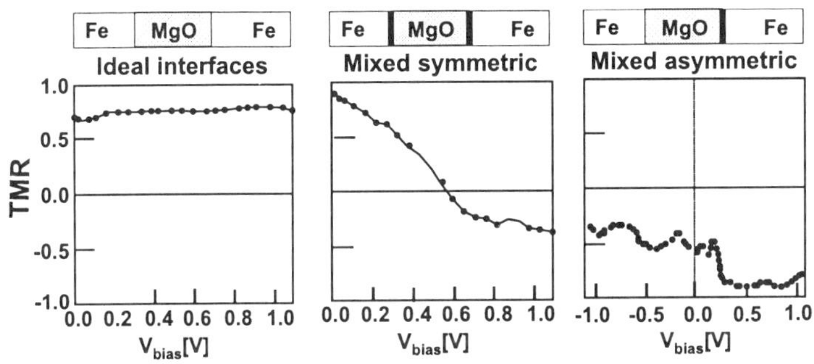

Fig. 11. Dependence of TMR on the external bias voltage for three different kinds of interfaces in Fe/MgO/Fe MTJ (after Ref. [114]).

sidered range of bias voltage, whereas TMR is negative for the asymmetric junction (with one ideal and the second mixed interface) in the whole range of external voltage. In the case of symmetric structure with two mixed FeO interfaces TMR changes sign for bias larger than roughly 0.5 V. These results clearly demonstrate that the bias characteristic is strongly influenced by the interface geometry. The modification of the interfaces allows the generation of positive and negative TMR ratios. These studies shed light on the physics behind the experimental findings and emphasize the importance of *ab initio* calculations for quantitative description of TMR in FM/I/FM type of MTJs. We close this subsection just mentioning another calculations for Fe/MgO/Fe[115] and Co/$SrTiO_3$/Co[88] MTJs.

3.5. *Models for disordered structures*

Actual tunnel junctions contain large amounts of disorder in the electrodes, in the barrier, and at the electrode/barrier interfaces. This disorder may represent interdiffusion at the interfaces, interface roughness, impurities, and defects such as grain boundaries, stacking faults, and vacancies. These factors critically influence TMR. It has been demonstrated in many papers and here we would like to mention only few of them. Interdiffusion dramatically changes the electronic and atomic structure, which affects TMR in a critical way (see for example Refs. [116–118]). Interface roughness leads to fluctuations in the barrier thickness that strongly alter the tunneling conductance.[119] Impurities and defects in the barrier introduce complex mechanisms that assist tunneling.[120–125] This is especially important in the case of amorphous barriers, although even in epitaxially grown tunnel junctions the effects of disorder might be decisive. Disorder in the electrodes mixes bulk and interface states and thereby influences TMR.[126–128] The nice review of the theoretical works can be found in Ref. [52]. These works include also first-principles calculations of the role of disorder.[129,130]

Up to now, we discussed theoretical approaches that employed coherent transport approach for MTJs. However, scattering processes may also cause that diffusive transport plays also a role for TMR ratio. Few years ago, scattering from magnons at the electrode-insulator interface was proposed as the mechanism for randomizing the tunneling process and opening the spin-flip channels that reduce the magnetoresistance.[131] These scattering effects has been included in the analysis of the spin-dependent transmission coefficients for MTJs in Ref. [132].

4. Tunneling magnetoresistance in junctions based on ferromagnetic semiconductors

The metallic MTJs discussed in Secs. 2 and 3 are important building blocks of the already existing spintronic devices but are plagued by the problem of achieving good quality of both FM/I interfaces. This is caused by the technology of the growth process. On the other hand, all-semiconductor FS/S/FS MTJs, where the electrodes consist of ferromagnetic semiconductor, offer potential for precise control of interfaces and barrier properties, particularly in the case of III-V compounds, for which epitaxial growth of complex heterostructures containing ferromagnetic (Ga,Mn)As or (In,Mn)As layers is particularly advanced.[17,133] Extensive studies of MTJs with (Ga,Mn)As ferromagnetic contacts carried out by various groups resulted in an increase of the observed TMR ratio from about 70% reported by Tanaka and Higo[38] to values higher than 250%.[39–41,134] These experiments demonstrated also that the TMR decreases rapidly with the increase of the applied bias – a phenomenon observed also in the metallic TMR structures, and still not fully understood (see Secs. 2) and 3). Finally, the experiments suggest that TMR is sensitive to the direction of the applied magnetic field in respect to the direction of current and crystallographic axes. This so-called tunnel anisotropic magnetoresistance (TAMR) effect was observed in structures containing a single ferromagnetic electrode[135,136] as well as in typical TMR MTJ with two ferromagnetic contacts.[137,138]

These challenging experimental findings call for a theory that would describe the tunneling in semiconductor MTJs and would indicate the ways for optimized design of the devices. Since the ferromagnetic coupling in (Ga,Mn)As is mediated by the holes,[36] a meaningful theory should in principle to take into account the entire complexity of the valence band, including the spin-orbit interaction. However, some qualitative physics can be gained also from simple two band model as described in Sec. 3.1, which neglect the spin-orbit coupling. Such models have been employed to the calculation of tunneling currents in the FS/S/FS MTJs[139,140] and their results will be summarized in the following subsection. Also TMR in the GaMnAs/GaAlAs/GaMnAs tunneling structure has been studied within the 6×6 *kp* theory.[141]

4.1. *Free carrier models of FS/S/FS tunnel junctions*

The TMR in the (Ga,Mn)As/AlAs/(Ga,Mn)As DMS tunnel junctions has been studied within the free carrier model in Refs. [139,140]. In Ref. [139] the dependence of the TMR ratio on the barrier width has been studied. The valence bands of (Ga,Mn)As and AlAs were treated as parabolic bands with effective masses of heavy hole bands equal to $m^*_{\mathrm{hh}} = 0.45\ m_{\mathrm{e}}$ and $m^*_{\mathrm{hh}} = 0.75\ m_{\mathrm{e}}$, with m_{e} being the electron mass, respectively. The effective masses for light hole bands were taken as $m^*_{\mathrm{lh}} = 0.08\ m_{\mathrm{e}}$ and $m^*_{\mathrm{lh}} = 0.143\ m_{\mathrm{e}}$, for (Ga,Mn)As and AlAs, respectively. The calculated TMR at T = 8 K and external bias V = 1 mV as a function of the barrier thickness is depicted in Fig. 12. As can be seen, the TMR first increases with barrier thickness then decreases nearly exponentially. In Ref. [140], the author studied the relative conductance ratio for magnetization orientations in the electrodes being at angle θ, i.e., $R(\theta) = (G(0) - G(\theta))/G(0)$. They found very simple dependence of $R(\theta)$ on the angle θ, with $R(\theta)$ being proportional to $\sin^2(\theta/2)$.[140] The simple model gives very weak dependence of $R(\theta)$ on the temperature up to temperature threshold value approximately equal to 0.9 T_c. For temperatures higher than 0.9 T_c, the $R(\theta)$ goes very quickly to zero. This result just indicates the simple possibility of magnetoresistance tuning. Both papers have found that the Julliere's formula is generally not a good approximation for this system.

4.2. *Full band model for all-semiconductor MTJs*

The theoretical approach that describes vertical spin dependent transport in modulated structures of magnetic semiconductors taking into account details of the semiconductor band structure, including spin orbit-interaction, has been proposed recently.[143,144] The scheme combines the two-terminal

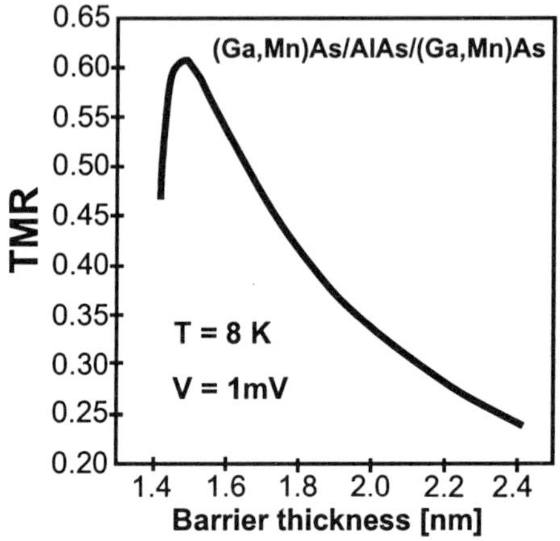

Fig. 12. Dependence of TMR in the (Ga,Mn)As/AlAs/(Ga,Mn)As MTJ s on the AlAs barrier thickness at 8 K and bias of 50meV (after Ref. [139]).

Landauer-Büttiker formalism with the empirical multi-orbital tight-binding description of the semiconductor band structure. In this way, the quantum character of spin transport over the length scale relevant for the devices in question is taken into account. Furthermore, the tight-binding approach, in contrast to free carrier models[139,140] or kp models[141,142] employed so-far, allows for a proper description of effects crucial for spin transport in heterostructures such as atomic structure of interfaces, spin-orbit coupling caused effects of Rashba and Dresselhaus terms as well as tunneling involving $\mathbf{k}$ states away from the center of the Brillouin zone. This model has recently been applied to describe TMR devices,[143,145] selected features of Zener-Esaki diodes[145,146] as well as it was adopted to examine an intrinsic domain-wall resistance in (Ga,Mn)As.[147,148]

In contrary to the FM/I/FM MTJ structures considered in Sec. 3, in FS/S/FS MTJs the spin-orbit coupling plays an important role. In the presence of spin-orbit coupling, the spin is not a good quantum number. The only preserved quantities in tunneling are the energy E and, due to spatial in plane symmetry of our structures, the in-plane wave vector $\mathbf{k}_{\|}$. For a given energy E and in-plane wave-vector $k_{\|}$, the Bloch states in the left L and right R leads (i and j, respectively) are characterized by the wave vector component $k_{\perp}$ perpendicular to the layers and are denoted by $|L, k_{L,\perp,i}\rangle$ and $|R, k_{R,\perp,j}\rangle$, respectively. The indices i and j indicate all possible pairs for 40 bands described by the tight-binding Hamiltonian. The transmission probability $T_{L,k_{L,\perp,i}\rightarrow R,k_{R,\perp,j}}$ is a function of the transmission amplitude $t_{L,k_{L,\perp,i}\to R,k_{R,\perp,j}}(E,\mathbf{k}_{\|})$ and group velocities in the left and right lead, $v_{L,\perp,i}$ and $v_{R,\perp,j}$:

$$\begin{aligned} T_{L,k_{L,\perp,i}\to R,k_{R,\perp,j}}(E,\mathbf{k}_{\|}) &= \\ &= \left|t_{L,k_{L,\perp,i}\to R,k_{R,\perp,j}}(E,\mathbf{k}_{\|})\right|^2 \frac{v_{R,\perp,j}}{v_{L,\perp,i}}. \end{aligned} \tag{20}$$

The current flowing in the right direction can now be written as:[149]

$$\begin{aligned} j_{L\to R} = \frac{-e}{(2\pi)^3\hbar}\int_{BZ} d^2k_{\|}dEf_L(E) \\ \sum_{\substack{k_{L,\perp,i},k_{R,\perp,j} \\ v_{L,\perp,i},v_{R,\perp,j}>0}} T_{L,k_{L,\perp,i}\to R,k_{R,\perp,j}}(E,\mathbf{k}_{\|}), \end{aligned} \tag{21}$$

where f_L or respectively f_R are the electron Fermi distributions in the left and right interface and i,j number the corresponding Bloch states. Plugging in the expression given in Eq. (20) and using the time reversal symmetry:

$$\begin{aligned} T_{L,k_{L,\perp,i}\to R,k_{R,\perp,j}}(E,\mathbf{k}_{\|}) &= \\ &= T_{L,-k_{R,\perp,j}\to R,-k_{L,\perp,i}}(E,\mathbf{k}_{\|}) \end{aligned} \tag{22}$$

we get

$$\begin{aligned} j = &\frac{-e}{(2\pi)^3\hbar}\int_{BZ} d^2k_{\|}dE\left[f_L(E)-f_R(E)\right] \\ &\sum_{\substack{k_{L,\perp,i},k_{R,\perp,j} \\ v_{L,\perp,i},v_{R,\perp,j}>0}} \left|t_{L,k_{L,\perp,i}\to R,k_{R,\perp,j}}(E,\mathbf{k}_{\|})\right|^2 \frac{v_{R,\perp,j}}{v_{L,\perp,i}}. \end{aligned} \tag{23}$$

To calculate the current one has to determine the transmission probability, thus the transmission amplitude $t_{L,k_{L,\perp,i}\to k_{R,R,\perp,j}}(E,\mathbf{k}_{\|})$ and the group velocities $v_{L,\perp,j}$ of the ingoing and $v_{R,\perp,j}$ of outgoing states. These can be obtained by solving the Schrödinger equation for the structure with the appropriate scattering boundary conditions. The number of considered scattering states doubles in comparison to the case with spin degeneracy. This considerably increases computational burden, however, the transmission amplitudes between spin mixed states and the group velocities can be determined using exactly the same procedures as for the spin polarized case.[143]

To calculate TMR ratio in (Ga,Mn)As/AlGaAs/(Ga,Mn)As TMJs one has to determine Hamiltonian of the system. In Ref. [143] the rather sofisticated tight-binding Hamiltonian for the magnetic (Ga,Mn)As and nonmagnetic GaAs and AlAs semiconductors have been employed.[150] This Hamiltonian involves nearest neighbors couplings between $sp^3d^5s^*$ atomic orbitals and reproduces excellently the effective masses and the band structure of GaAs and AlAs in the whole Brillouin zone. The presence of Mn ions in (Ga,Mn)As is taken into account by including the *sp-d* exchange interactions within the virtual crystal and mean-field approximations. In the spirit of the tight-binding method, the effects of an external interaction are included in the on-site diagonal matrix elements of the tight-binding Hamiltonian. The shifts of on-site energies caused by the *sp-d* exchange interaction are parameterized in such a way that they reproduce experimentally obtained spin splitting: $N_0\alpha = 0.2$ eV of the conduction band and $N_0\beta = -1.2$ eV of the valence band.[151] The other parameters of the model for the (Ga,Mn)As material and for the NN interactions between GaAs and (Ga,Mn)As are taken to be the same as for GaAs. This is well motivated because the valence-band structure of (Ga,Mn)As with small fraction of Mn has been shown to be quite similar to that of GaAs.[151] Consequently, the valence band offset between (Ga,Mn)As and GaAs originates only from the spin splitting of the bands in (Ga,Mn)As.

4.3. *Bias dependence of TMR in FS/S/FS TMJs*

The model described in Sec. 4.2 has been applied to study the TMR ratio in (Ga,Mn)As/AlGaAs/(Ga,Mn)As TMJs.[143,145] Such structures can be obtained for various concentration of Mn in the electrodes and various concentration of Al in barrier alloy. In Fig. 13 we depict TMR ratio for the structure with GaAs barrier and different concentrations of Mn at constant typical hole concentration in (Ga,Mn) As, which is equal to $p = 3.5 \times 10^{20}$ cm^{-3}.[152] Theoretically calculated TMR ratio increases with the content of the magnetic ions and reproduces excellently experimental data giving TMR equal to 250% [40] and 60% [38] for Mn concentrations of 8% and 4%, respectively. On the other hand, the theory predicts the TMR to decrease with the increasing hole concentration when the concentration of Mn ions in (Ga,Mn)As is fixed (see Fig. 13). Therefore, the calculations seem to suggest that for obtaining high TMR, large exchange splittings, i.e., high content of magnetic ions is needed. Unfortunately, the presented in Fig. 13 dependence suggests that the attempts to increase the hole concentration in (Ga,Mn)As, in order to obtain higher Curie temperature, may result in much smaller TMR in this structure.

As expected, the theory predicts also decay of the TMR ratio with external bias. In Fig. 14 the obtained TMR values for a set of different hole

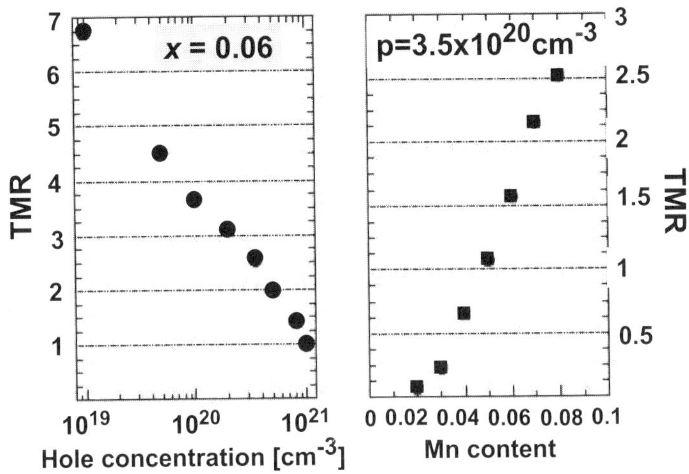

Fig. 13. TMR in $Ga_{1-x}Mn_xAs/GaAs/Ga_{1-x}Mn_xAs$ (left) as a function of hole concentration p for fixed Mn content $x = 0.08$; (right) as a function of Mn content for fixed hole concentration $p = 3.5 \times 10^{20}$ cm^{-3}. The bias applied to the structure is $V = 0.05V$.

concentrations in the (Ga,Mn)As layers, are plotted as a function of the applied bias. As can be easily seen from this figure, TMR depends strongly on the hole concentration and decays with the external bias for each hole concentration. As TMR is determined primarily by the spin polarization of the carriers at the Fermi level,[145] this agrees with the statement of Ref. [153] that the higher the hole concentration the smaller is the polarization at the Fermi level. This suggests that the dependence of TMR on the applied bias results predominantly from the band structure effects in this case. It is worth to stress that in the range of calculated biases TMR remains positive and shows no oscillations observed in the metallic TMJs (see Secs. 2 and 3). In Fig. 15 TMR as a function of external bias is shown for two different barriers, AlAs and GaAs, of the same width. The valence band offsets (VBO) are equal to 0.81 eV and 0.26eV for (Ga,Mn)As/AlAs and (Ga,Mn)As/GaAs interfaces respectively. It turns out that TMR is in the whole range of external voltages higher for material with larger VBO. Also the TMR magnitude decreases with the bias much slower for the junction with AlAs barrier than in the case of GaAs. In TMJs with both barriers AlAs and GaAs the TMR decreases with the width of the barriers in the whole bias range and for all hole concentrations and Mn concentrations.

The calculated dependence of TMR on the in-plane direction of the magnetization vector shows anisotropy of tunneling magnetoresistance be-

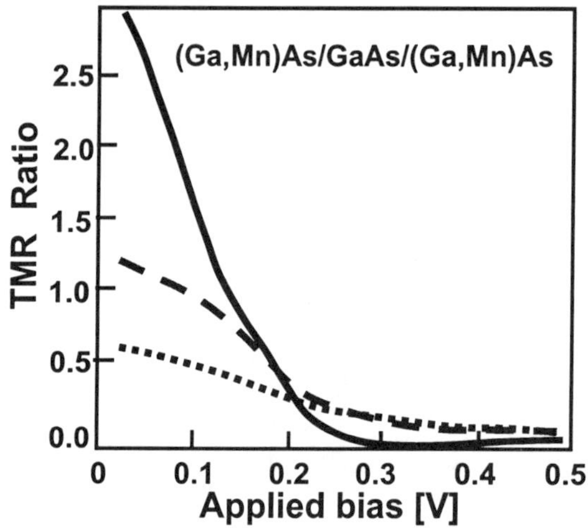

Fig. 14. TMR for the (Ga,Mn)As/GaAs/(Ga,Mn)As tunneling structure as a function of external bias for various hole concentrations in ferromagnetic leads: p=10^{20} cm^{-3} (solid line), p=5 × 10^{20} cm^{-3} (dashed line), and p= 10^{21} cm^{-3} (dotted line). Mn concentration of 6.5% has been used for these calculations.

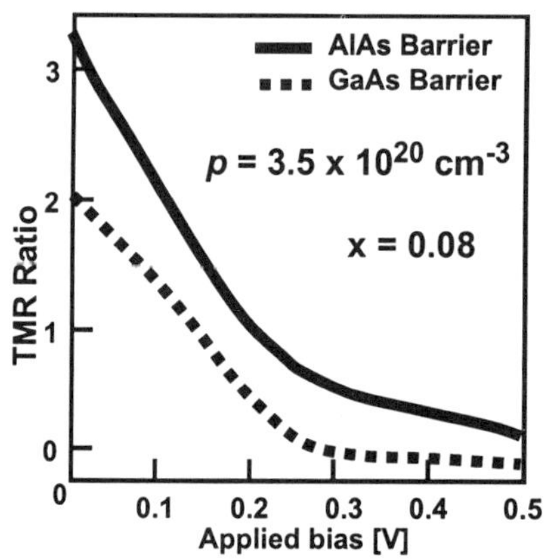

Fig. 15. TMR as a function of external bias for (Ga,Mn)As/GaAs/(Ga,Mn)As (dotted line) and (Ga,Mn)As/AlAs/(Ga,Mn)As (continous line) tunneling structures. The width of the barrier is in both cases equals to 4 atomic double layers. Hole concentrations in ferromagnetic leads equals p=3.5^{20} cm^{-3} Mn concentration of 8%

low 10% for typical hole concentrations. This anisotropy corresponds to the D_{2d} symmetry and comes mostly from the antiparallel orientations of magnetization vectors in the (Ga,Mn)As electrodes.

5. Conclusions

The TMR effect is essential for the field of spintronics. In this review, we have shown the rich physics of this effect and presented theoretical approaches that allow, on one hand, for insight into the physical mechanisms and, on the other hand, for qualitative predictions that are necessary for reasonable design of novel structures for functional devices based on the TMR effect. The enormous progress has been achieved in FM/I/FM TMJs that are already utilized in commercial devices. However, the search for the best ferromagnetic electrode material is going on and the systems utilizing manganites are very promising. Still, in the metallic TMJs the difficult problem of the interfaces is essential and definitely requires more experimental and theoretical effort. The all-semiconductor tunnel junctions are very promising systems that exhibit similar TMR ratios as FM/I/FM TMJs and similar behavior with the external bias. However, the FS/S/FS TMJs do not show the change of the TMR sign or oscillation with bias as FM/I/FM TMJs sometimes do. The problem is that the Curie temperature of the homogeneous ferromagnetic semiconductors is lower than 300 K. On the other hand, room temperature wide-gap semiconductors can be also used in TMJs. However, the ferromagnetism in these semiconductors originates from spinodal decomposition. It remains unclear how the introduced inhomogeneity in the ferromagnetic electrode will influence the atomic structure of the interface to the non-magnetic semiconductor and further the tunneling current and TMR.

Acknowledgments

This work was partly supported by the EC project NANOSPIN (FP6-2002-IST-015728) and by the research grant of the Polish Ministry of Science and Education (N20202632).

References

1. S. A. Wolf, D. D. Awschalom, R. A. Buhrman, J. M. Daughton, S. von Molnar, M. L. Roukes, A. Y. Chtchelkanova and D. M. Treger, *Science* **294**, 1488 (2001).
2. Dante Gatteschi, Roberta Sessoli and Jacques Villain, *Molecular Nanomagnets* (Oxford University Press, 2006).
3. M. Oestreich, M. Brender, J. Hübner, D. H. W. W. Rühle, T. H. P. J. Klar, W. Heimbrodt, M. Lampalzer, K. Voltz and W. Stolz, Semicond. Sci. Technol. **17**, 285 (2002).
4. I. Zutic, J. Fabian and S. Das Sarma, Rev. Mod. Phys. **76**, 323 (2004) and references therein.

5. C. Felser, G. H. Fecher and B. Balke, Angew. Chem. Int. Ed. **46**, 668 (2007).
6. R. Hanson, L. P. Kouwenhoven, J. R. Petta, S. Tarucha, L. M. K. Vandersypen, Rev. Mod. Phys. **79**, 1217 (2007).
7. *Magnetic Multilayers and Giant Magnetoresistance*, Ed. U. Hartman (Springer-Verlag, Berlin, 2000).
8. E. Hirota, H. Sakakima and K. Inomata, *Giant Magneto-Resistance Devices* (Springer-Verlag, Berlin, 2002).
9. *Spin Dependent Transport in Magnetic Nanostructures*, Eds. Sadamichi Maekawa and Teruya Shinjo (CRS Press, 2002).
10. *Magnetic Interactions and Spin Transport*, Eds. A. Chtchelkanova, A. Stuart, Y. Wolf and I. Yves, (Springer-Verlag, Berlin, 2003).
11. *Spin Electronics*, Eds. D. D. Avschalom, R. A. Buhrman, J. D. Daughton, S. von Molnar and M. L. Roukes, (Kluwer Academic Publishers, Dortrecht, 2004).
12. *Concepts in Spin Electronics*, Ed. Sadamichi Maekawa, (Oxford University Press, 2006).
13. *Spintronic Materials and Technology*, Eds. Y.B.Yu and S.M. Thompson (Taylor & Francis, Boca Raton, 2007).
14. *Local-Moment Ferromagnets - Unique Properties for Modern Applications*, Eds. M. Donath and W. Nolting (Springer-Verlag, Berlin, 2005).
15. Mathias Getzlaff, *Fundamentals of Magnetism* (Springer-Verlag, Berlin, 2008).
16. S. Datta and B. Das, Appl. Phys. Lett. **56**, 665 (1990).
17. H. Ohno, Science **281**, 951 (1998).
18. T. Dietl, H. Ohno, F. Matsukura, J. Cibert and D. Ferrand, Science **287**, 1019 (2000).
19. T. Dietl, Semicond. Sci. Technol. **17**, 377 (2002).
20. S.J. Pearton, C.R. Abernathy, M.E. Overberg, G.T. Thaler, D.P. Norton, N. Theodoropoulou, A.F. Hebard, Y.D. Park, F. Ren, J. Kim and L.A. Boatner, J. Appl. Phys. **93**, 1 (2003).
21. T. Jungwirth, J. Sinova, J. Masek, J. Kucera and A.H. MacDonald, Rev. Mod. Phys. **78**, 809 (2006).
22. V. Dediu, M. Murgia, F.C. Matacotta, C. Taliani and S. Barbanera, Solid State Commun. **122**, 181 (2002).
23. D.A. Pejakovic, C. Kitamura, J.S. Miller and A.J. Epstein, Phys. Rev. Lett. **88**, 057202 (2002).
24. A.J. Epstein MRS Bull. **28**, 492 (2003).
25. A.M. Goldman, K. Nikolaev, P. Kraus, V. Vasko, A. Bhattacharya and W. Cooley, J. Supercond. **14**, 283 (2001).
26. K. Tsukagoshi, B.W. Alphenaar and H. Ago, Nature (London) **401**, 572 (1999).
27. B. Zhao, I. Mönch, H. Vinzelberg, T. Mühl and C.M. Schneider, Appl. Phys. Lett. **80**, 3144 (2002).
28. see for example, http://www.intel.com/.
29. G. Prinz, Science **282**, 1660 (1998).

30. Jian-Gang (Jimmy) Zhu and Chando Park, Materials Today **9**, 36 (2006).
31. G. Grynkewich, J. Akerman, P. Brown, B. Butcher, R.W. Dave, M. DeHerrera, M. Durlam, B.N. Engel, J. Janesky, S. Pietambaram, N.D. Rizzo, J.M. Slaughter, K. Smith, J.J. Sun and S. Tehrani, MRS Bulletin **29**, No. 11 (2004).
32. A. Ney, C. Pampuch, R. Koch and K.H. Ploog, Nature **425**, 485 (2003).
33. J.S. Moodera, L.R. Kinder, T.M. Wong and R. Meservey, Phys. Rev. Lett. **74**, 3273 (1995).
34. H. Ohno, H. Munekata, T. Penney, S. von Molnar and L.L. Chang, Phys. Rev. Lett. **68**, 2664 (1992).
35. H. Ohno, A. Shen, F. Matsukura, A. Oiwa, A. Endo, S. Katsumoto and Y. Iye, Appl. Phys. Lett. **69**, 363 (1996).
36. T. Dietl and H. Ohno, MRS Bulletin **28**, 714 (2003).
37. T. Dietl, Physica **E35**, 293 (2006).
38. M. Tanaka and Y. Higo, Phys. Rev. Lett. **87**, 026602 (2001).
39. R. Mattana, J.-M. George, H. Jaffrés, F. Nguyen Van+Dau, A. Fert, B. Lépine, A. Guivarh and G. Jézéquel, Phys. Rev. Lett. **90**, 166601 (2003).
40. D. Chiba, F. Matsukura and H. Ohno, Physica **E21**, 966 (2004).
41. M. Elsen, O. Boulle, J.-M. George, H. Jaffrés, R. Mattana, V. Cros, A. Fert, A. Lemaître, R. Giraud and G. Faini, Phys. Rev. **B73**, 035303 (2006).
42. T. Dietl, H. Ohno and F. Matsukura, Phys. Rev. **B63**, 195205 (2001).
43. Y. Ohno, D.K. Young, B. Beschoten, F. Matsukura, H. Ohno and D.D. Awschalom, Nature **402**, 790 (1999).
44. M. Kohda, Y. Ohno, K. Takamura, F. Matsukura and H. Ohno, Jap. J. Appl. Phys. **40**, (Part 2), L1274 (2001).
45. M. Kohda, Y. Ohno, F. Matsukura and H. Ohno, Physica **E32**, 438 (2006).
46. P. Van Dorpe, Z. Liu, W. Van Roy, V.F. Motsnyi, M. Sawicki, G. Borghs and J. De Boeck, Appl. Phys. Lett. **84**, 3495 (2004).
47. M. Julliere, Phys. Lett. **A54**, 225 (1975).
48. P.M. Tedrow and R. Meservey Phys. Rev. Lett. **26**, 192 (1970).
49. S. Maekawa and U. Gäfvert, IEEE Trans. Magn. **18**, 707 (1982).
50. T.S. Plaskett, P.P. Freitas, N.P. Barradas, M.F. da Silva and J.C. Soares J. Appl. Phys. **76**, 6104 (1994).
51. J. Nowak and J. Rauluszkiewicz, J. Magn. Magn. Mater. **109**, 79 (1992).
52. E.Y. Tsymbal, O.N. Mryasov and P.R. LeClair, J. Phys.: Condens. Matter **15**, R109 (2003).
53. Xin Jiang and Stuart Parkin, in *Concepts in Spin Electronics* ed. Sadamichi Maekawa, p 239 (Oxford University Press, 2006).
54. T. Miyazaki and N.J. Tezuka, J. Magn. Magn. Mater. **139**, L231 (1995).
55. D. Wang, C. Nordman, J.M. Daughton, Z. Qian and J. Fink, IEEE Trans. Magn. , **40**, 2269 (2004).
56. S. Yuasa, T. Sato, E. Tamura, Y. Suzuki, H. Yamamori, K. Ando and T. Katayama, Europhys. Lett. **52**, 344 (2000).
57. H. Boeve, E. Girgis, J. Schelten, J. D. Boeck and G. Borghs, Appl. Phys. Lett. **76**, 1048 (2000).

58. J.S. Moodera, J. Nowak and R.J.M. van de Veerdonk, Phys. Rev. Lett. **80**, 2941 (1998).
59. W. Wulfhekel, H.F. Ding and J. Kirschner J. Magn. Magn. Mater. **242 – 245**, 47 (2002).
60. J. Zhang and R. White, J. Appl. Phys. **83**, 6512 (1998).
61. R. Jansen and J.S. Moodera, J. Appl. Phys. **83**, 8883 (1998).
62. R. Jansen and J.S. Moodera, Phys. Rev. **61**, 9047 (2000).
63. R. Jansen and J.S. Moodera, Appl. Phys. Lett. **75**, 400 (1999).
64. P. LeClair, J.T. Kohlhepp, C.H. van de Vin, H. Wieldraaijer, H.J.M. Swagten and W.J. M. de Jonge, Phys. Rev. Lett. **88**, 107201 (2002).
65. M. Sharma, S.X. Wang and J.H. Nickel, Phys. Rev. Lett. **82**, 616 (1999).
66. J.M. de Teresa, A. Barthelemy, A. Fert, J.P. Contour, R. Lyonnet, F. Montaigne, P. Seneor and A. Vaurès, Phys. Rev. Lett. **82**, 4288 (1999).
67. W.E. Pickett and J.S. Moodera, Phys. Today **5**, 39 (2001).
68. J.H. Park, E. Vescovo, H.J. Kim, C. Kwon, R. Ramesh and T. Venkatesan, Nature **392**, 794 (1998).
69. D. Ristoiu, J.P. Nozières, C.N. Borca, T. Komesu, H-K. Jeong and P.A. Dowben, Europhys. Lett. **49**, 624 (2000).
70. Y. Ji, G.J. Strijkers, F.Y. Yang, C.L. Chien, J.M. Byers, A. Anguelouch, G. Xiao and A. Gupta, Phys. Rev. Lett. **86**, 5585 (2001).
71. J.S. Parker, S.M. Watts, P.G. Ivanov and P. Xiong, Phys. Rev. Lett. **88**, 196601 (2002).
72. Y. Lu, X.W. Li, G.Q. Gong, G. Xiao, A. Gupta, P. Lecoeur, J.Z. Sun, Y.Y. Wang and V.P. Dravid, Phys. Rev. **B54**, R8357 (1996).
73. M. Viret, M. Drouet, J. Nassar, J.P. Contour, C. Fermon and A. Fert, Europhys. Lett. **39**, 545 (1997).
74. D.C. Worledge and T.H. Geballe, Appl. Phys. Lett. **76**, 900 (2000).
75. M. Ziese, Rep. Prog. Phys. **65**, 143 (2002).
76. M. Bowen, J-L. Maurice, A. Barthélémy, M. Bibes, D. Imhoff, V. Bellini, R. Bertacco, D. Wortmann, P. Seneor, E. Jacquet, A. Vaurès, J. Humbert, J-P. Contour, C. Colliex, S. Blügel and P.H. Dederichs, J. Phys.: Condens. Matter **19**, 315208 (2007).
77. J.-N. Chazalviel and Y. Yafet, Phys. Rev. **B15**, 1062 (1977).
78. S. Yuasa, T. Nagahama, A. Fukushima, Y. Suzuki and K. Ando, Nature Materials **3**, 868 (2004).
79. S.S.P. Parkin, C. Kaiser, A. Panchula, M. Rice, B. Hughes, M. Samant and See-Hun Yang, Nature Materials **3**, 862 (2004).
80. W.H. Butler, X.-G. Zhang, T.C. Schulthess and J.M. Maclaren, Phys. Rev. **B63**, 054416 (2001).
81. J. Mathon and A. Umerski, Phys. Rev. **B63**, 220403R (2001).
82. S. Yuasa, A. Fukushima, H. Kubota, Y. Suzuki and K. Ando, Appl. Phys. Lett. **89**, 042505 (2006).
83. J. Hayakawa, S. Ikeda, Y.M. Lee, F. Matsukura and H. Ohno, arXiv:cond-mat/0610526, unpublished.

84. M. Sugiyama, J. Hayakawa, K. Itou, H. Asano, M. Matsui, A. Sakuma and M. Ichimura, J. Magn. Soc. Japan **25**, 795 (2001).
85. J.Z. Sun, K.P. Roche and S.S.P. Parkin, Phys. Rev. **B61**, 11244 (2000).
86. Yuan Ren, Zheng-zhong Li, Ming-wen Xiao and An Hu, J. Phys.: Condens. Matter **17**, 412 (2005).
87. P.X. Xu, K. Xia, M. Zwierzycki, M. Talanana and P.J. Kelly, Phys. Rev. Lett. **96**, 176602 (2006).
88. J.P. Velev, K.D. Belashchenko, D.A. Stewart, M. van Schilfgaarde, S.S. Jaswal and E.Y. Tsymbal, Phys. Rev. Lett. **95**, 216601 (2005).
89. C. Tiusan, J. Faure-Vincent, C. Bellouard, M. Hehn, E. Jouguelet and A. Schuhl, Phys. Rev. Lett. **93**, 106602 (2004).
90. C. Tiusan, M. Sicot, J. Faure-Vincent, M. Hehn, C. Bellouard, F. Montaigne, S. Andrieu and A. Schuhl, J. Phys.: Condens. Matter **18**, 941 (2006).
91. J. Moser, M. Zenger, C. Gerl, D. Schuh, R. Meier, P. Chen, G. Bayreuther, W. Wegscheider, D. Weiss, C.-H. Lai, R.-T. Huang, M. Kosuth and H. Ebert, Appl. Phys. Lett. **89**, 162106 (2006).
92. J.S. Moodera, M.E. Taylor and R. Meservey, Phys. Rev. **B40**, R11980 (1989).
93. P. LeClair, J.T. Kohlhepp, H.J.M. Swagten and W.J.M. de Jonge, Phys. Rev. Lett. **86**, 1066 (2001).
94. S. Yuasa, T. Nagahama and Y. Suzuki, Science **297**, 234 (2002).
95. M. Bibes, K. Bouzehouane, A. Barthélémy, M. Besse, S. Fusil, M. Bowen, P. Seneor, J. Carrey, V. Cros, A. Vaurès, J.-P. Contour and A. Fert, Appl. Phys. Lett. **83**, 2629 (2003).
96. Y. Sakuraba, M. Hattori, M. Oogane, Y. Ando, H. Kato, A. Sakuma, T. Miyazaki and H. Kubota, Appl. Phys. Lett. **88**, 192508 (2006).
97. C.H. Shang, J. Nowak, R. Jansen and J.S. Moodera, Phys. Rev. **B58**, R2917 (1998).
98. X-G. Zhang and W.H. Butler, J. Phys.: Condens. Matter **15**, R1603 (2003).
99. A.H. Davis and J.M. MacLaren, J. Phys.: Condens. Matter **14**, 4365 (2002).
100. J. Mathon and A. Umerski, Phys. Rev. **B63**, 220403(R) (2001).
101. J.C. Slonczewski, Phys. Rev. **B39**, 6995 (1989).
102. N.N. Beletskii, G.P. Berman, A.R. Bishop, S.A. Borysenko and V.M. Yakovlenko, Phys. Rev. **B75**, 174418 (2007).
103. Zhiming Zheng, Yunong Qi, D. Y. Xing and Jinming Dong, Phys. Rev. **B59**, 14505 (1999).
104. C.B. Duke, *Tunneling in solids* (New York, Academic, 1969).
105. A. Van Esch, L. Van Bockstal, J. De Boeck, G. Verbanck, A.S. van Steenbergen, P.J. Wellmann, B. Grietens, R. Bogaerts, F. Herlach and G. Borghs, Phys. Rev. **B56**, 13103 (1997).
106. J.S. Moodera and L.R. Kinder, J. Appl. Phys. **79**, 4724 (1996).
107. J. Mathon, A. Umerski and M.A. Villeret, Phys. Rev. **B55**, 14378 (1997).
108. J. Mathon, Phys. Rev. **B56**, 11 810 (1997).
109. J. Mathon and A. Umerski, Phys. Rev. **B60**, 1117 (1999).
110. J.M. MacLaren, X-G. Zhang, W.H. Butler and X. Wang, Phys. Rev. **B59**, 5470 (1999).

111. H.L. Meyerheim, R. Popescu, N. Jedrecy, M. Vedpathak, M. Sauvage-Simkin, R. Pinchaux, B. Heinrich and J. Kirschner, Phys. Rev. **B65**, 144433 (2002).
112. X.-G. Zhang, W.H. Butler and A. Bandyopadhyay, Phys. Rev. **B68**, 092402 (2003).
113. C. Zhang, X.-G. Zhang, P.S. Krstic, Hai-pingCheng, W.H. Butler and J.M. MacLaren, Phys. Rev. **B69**, 134406 (2004).
114. Ch. Heiliger, P. Zahn, B.Yu. Yavorsky and I. Mertig, Phys. Rev. **B72**, 180406(R) (2005).
115. Derek Waldron, Lei Liu and Hong Guo, Nanotechnology **18**, 424026 (2007).
116. S. Cardosa, P.P. Freitas, C. de Jesus, P. Wei and J.C. Soares, Appl. Phys. Lett. **76**, 610 (2000).
117. J.J. Sun, K. Shimazawa, N. Kasahara, K. Sato, S. Saruki, T. Kagami, O. Redon, S. Araki, H. Morita and M. Matsuzaki, Appl. Phys. Lett. **76**, 2424 (2000).
118. J. Schmalhorst, H. Bruckl, M. Justus, A. Thomas, G. Reiss, M. Vieth, G. Gieres and J. Wecker, J. Appl. Phys. **89**, 586 (2001).
119. V. Da Costa, C. Tiusan, T. Dimopoulos and K. Ounadjela, Phys. Rev. Lett. **85**, 876 (2000).
120. A.M. Bratkovsky, Phys. Rev. **B56**, 2344 (1997).
121. A.M. Bratkovsky, Appl. Phys. Lett. **72**, 2334 (1998).
122. H. Itoh, J. Inoue, S. Maekawa and P. Bruno, J. Magn. Magn. Mater. **199**, 545 (1999).
123. E.Y. Tsymbal and D.G. Pettifor, Phys. Rev. **B58**, 432 (1998); E.Y. Tsymbal and D.G. Pettifor, J. Magn. Magn. Mater. **199**, 146 (1999); E.Y. Tsymbal and D.G. Pettifor, J. Appl. Phys. **85**, 5801 (1999).
124. R. Jansen and J.C. Lodder, Phys. Rev. **B61**, 5860 (2000).
125. J. Inoue, N. Nishimura and H. Itoh, Phys. Rev. **B65**, 104433 (2002).
126. P.M. Levy, K.S. Wang, P.H. Dederichs, C. Heide, S.F. Zhang and L. Szunyogh, Phil. Mag. **B82**, 763 (2002).
127. C. Uiberacker and P.M. Levy, Phys. Rev. **B65**, 169904 (2002).
128. D. Bagrets, A. Bagrets, A. Vedyayev and B. Dieny, Phys. Rev. **B65**, 064430 (2002).
129. Yoshio Miura, Kazutaka Nagao and Masafumi Shirai, Phys. Rev. **B69**, 144413 (2004).
130. P.X. Xu, V.M. Karpan, K. Xia, M. Zwierzycki, I. Marushchenko and P.J. Kelly, Phys. Rev. **B73**, 180402(R) (2006).
131. S. Zhang, P.M. Levy, A.C. Marley and S.S.P. Parkin, Phys. Rev. Lett. **79**, 3744 (1997).
132. C.A. Dartora and G.G. Cabrera, Phys. Rev. **B72**, 014424 (2005).
133. H. Munekata, H. Ohno, S. von Molnar, A. Segmüller, L.L. Chang, L. Esaki and H. Ohno, Phys. Rev. Lett. **63**, 1849 (1989).
134. W.G. Wang, C. Ni, T. Moriyama, J. Wan, E. Nowak and J.Q. Xiao, Appl. Phys. Lett. **88**, 202501 (2006).

135. C. Gould, C. Rüster, T. Jungwirth, E. Girgis, G.M. Schott, R. Giraud, K. Brunner, G. Schmidt and L.W. Molenkamp, Phys. Rev. Lett. **93**, 117203 (2004).
136. R. Giraud, M. Gryglas, L. Thevenard, A. Lemaître and G. Faini, Appl. Phys. Lett. **87**, 242505 (2005).
137. C. Ruster, C. Gould, T. Jungwirth, E. Girgis, G.M. Schott, R. Giraud, K. Brunner, G. Schmidt and L.W. Molenkamp, Phys. Rev. Lett. **94**, 027203 (2005).
138. A. Giddings, M. Khalid, T. Jungwirth, J. Wunderlich, S. Yasin, R. Campion, K. Edmonds, J. Sinova, K. Ito, K.Y. Wang, D. Williams, B. Gallagher and C.T. Foxon, Phys. Rev. Lett. **94**, 127202 (2005).
139. Y.C. Tao, J. G. Hu and H. Liu, J. Appl. Phys. **96**, 498 (2004).
140. J.E. Bunder, Appl. Phys. Lett. **91**, 092111 (2007).
141. L. Brey, C. Tejedor and J. Fernàndez-Rossier, Appl. Phys. Lett. **85**, 1996 (2004).
142. A.G. Petukhov, A.N. Chantis and D.O. Demchenko, Phys. Rev. Lett. **89**, 107205 (2002).
143. P. Sankowski, P. Kacman, J.A. Majewski and T. Dietl, Phys. Rev. **B75**, 045306 (2007).
144. P. Sankowski, P. Kacman, J.A. Majewski and T. Dietl, Low Temp. Phys. **33**, 187 (2007).
145. P. Sankowski, P. Kacman, J.A. Majewski and T. Dietl, Physica **E32**, 375 (2006).
146. P. Van Dorpe, W. Van Roy, J. De Boeck, G. Borghs, P. Sankowski, P. Kacman, J.A. Majewski and T. Dietl, Phys. Rev. **B72**, 205322 (2005).
147. R. Oszwałdowski, J.A. Majewski and T. Dietl, Phys. Rev. **B74**, 153310 (2006).
148. P. Sankowski, R. Oszwaldowski, P. Kacman, J.A. Majewski and T. Dietl, phys.stat.sol. (c) **4**, 2399 (2007).
149. Aldo di Carlo, Semicond. Sci. Technol. **18**, R1 (2003).
150. J.-M. Jancu, R. Scholz, F. Beltram and F. Bassani, Phys. Rev. **B57**, 6493 (1998).
151. J. Okabayashi, A. Kimura, O. Rader, T. Mizokawa, A. Fujimori, T. Hayashi and M. Tanaka, Physica **E10**, 192 (2001).
152. K.M. Yu, W. Walukiewicz, T. Wojtowicz, W.L. Lim, X. Liu, Y. Sasaki, M. Dobrowolska and J.K. Furdyna, Appl. Phys. Lett. **81**, 844 (2002).
153. T. Dietl, H. Ohno and F. Matsukura, Phys. Rev. **B63**, 195205 (2001).

PHYSICS OF CARBON NANOSTRUCTURES

V.A. OSIPOV

Joint Institute for Nuclear Research,
Bogoliubov Laboratory of Theoretical Physics,
141980 Dubna, Moscow region, Russia
osipov@theor.jinr.ru

Carbon is a prominent element that appears in various structures with new promising technological applications. The physics of carbon nanostructures is one of the hot topics in modern condensed matter theory. This is a brief introduction into the theory of variously shaped carbon nanostructures with paying special attention to the continuum field-theory models. The review includes a brief historical excursus, the most interesting experimental observations, the formulation of generic field-theory models for the description of electronic states in carbon nanoparticles, and some possible applications.

1. Introduction

One of the most prominent advances of the last century was the discovery of the third form of carbon. Doubtless, this finding opened the "nanocosm" in condensed matter physics with variously shaped carbon nanoparticles serving as the building blocks. Over a long period there were known two main structures of carbon lattices. The first one is a tetrahedral network of two interpenetrating face centered cubic lattices. Four outer electrons of each carbon atom are 'localised' between the atoms in covalent bonding having sp^3 hybridization. Therefore, the movement of electrons is restricted and, as a result, this material does not conduct an electric current. Certainly, this is a diamond, the hardest known natural substance. There is a principally different carbon lattice having a layered structure. In this case, each carbon atom is covalently bonded (sp^2 hybridization occurs) to three others in the same plane. The bond angle is equal to 120°, so that carbon atoms form six-membered rings (the hexagonal structure). The fourth electrons form much weaker π bonds. There are also weak van der Waals forces between the layers. This material can conduct electricity along the planes and does not conduct in a direction normal to planes. This is the well-known graphite. Everybody met graphite already in the early childhood inasmuch as it serves as leads in pencils.

There were many assumptions that carbon could form stable ordered structures other than graphite and diamond. Let us mention the most important ideas. In 1970 Osawa suggested that a molecule made up of sp^2 hybridized carbons could be stable and have a soccer structure.[1] In 1973 Bochvar and Gal'pern published quantum chemical calculations of both C_{60} and C_{20} structures and described some of their properties.[2] However, the crucial year was 1985 when Kroto suggested the idea of the experiment with vaporized carbon.[3] To produce the carbon nanoparticles he proposed to use the cluster beam technique, that is the same apparatus as for production of metal clusters. The only differences are the utilization of graphite disk and the selection of appropriate conditions. This idea was realized by using of Smalley's graphite vaporization apparatus consisting of the rotating graphite disk, focused vaporization laser and a pulsed jet of helium which causes clustering and cooling. A cluster beam deposition was then analysed by a mass spectrometer to ascertain the cluster mass distributions. Using of this apparatus gave in 1985 the first evidence of spherical particles called fullerenes. It was found that the obtained spectra markedly depend on the effective helium density. At low helium pressure (less than 10 Torr) a nearly uniform distribution of even-numbered clusters was observed. Under certain (optimized) clustering conditions the relative abundance of C_{60} was greatly enhanced. A similar but less pronounced effect was found to take place for C_{70}.

The breakthrough in physics of carbon nanostructures was made in 1990 when Krätschmer and Huffman suggested the principally different way of graphite vaporization[4,5] based on the using of an electric arc discharge. The reaction kettle consists of two electrodes with a current source, a graphite rod, and a graphite base. One may also use two graphite rods. The electric arc is between points of the graphite rods and the process of vaporization occurs in a helium environment at large pressure. This apparatus allowed them to produce macroscopic quantities of solid fullerenes which made it possible to perform a more complete analysis of C_{60} properties. The analyses led to the first solid evidence that C_{60} was a closed sphere. It should be noted that the absence of a laser made this apparatus available in practically all laboratories. Notice also that namely Krätschmar-Huffman-based evaporators are currently used in industrial production of fullerenes as well as other forms of carbon nanoparticles.

In 1991 Iijima discovered carbon nanotubes and some other nanostructures.[6] The first discovered nanotubes were multi-walled, had a cylindrical lengthy form and were closed at both ends. The outer diameter varied from about 3 nm to 30 nm. In 1993, there were observed a new class of carbon nanotubes having solely a single layer.[7] These single-walled nanotubes are generally narrower than the multiwalled tubes, with diameters typically in the range of 1-2 nm, and tend to be curved rather than straight. It was soon established that these new fibres had a range of exceptional properties (see below). After that a great variety of different carbon nanoparticles was synthesized: nanocones, nanohorns, onions, endohedral fullerenes (having additional atoms, ions, or clusters inside a cave), etc. As the recent prominent result, let us note an experimental observation of graphene (a single atomic layer of graphite).[8]

2. Defects, geometry, electronic structure

In this section, I would like to discuss the problem of geometry and defects in carbon nanostructures, their influence on the electronic structure as well as some important experiments and the results of numerical (tight-binding) calculations.

Fig. 1 shows the football and the fullerene molecule. There are three evi-

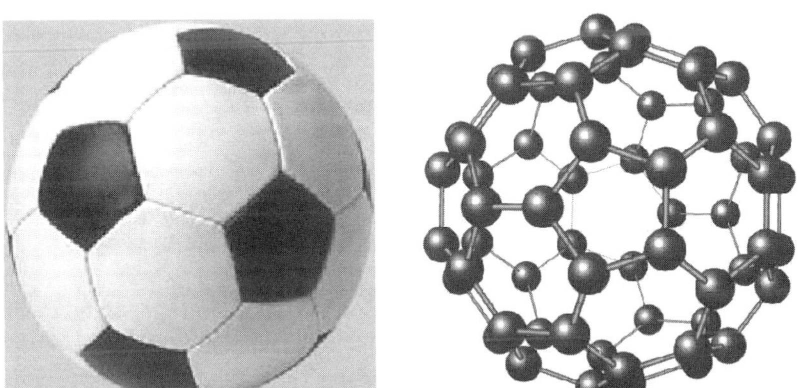

(a) (b)

Fig. 1. (a) Football. (b) Fullerene C_{60}.

dent similarities between these pictures. First of all, one can see the hexagonal cells. Secondly, both objects contain pentagons (fivefold rings) and, finally, they have spherical shapes. Notice that if we count all pentagons on the surface the answer will be exactly twelve. A reasonable question arises: why twelve? The simplest naive answer is that this number could be responsible for the spherical form. This is not true. In fact, when you try to build from hexagons either a rugby ball or non-spherical fullerenes like, for example, C_{70} and C_{40}, you will also need for exactly twelve pentagons. Moreover, the same number of pentagons is required to get a closed nanotube. In this case, the pentagons will be situated in the cap regions. Actually, the right answer gives the Euler's theorem which relates the number of vertices, edges and faces of a polyhedron. For the hexagonal carbon lattice it can be written in the form (see, e.g., Ref. [9])

$$...2n_4 + n_5 - n_7 - 2n_8... = \sum_x (6-x)n_x = \chi = 12(1-g), \tag{1}$$

where n_x is the number of polygons having x sides, χ is the Euler characteristic (a surface integral of Gaussian curvature) which is a geometrical invariant related to the topology of the structure, and g is the genus or a number of handles of an arrangement (the maximal number of independent closed cuts that leave the surface in one piece). According to (1) there is no contribution to Gaussian curvature for $x = 6$. This means that two-dimensional carbon lattice consisting only of hexagons is flat and one has to introduce some additional polygons to obtain a nontrivial shape. A similar picture takes place for closed structures with $g = 1$. For example, the torus can be created from hexagons only and no other polygons are required. The situation changes drastically for a sphere where $g = 0$ and, therefore, according to (1) the minimal number of necessary pentagons is exactly $n_5 = 12$. This means the presence of twelve 60° disclinations on the closed hexatic surface. Thus, the closed hexagonal microcrystals can only be formed by having a total disclination of 4π. As an important conclusion, let us mention that disclinations are *generic* defects in closed carbon structures.

To gain a better insight into the relation between pentagons and disclinations, let us look at Fig. 2 where the planar hexagonal lattice is shown. Imagine, we wish to create a fivefold ring near the point α. How to perform this task? One can easily see that the only possible way to remove a single atom from its position is to cut the whole sector $\beta\alpha\gamma$ and then glue the borders. It should be noted that this cut-and-glue procedure is well known in solid state physics and describes the creation of a rotational topological defect (a disclination). The power of the disclination (the Frank index) is determined by the sector angle. In our case, the sector angle is 60°, so that one gets a 60° disclination with the Frank index 1/6. The creation of the disclination will inevitably be accompanied by a buckling into the third di-

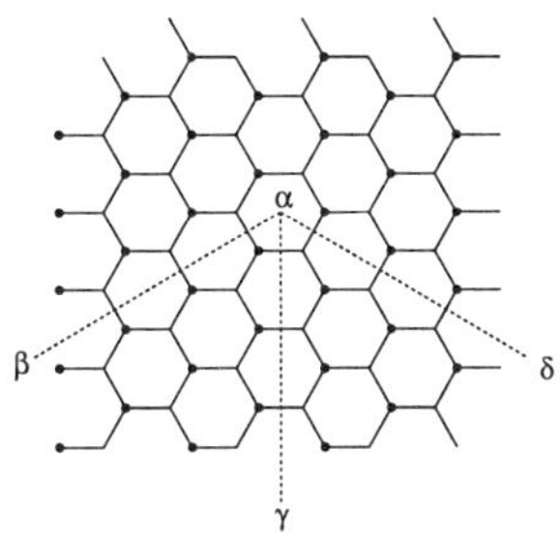

Fig. 2. Two-dimensional graphite lattice

mension (see Fig. 3) due to a tendency of an elastic surface to screen elastic strains caused by the defect. So, the final structure has a cone-like form. In the case of a rigid membrane, the ordinary cone will be created.

Another possible defect is a heptagon. To produce the heptagon one has to insert a sector in the graphite plane. In this case, the negative -60° disclination appears (see Fig. 3). In the presence of a heptagon the surface is

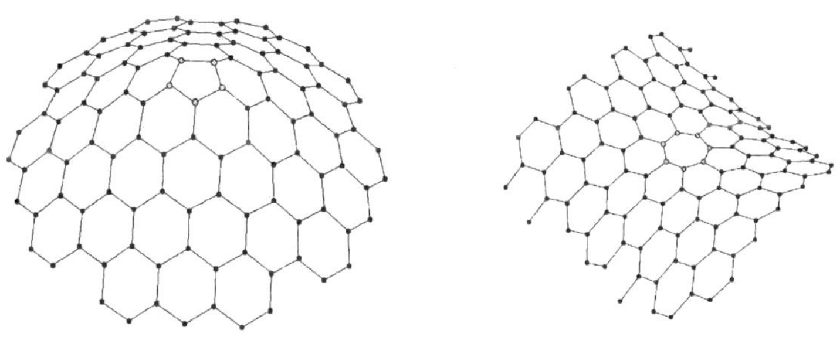

(a) (b)

Fig. 3. The pentagon (left) and the heptagon (right) in the hexagonal graphite lattice.

also buckled but geometry is different: the saddle-like point appears. Summarizing, the pentagon gives a cone-like structure with positive curvature and the heptagon results in negative curvature in the hexagonal graphite lattice.

Fig. 4 shows the deformation of a nanotube in the presence of two symmetrically situated heptagons. As is seen, a shape of the one-sheet hyperboloid appears. This geometry becomes more pronounced for a ring of heptagons.

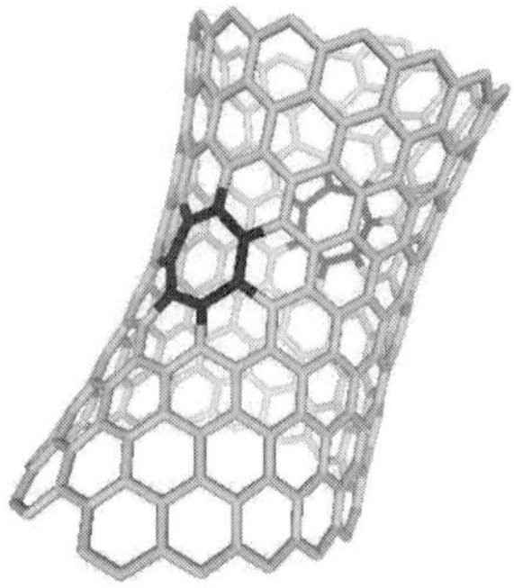

Fig. 4. A carbon nanotube with two inserted sevenfolds (marked). This surface has the negative gaussian curvature and can be approximated by one-sheet hyperboloid.

Until now we have discussed the pentagons and heptagons from a theoretical (geometrical) point of view. Are there any experimental evidences in favor of these defects in carbon nanostructures? The first experimental observation of a pentagon at the apex of a cone was presented by An et al. in Ref. [10]. They used the scanning tunneling microscope to study the structure of a conical protuberance and found five bright spots at the apex of the nanocone. This was the clear evidence that the pentagon is located at the apex. In addition, the bright spots indicate the enhanced charge density localized at each carbon atom in the pentagon which means an increase in the electronic density of states (DOS).

This finding was approved by the numerical tight-binding calculations for nanocones with different number of pentagons at the apex.[11] In particular, it was found that the DOS behaves smoothly for a single layer of graphite (graphene). For one pentagon, there appear some additional peaks. For two pentagons at the apex, peaks are also present while their strength and positions are changed and depend on the relative positions of pentagons. The same tendencies were found to take place for three, four and five pentagons. This study shows the crucial influence of pentagonal rings on the electronic structure of the graphite surface.

A similar picture was observed in capped single-walled nanotubes both theoretically and experimentally.[12] Namely, the experimental tunneling spectra from the end, near the end and far from the end of the nanotube show additional peaks in the cap region. The tight-binding calculations are in agreement with the experimental data. The numerical calculations were performed for two different morphologies of six pentagons at the cap. It was established a good agreement with the experimental curves for one of the suggested morphologies. Notice that the peaks disappear far from the cap. So, pentagons markedly influence the electronic structure in the cap region and have no effect at large distances.

The tight-binding calculations for nanohorns were performed in Ref. [13]. Nanohorns are the cone-like structures with exactly five pentagons at the apex (see Fig. 5). It was found that both the presence of

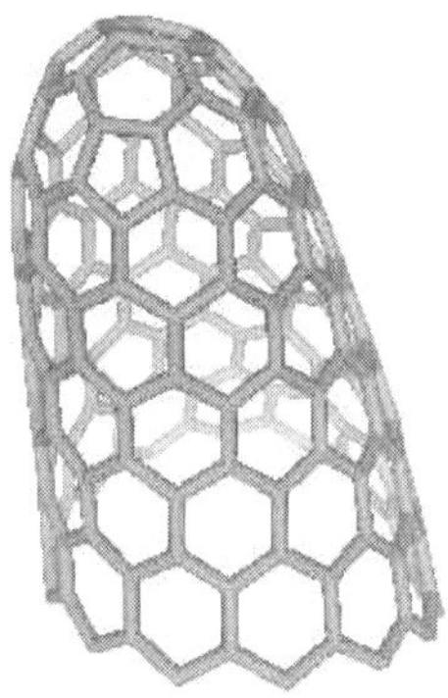

Fig. 5. A carbon nanohorn. One of the possible morphologies is shown.

pentagons and different possible morphologies of pentagons at the terminating cap influence the DOS. Therefore, one can finally conclude that pentagons/heptagons have a marked impact on the electronic structure of carbon nanoparticles.

3. Continuum field-theory model

As it was shown in the previous section, the carbon nanoparticles are variously shaped and contain topological defects as the basic elements of their structure. Both these factors should be taken into account in theoretical description of carbon nanostructures. This means that one has to formulate a model for the description of electronic states on curved surfaces with topological defects taken into account. In addition, the specific properties of a two-dimensional carbon lattice should be properly taken into account. Of course, it is unlikely to expect a simple theoretical formulation of this problem. There is, however, a good basis to start this study. Namely, the case of planar graphite lattice was considered and properly described in 1947 by Wallace[14] (see also Ref. [15]). Two important findings should be mentioned. First, the unit cell contains two carbon atoms, so that the lattice becomes divided in two sublattices. Second, the ground state of graphite layer is twofold degenerate. The electronic spectrum has a fixed point (K point) where the conduction and valence bands are touched, i.e. there is no gap in the spectrum. This fact allows us to classify the graphite layer as a semimetal. In fact, K is the Fermi point with a linear spectrum around it. In addition, the K and the -K points are equivalent. Notice, however, that this is valid only for the defect free case. As it was shown in Ref. [14], the effective mass approximation is equivalent to the $\vec{k}\cdot\vec{p}$ expansion of the graphite energy bands around the $\vec{K}$ point in the Brillouin zone. Since there are two degenerate Bloch eigenstates at $\vec{K}$, the wave function can be approximated by

$$\Psi(\vec{k},\vec{r}) = f_1(\vec{\kappa})e^{i\vec{\kappa}\vec{r}}\Psi_1^S(\vec{K},\vec{r}) + f_2(\vec{\kappa})e^{i\vec{\kappa}\vec{r}}\Psi_2^S(\vec{K},\vec{r}), \tag{2}$$

where $\vec{k} = \vec{K} + \vec{\kappa}$. Keeping terms of order $\vec{\kappa}$ in the Schrödinger equation gives the secular equation for amplitudes $f_{1,2}(\vec{\kappa})$ and after diagonalization one finally gets the two-dimensional Dirac equation (see Ref. [16] for details)

$$i\sigma^\mu\partial_\mu\psi(\vec{r}) = E\psi(\vec{r}), \tag{3}$$

where σ^μ are the conventional Pauli matrices ($\mu = 1, 2$), the energy E is counted from the Fermi energy, the Fermi velocity V_F is taken to be one, and the two-component wave function ψ represents two graphite sublattices (A and B).

The continuum field-theory model for the description of low-energy electronic states in carbon nanostructures of arbitrary geometry was formulated in Ref. [17] in the form

$$-i\sigma^\alpha e_\alpha{}^\mu(\nabla_\mu - ia_\mu^k - iW_\mu)\psi^k = E\psi^k. \tag{4}$$

By using the index k in (4) we take into account electronic states at two independent Fermi wave vectors in carbon lattice (so-called "K-spin up" ($K^\uparrow$) and "K-spin down" ($K^\downarrow$) states). Geometry is involved via zweibeins e_μ^α, $\nabla_\mu = \partial_\mu + \Omega_\mu$ where the spin connection term $\Omega_\mu = (1/8)\omega^\alpha_\mu{}^\beta[\sigma_\alpha, \sigma_\beta]$, and $(\omega^\mu)^{\alpha\beta}$ are the spin connection coefficients:

$$(\omega_\mu)^{ab} = e_\nu^a(\partial_\mu e^{\nu b} - \Gamma^\nu_{\mu\chi}e^{\chi b}) = -(\omega_\mu)^{ba}, \tag{5}$$

where

$$\Gamma^\nu_{\mu\lambda} = \frac{g^{\nu\chi}}{2}(\partial_\mu g_{\chi\lambda} + \partial_\lambda g_{\chi\mu} - \partial_\chi g_{\mu\lambda}) \tag{6}$$

are the metric connection coefficients.

In order to take into account disclinations, two compensating gauge fields a_μ and W_μ are introduced in (4). The Abelian field W_μ is responsible for the elastic flux due to a disclination.[18,19] Its circulation around a single disclination is found to be exactly $2\pi/6$ (the appearance of a pentagon in the honeycomb lattice is equivalent to the creation of 60° disclination)

$$\oint W_\mu dx^\mu = \frac{2\pi}{6}. \tag{7}$$

The non-Abelian field $\vec{a}$ allows us to take into account the exchange between A and B sublattices in the presence of a pentagon.[17,20,21] Namely, this exchange can be described by using an appropriate boundary condition for the K spin part of the four-component spinor wavefunction $\psi = (\psi^\uparrow \psi^\downarrow)^T$ in the form $\psi(\varphi + 2\pi) = -T\psi(\varphi)$. Here the holonomy operator T is composed as a product of two operators in the form $\exp(i\Phi\tau_i)$, i=2,3 where the isospin Pauli matrices τ act on the K part of the spinor components. The general consideration takes proper account of the relative placement of pentagons. For even number of defects one has (see Ref. [21])

$$\oint a_\mu^k dx^\mu = \pm(N\frac{2\pi}{4} + M\frac{2\pi}{3}). \tag{8}$$

Here the sign plus (minus) is taken for $k = K^\uparrow$ ($k = K^\downarrow$), respectively, N is a number of defects and M ($M = n - m \pmod 3$) depends on the arrangement of pentagons, n and m are the numbers of steps in positive and negative directions, respectively. The directions rotated by $2\pi/3$ are considered to be identical. $M = 0$ for an odd number of pentagons. Let us apply this model to carbon nanoparticles of different geometries.

4. Carbon nanostructures of different geometries

4.1. *Nanocones*

Generally, there are two possible scenarios for modelling carbon nanocones. First, one can accomplish a cut-and-glue procedure in which the pentagon in the hexagonal network is constructed by cutting out a 60° sector from a graphene sheet. In this case, pentagonal defects in cones can be considered as apical disclinations and the opening angle is directly connected to the Frank index of a disclination. Due to the symmetry of a graphite lattice only five types of cones can be created from a continuous sheet of graphite. The total disclinations of all these cones are multiplies of 60°, corresponding to the presence of a given number (n) of pentagons at the apices. It is important to mention that carbon nanocones with cone angles of 19°,

39°, 60°, 85°, and 113° have been observed in a carbon sample.[22] These angles might correspond to 300°, 240°, 180°, 120°, and 60° disclinations in graphite, respectively. Disks (n=0) and one-open-end nanotube (n=6) have also been observed in the same sample.[22] This case has been theoretically studied in Refs. [11,23,24]. Notice that the morphology of pentagons at the apex was ignored within the continuum description in Refs. [23,24]. More delicate consideration was presented in Ref. [21].

Second, a single disclination on a finite graphite sheet is known to be buckled to screen its energy thus leading to curved hexagonal network.[25] In this context, the pentagon in a graphene can result in a curved cone-like structure. The most appropriate cone-like figure is a hyperboloid. Let us consider here this scenario. An upper half of a two-sheet hyperboloid can be regarded as an embedding

$$(\chi, \varphi) \to (a \sinh\chi \cos\varphi, a \sinh\chi \sin\varphi, c \cosh\chi).$$

From this one can easily find the metrical tensor and nonvanishing connection coefficients. In a rotating $SO(2)$ frame the zweibeins become

$$\begin{aligned} {e^1}_\chi &= \sqrt{g_{\chi\chi}} \cos\varphi, \; {e^2}_\chi = \sqrt{g_{\chi\chi}} \sin\varphi, \\ {e^1}_\varphi &= -a \sinh\chi \sin\varphi, \; {e^2}_\varphi = a \sinh\chi \cos\varphi, \end{aligned} \tag{9}$$

which in view of Eq. (5) gives for the spin connection coefficients

$$\omega_\chi^{12} = \omega_\chi^{21} = 0, \quad \omega_\varphi^{12} = -\omega_\varphi^{21} = \frac{1}{2}\left[1 - \frac{a \cosh\chi}{\sqrt{g_{\chi\chi}}}\right] = \omega, \tag{10}$$

so that the spin connection $\Omega_\varphi = i\omega\sigma^3$.

External gauge potential in this case is found to be $W_\chi = a_\chi = 0$, $W_\varphi = \nu$, and the Dirac operator on the hyperboloid takes the form[26]

$$\hat{\mathcal{D}} = \begin{bmatrix} 0 & e^{-i\varphi}\left(-\frac{\partial_\chi}{\sqrt{g_{\chi\chi}}} + \frac{1}{a \sinh\chi}(i\partial_\varphi + \nu + a_\varphi + \omega)\right) \\ e^{i\varphi}\left(\frac{\partial_\chi}{\sqrt{g_{\chi\chi}}} + \frac{1}{a \sinh\chi}(i\partial_\varphi + \nu + a_\varphi - \omega)\right) & 0 \end{bmatrix}.$$

The substitutions $\psi = \begin{pmatrix} u \\ v \end{pmatrix} e^{ij\varphi}$ and $\tilde{\psi} = \psi\sqrt{\sinh\chi}$ reduce the eigenvalue problem (4) (after diagonalization by the K-spin part) to

$$\begin{aligned} \partial_\chi \tilde{u} - \sqrt{\coth^2\chi + b^2}\, \Phi \tilde{u} &= \tilde{E}\tilde{v}, \\ -\partial_\chi \tilde{v} - \sqrt{\coth^2\chi + b^2}\, \Phi \tilde{v} &= \tilde{E}\tilde{u}, \end{aligned} \tag{11}$$

where $\tilde{E} = \sqrt{g_{\chi\chi}}\, E$, $b = c/a$, and $\Phi = j - n_\Omega/6 + 1/2 \mp (n_\Omega/4 - M/3)$ with $M = -1, 0, 1$ being a factor depending on the morphology of defects. For odd number of defects (and when all the defects are located at the same point), one has $M = 0$ as stated above.

The most interesting point is the existence of the zero-energy modes in this case. Namely, for $E = 0$ the general solution to (11) reads[27]

$$\tilde{u}_0(\chi) = A\left[(k\cosh\chi + \Delta)^{2k}\,\frac{\Delta - \cosh\chi}{\Delta + \cosh\chi}\right]^{\frac{\Phi}{2}},$$
$$\tilde{v}_0(\chi) = A\left[(k\cosh\chi + \Delta)^{2k}\,\frac{\Delta - \cosh\chi}{\Delta + \cosh\chi}\right]^{-\frac{\Phi}{2}}, \qquad (12)$$

where $k = \sqrt{1+b^2}$, $\Delta = \Delta(\chi) = \sqrt{1 + k^2\sinh^2\chi}$, and A is a normalization factor. Due to the cone asymptotic, it is quite reasonable to assume a correspondence between the parameter k and the Frank index ν. Namely, like for a cone one can specify $k = 1/(1-\nu)$ (see, for example, Ref. [28]). In this case, only $\tilde{v}_0$ mode becomes normalized and only for $j = 2$ and $4 < n_\Omega < 6$. Therefore, a true zero-mode solution exists for exactly five disclinations. It should be noted that this is the typical nanohorn configuration (see Fig. 5).

In the general case, Eq. (11) was integrated numerically in Ref. [27]. The results are shown in Fig. 6. As is seen, the wave functions have an oscilla-

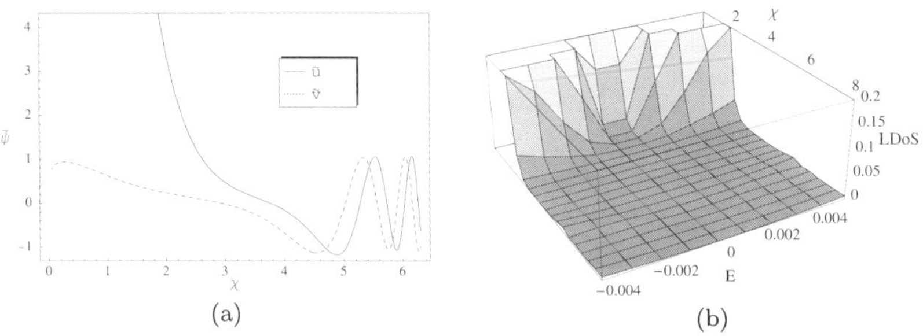

(a) (b)

Fig. 6. (a) The solution to (11) with a set of parameters $a = 1$, $c = 4$, $j = 0$, $E = 0.01$, and $\nu = 5/6$. (b) Local density of states (per area) in arbitrary units as a function of χ and E.

tory behavior at large distances. With ξ decreasing, $\tilde{u}(\chi)$ rapidly increases while $\tilde{v}(\chi)$ decreases. Fig. 6(b) shows the calculated LDoS (per area) as a function of ξ and energy. One can see a remarkable increase of the LDoS near the defect. There is a finite density of electronic states at the Fermi level slowly growing with E. This finding has an important physical consequence. The increase of the LDoS near the defect means that five pentagons serve as active emission centers thus resulting in the enhanced field emission properties of nanohorns. This agrees well with the experimental results in Ref. [29] where field emission properties of carbon nanohorn films were observed and compared with the best nanotube emitters. For nanotubes, the number of emission centers per unit cap is larger (six pentagons) and, therefore, the emission is better.

4.2. *One-sheet hyperboloids*

It was observed in Ref. [30] that there exist single-walled nanotube-based structures with the negative curvature. As discussed above, a possible origin of such curvature is the presence of heptagons. Let us consider two or more negative disclinations of fixed power (heptagons) situated symmetrically along the ring. In this case, a one-sheet hyperboloid geometry can be realized. Geometrically, the one-sheet hyperboloid can be regarded as an embedding

$$(\chi, \varphi) \to (a \cosh \chi \cos \varphi, a \cosh \chi \sin \varphi, c \sinh \chi).$$

The field equations for the one-sheet hyperboloid are similar to Eq.(11) with the only replacement sinh $\Leftrightarrow$ cosh (see Ref. [26])

$$\begin{aligned} \partial_\chi \tilde{u} - \sqrt{\tanh^2 \chi + b^2}\, \Phi \tilde{u} &= \tilde{E} \tilde{v}, \\ -\partial_\chi \tilde{v} - \sqrt{\tanh^2 \chi + b^2}\, \Phi \tilde{v} &= \tilde{E} \tilde{u} \end{aligned} \tag{13}$$

where $\tilde{E} = \sqrt{g_{\chi\chi}}\, E$, $b = c/a$, and $\Phi = \Phi(\text{sign}\chi) = j + 1/2 + n_\Omega/12\, \text{sign}\,(\chi) \mp ((6 - n_\Omega/2)/4 + M(\text{sign}\chi)/3)\text{sign}\chi$. Here the morphology factor M depends on signχ, however, as for the two-sheet hyperboloid, when the value $n_\Omega/2$ is odd one has $M = 0$.

Fig. 7 shows the LDoS as a function of the energy and a position with $\nu = -1/3$ (two heptagons in a ring) and $\nu = -1$ when there are six heptagons in a ring. As is seen, for $\nu = -1/3$ the LDoS is a symmetric function of both arguments. It has a distinct minimum near both $\chi = 0$ and the Fermi energy. Near $\chi = 0$ the LDoS has a power-like dependence on energy while at large distances it tends to a 'vacuum'-like behavior which is linear in energy. For $\nu = -1$ the LDoS reaches the highest possible value in the region near disclinations (which is about a half of its value at infinity). As is shown in Fig. 7(b), in this case the LDoS is linear in energy almost everywhere. Thus, summarizing, the LDoS markedly decreases near the negative disclinations located on the one-sheet hyperboloid.

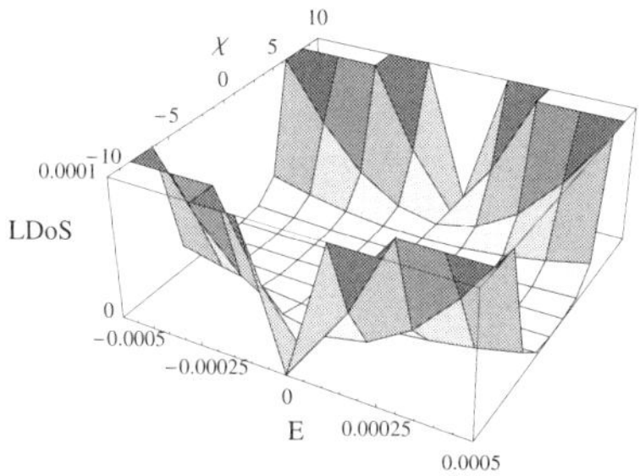

(a)

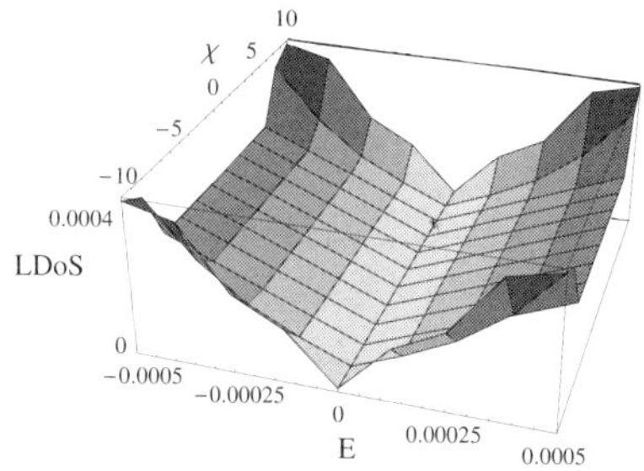

(b)

Fig. 7. Local density of states (per area) in arbitrary units as a function of χ and E: (a) for $\nu = -1/3$, (b) for $\nu = -1$

4.3. *Nanotubes*

Let us propose an appropriate geometry for the capped tube: a single manifold Σ which reproduces the sphere in the cap region and the tube outside

$$\vec{R}(\rho(z)\cos\varphi, \rho(z)\sin\varphi, z), \ \rho(z) = R_t\sqrt{1-\exp(-2\Lambda)}, \tag{14}$$

with $\Lambda = (z + R_f)/R_f$, $\alpha = R_t/R_f$, $z \geq -R_f$, $0 \leq \varphi < 2\pi$, where R_f characterizes the capped area and R_t is the tube radius. It is taken into account that R_f can generally be different from R_t. To simplify the problem, one can approximate $S = 2\pi\Lambda$ as if the surface would be the hemisphere with the radius R_f. The metric tensor is written as

$$g_{zz} = \alpha^2\frac{e^{-4\Lambda}}{1-e^{-2\Lambda}} + 1, \quad g_{\varphi\varphi} = \rho^2(z), \quad g_{z\varphi} = 0. \tag{15}$$

The nonzero metric connection coefficients are found to be

$$\Gamma^z_{zz} = \frac{-1}{R_f g_{zz}}(2(g_{zz}-1) + \frac{e^{2\Lambda}}{\alpha^2}(g_{zz}-1)^2),$$
$$\Gamma^z_{\varphi\varphi} = -\frac{R_t^2 e^{-2\Lambda}}{R_f g_{zz}}, \quad \Gamma^\varphi_{z\varphi} = \Gamma^\varphi_{\varphi z} = \frac{R_t^2 e^{-2\Lambda}}{R_f g_{\varphi\varphi}}. \tag{16}$$

The zweibeins are $e^1_z = \sqrt{g_{zz}}$, $e^2_\varphi = \rho(z)$, and the spin connection coefficients are $\omega^{12}_\varphi = -\omega^{21}_\varphi = \alpha R_t e^{-2\Lambda}/(\rho(z)\sqrt{g_{zz}})$, so that Eq. (4) includes the spin connection term

$$\Omega_\varphi = \frac{i\sigma_3\alpha R_t e^{-2\Lambda}}{2\rho(z)\sqrt{g_{zz}}}. \tag{17}$$

Finally, the Dirac equation (4) takes the form (see Ref. [31])

$$
\begin{aligned}
-i(\frac{\partial_z}{\sqrt{g_{zz}}} + \frac{1}{\rho(z)}(j - \frac{\alpha R_t e^{-2\Lambda}}{\rho(z)\sqrt{g_{zz}}} - W_\varphi - a_\varphi))v = Eu, \\
-i(\frac{\partial_z}{\sqrt{g_{zz}}} - \frac{1}{\rho(z)}(j + \frac{\alpha R_t e^{-2\Lambda}}{\rho(z)\sqrt{g_{zz}}} - W_\varphi - a_\varphi))u = Ev.
\end{aligned} \tag{18}
$$

Notice that the momentum j takes half-integer values and enters (18) only by the combination $j - a_\varphi$. In the tube region this model is identical to that proposed in Ref. [32]. In the general case, the system (18) was solved numerically.[31] Fig. 8 shows the local density of electronic states (LDoS) as

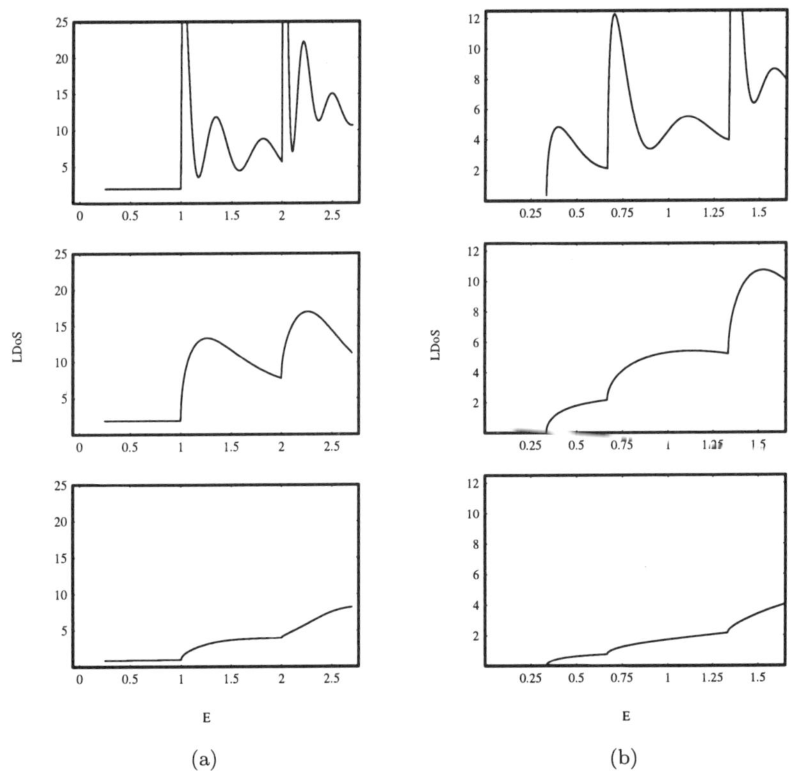

Fig. 8. The local density of states (per unit area, in arbitrary units) as a function of the energy in the cap region (bottom), near the cap (middle) and far from the cap (top). The case of metallic (left) and semiconducting (right) tubes is presented. We take $1/\alpha$=0.9, and the energy E is measured in the units of $\hbar V_F/R_t$.

a function of the energy in three regions: far from the cap, near the cap and on the cap for both metallic and semiconducting cases. As is seen, the peaks appear at the energies higher than the threshold energy m, where $mR_t = 0, \pm 1, \pm 2, ..$ for metallic and $mR_t = \pm 1/3, \pm 2/3, \pm 4/3, ..$ for semiconducting case. Notice the appearance of secondary (less pronounced) peaks far from the cap. For metallic tube, the constant LDoS below the threshold energy is found. The most interesting finding is the finite values of the LDoS near points $E = m$ where it behaves linearly in $E - m$. Moreover, as is seen from Fig. 8, the smoothed VHS (van Hove singularities) peaks are shifted to higher energies. Within the model the dispersion relation has the form $E = \sqrt{k^2 + m^2}$ and, therefore, $DoS_0(E)$ diverge like $(E - m)^{-1/2}$ when the energy approaches the threshold m. Accordingly, the LDoS should be singular at the threshold points but this was not observed. This was explained in Ref. [31] by the fact that the phase of the wavefunction in the tube region is determined by the asymptotic solution near the cap. As it was analytically shown, in the cap region the mass factor markedly increases due to geometry. This results in a fixation of the wave-function phase which, in turn, provides the smoothing of VHS.

4.4. *Icosahedral fullerenes*

According to the Euler's theorem, the fullerene molecule consists of exactly twelve disclinations. Generally, it is difficult to take into account properly all the disclinations. There are two ways to simplify the problem. First, one can consider a situation near a single defect (similar to Ref. [28]) taking into account that each defect in the fullerene can be simulated by two fluxes: the K-spin flux in Eq. (8) and the elastic flux in Eq. (7). In the case of a sphere, however, the most appropriate approximation is to introduce the effective field replacing the fields of twelve disclinations by the field of the magnetic 't Hooft-Polyakov monopole with a constant flux density and the half-integer charge A.[20] In this case, the total flux of the monopole $4\pi A$ is equal to the sum of fluxes from all disclinations.

The procedure of summing up non-abelian fictitious fluxes from apical defects placed at different points of the graphite cones was presented in Ref. [21] (see Eq. (8)). This gives a natural (n, m) classification of two-pentagon lattices: those for which $n \equiv m \pmod 3$, and those for which $n \not\equiv m \pmod 3$. For (I) fullerenes $M = M_2 = \pm 1$ or 0 for any two defects, $M_4 = 0$ for any four defects, and $M_6 = -M_2$ for six defects. For fullerenes with the full icosahedral symmetry (Ih) the combined flux turns out to be a sum of fluxes due to any pair of defects. Thus, the combined flux does not depend upon the arrangement of pentagons because any corresponding fragment of the lattice turns out to be of the above-mentioned class $(1,1)$. In other words, in icosahedral fullerenes $(n-m) \pmod 3 = 0$ due to the mirror symmetry of the lattice. Finally, the continuous fields take the form (see Ref. [17])

$$a_\theta^k = 0, \quad a_\varphi^k = \pm\frac{3}{2}\cos\theta, \qquad W_\theta = 0, \quad W_\varphi = -\cos\theta.$$

It should be noted that for the Dirac field in the external potential provided by a monopole, the effective "charge" A involves the "isospin" matrix τ' (see Ref. [33]). This matrix can be also diagonalized in the Dirac equation thus giving the additional sign $\pm$ to the whole charge. Finally, the total "charge" is written as $A = \pm(a_\varphi^k + W_\varphi)/\cos\theta = \pm 1/2, \pm 5/2$. Therefore, the Dirac operator in (4) takes the following form:

$$\hat{\mathcal{D}} = -i\sigma_x(\partial_\theta + \frac{\cot\theta}{2}) - i\frac{\sigma_y}{\sin\theta}(\partial_\varphi - iA\cos\theta).$$

The substitution

$$\begin{pmatrix}\psi_A\\ \psi_B\end{pmatrix} = \sum_j \frac{e^{ij\varphi}}{\sqrt{2\pi}}\begin{pmatrix}u_j(\theta)\\ v_j(\theta)\end{pmatrix}, j = 0, \pm 1, \pm 2, \ldots$$

leads to the equations for u_j and v_j

$$\begin{aligned} -i(\partial_\theta + [\frac{1}{2} - A]\cot\theta + \frac{j}{\sin\theta})v_j(\theta) &= Eu_j(\theta), \\ -i(\partial_\theta + [\frac{1}{2} + A]\cot\theta - \frac{j}{\sin\theta})u_j(\theta) &= Ev_j(\theta). \end{aligned} \tag{19}$$

The general solutions to (19) are as follows[17] (see also Ref. [34]): (i) for $|j| \geq ||A| + 1/2|$ one obtains the spectrum

$$E_n^2 = (n + |j| + 1/2)^2 - A^2, \tag{20}$$

and the eigenfunctions

$$\begin{aligned} u_j &= C_u(1-x)^\alpha(1+x)^\beta P_n^{2\alpha,2\beta}, \\ v_j &= C_v(1-x)^\gamma(1+x)^\delta P_n^{2\gamma,2\delta}, \end{aligned} \tag{21}$$

where P_n are the Jacobi polynomials, the unit of energy here is $\hbar V_F/R$ where R is the fullerene radius, and

$$\begin{aligned} \alpha &= \frac{1}{2}\left|j - A - \frac{1}{2}\right|, \quad \gamma = \frac{1}{2}\left|j - A + \frac{1}{2}\right| \\ \beta &= \frac{1}{2}\left|j + A + \frac{1}{2}\right|, \quad \delta = \frac{1}{2}\left|j + A - \frac{1}{2}\right| \end{aligned} \tag{22}$$

(ii) for $|j| \leq ||A| - 1/2|$ and $n = 0$, one gets exactly one zero-mode at fixed j and positive fixed A

$$u_0 = 0, \quad v_0 = C_v(1-x)^\gamma(1+x)^\delta. \tag{23}$$

Accordingly, if $A < 0$ there exists only one zero-mode solution u_0. Thus, for all possible values of j and all possible positive values of A there exists exactly six different zero-mode solutions v_0.

From Eqs. (20) and (23) one can calculate the energy spectrum. The possible "charges" are $A = -1/2, 5/2$, so that the first four levels are (in units of $\hbar V_F/R$) the following: $E = 0, 1.41, 2.45, 3.46$. Their degeneracies are $g = 6, 2, 6, 6$, respectively. It is interesting to note that both the existence of quasi-zero modes found for spherical fullerenes and their 6-fold degeneracy were established within the tight-binding calculations in Refs. [35–38]. There is also a good qualitative agreement in observed scaling of the energy gap between the highest occupied and lowest unoccupied energy levels with the size of the fullerene.

4.5. *Spheroidal fullerenes*

In this section I will briefly mention the case of slightly spheroidally deformed fullerenes and consider the role of a weak uniform magnetic field. The corresponding model and mathematical details can be found in our recent papers[39–41] where the problem of the low energy electronic states in spheroidal fullerenes[39] as well as the influence of a weak uniform external magnetic field pointed in the z and x directions[40,41] were studied. The main findings were a discovery of fine structure with a specific shift of the electronic levels upwards due to spheroidal deformation and the Zeeman splitting of electronic levels due to a weak uniform magnetic field. In addition, it was shown that the external magnetic field modifies the density of electronic states and does not change the number of zero modes.

Generally, the low energy electronic spectrum of spheroidal fullerenes was found to consist of three parts

$$E_{jn} = E^0_{jn} + E^{0B}_{jn} + E^\delta_{jn}. \tag{24}$$

Here the term E^δ_{jn} describes a shift of the energy levels due to spheroidal deformation and E^{0B}_{jn} comes from the uniform magnetic field and provides the Zeeman splitting. The z axis is defined as the rotational axis of the spheroid with maximal symmetry. As an illustration, Fig. 9 and Fig. 10 show schematically the structure of the first and the second levels for different field directions.

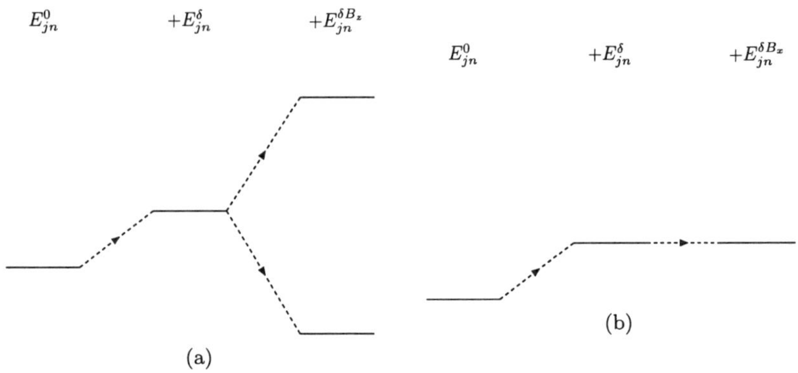

Fig. 9. The schematic picture of the first electronic level E_{jn} of spheroidal fullerenes in a weak uniform magnetic field pointed in the z (left) and x (right) directions.

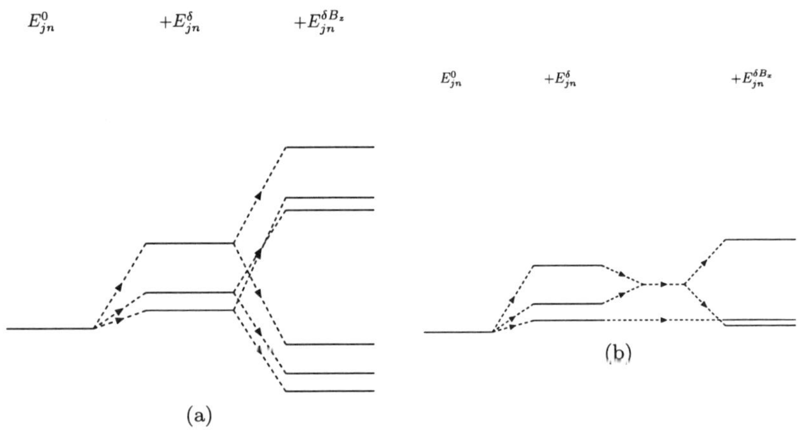

Fig. 10. The schematic picture of the second electronic level E_{jn} of spheroidal fullerenes in a weak uniform magnetic field pointed in the z (left) and x (right) directions.

As is seen, there is a marked difference between the behavior of the first and second energy levels in magnetic field. Indeed, in both cases the energy levels become shifted due to the spheroidal deformation. However, the uniform magnetic field does not influence the first energy level in the case of the x direction. For the second level, the initial (for $\delta = 0$) degeneracy of E^0_{jn} is equal to six (see the previous subsection). The spheroidal deformation provokes an appearance of three shifted double degenerate levels (fine structure). The magnetic field is responsible for the Zeeman splitting. The reason of such behavior is the proper symmetry of the spheroid which changes the role of the external magnetic field in comparison with the spherical case. This gives an additional possibility for the experimental study of the electronic structure of deformed fullerene molecules because there appears a possibility to change the structure of electronic levels in spheroidal fullerenes by altering the direction of the magnetic field.

5. Conclusions

In conclusion, I would like to discuss some important applications of carbon nanoparticles in modern technologies which are based on the unique properties of these structures. One of the important characteristics is the high flexibility of carbon. Notice that namely this property provides the production of variously shaped carbon nanoparticles including fullerenes, nanotubes, cones, toroids, graphitic onions, nanohorns, etc. Moreover, carbon nanotubes are both flexible and strong. They can be bent and straightened without damage unlike metals, which fracture when bent and restraightened. Compared to other materials multiwalled carbon nanotubes have the highest tensile strength.

Possible applications of carbon nanotubes (CNT) include field emission based flat panel displays, novel semiconducting devices, chemical sensors, and ultra-sensitive electromechanical sensors. Arrays of CNTs have proven to be very useful for electron emission due to their possibility to emit electrons from the tips at low electric field. This makes it possible to construct field emission displays (FED) using CNTs as field emission electron sources. The important difference of FEDs from the conventional television cathode ray tubes (CRT) is that each pixel has its own electron source. For this reason, FEDs are expected to be thick and still have the same picture quality as conventional CRTs. In addition, FEDs are more power saving in comparison with other kinds of flat panel display, such as LCD and Plasma. CNTs have a small diameter and a high aspect ratio, so that they can be used as tips in scanning probe microscopy, scanning tunneling microscopy, and atomic force microscopy allowing to enhance the resolution to nanoscale.

CNTs can be used as the conducting channels of a MOSFET due to a small diameter and high conductivity. Generally, nanoscale electronic devices made from CNTs, such as transistors and sensors, are much smaller and more versatile than those in conventional microelectronic chips. Recently, prototypes of memory devices based on carbon nanotube field-effect transistors (CNT-FETs) have been fabricated. These optoelectronic memory devices combine light sensitive polymers with CNTs on silicon wafers. The polymer layer converts photons to electric charge, which is stored by carbon nanotube transducer. The nanotubes also operate as electrodes to read and erase the charge stored. Thus, the functional CNT-FETs operate as optoelectronic memories, which are written optically and read and erased electrically. A probability of a memory element based on carbon nanopeapods in the nanometer ranges has also been recently reported.

Fullerenes can be effectively used for optoelectronics and photonics. These applications are particularly compelling with the observation of such promising properties as photoluminescence, electroluminescence, large non-resonant optical nonlinearity, and superconductivity. For example, nonlinear optical properties can be applied to high-speed integrated all-optical switching by using of fullerene thin films as the nonlinear medium. The results show many advantages of fullerenes and fullerene devices, including the simplicity of processing into guided wave structures for nonlinear integrated optics, large nonlinear coefficients, etc.

Among other applications let us mention that fullerenes have large gas-adsorption capacities (other atoms can be placed inside a cage) which is promising for their using in the fuel cells. If they are reacted with alkali metal ions, such as potassium, the resulting crystals exhibit superconducting behavior with $T_c \sim 30K$. Because of their affinity to organic molecules, fullerenes C_{60} and their derivatives can be coated onto piezoelectric crystal and applied as a chemical sensor for organic molecules. CNTs are expected to be one of the key materials to improve batteries performance. Of course, it is impossible to review all important applications of carbon nanostructures in this brief section. For example, I have passed over such important topic as using of carbon nanoparticles in medicine. I would like to refer interested readers to a great number of original publications where more details could be found.

Acknowledgments

I would like to thank the lecturers and participants of 43rd Karpacz Winter School of Theoretical Physics for fruitful discussions. Special thanks to Prof. J. Jędrzejewski for the kind invitation to present the lectures and useful discussions during the school. The main results presented in this brief review have been obtained in co-authorship with Drs. E. A. Kochetov, M. Pudlak, R. Pincak and D. V. Kolesnikov whom I am very grateful. This work has been supported by the Russian Foundation for Basic Research under grant No. 05-02-17721.

References

1. E. Osawa, Kagaku **25**, 854, (1970) (in Japanese).

2. D.A. Bochvar and E.G. Gal'pern, Dokl. Akad. Nauk SSSR **209**, 610 (1973) [Proc. Acad. Sci. USSR **209**, 239 (1973)].
3. H.W. Kroto, J.R. Heath, S.C. O'Brien, R.F. Curl and R.E. Smalley, Nature **318**, 162, (1985).
4. W. Krätschmer, K. Fostiropoulos and D.R. Huffman, Chem. Phys. Lett. **170**, 167 (1990).
5. W. Krätschmer, L.D. Lamb, K. Fostiropoulos and D.R. Huffman, Nature **347**, 354 (1990).
6. S. Iijima, Nature **354**, 56 (1991).
7. S. Iijima and T. Ichihashi, Nature **363**, 603 (1991).
8. K.S. Novoselov, A.K. Geim, S.V. Morozov, D. Jiang, Y. Zhang, S.V. Dubonos, I.V. Grigorieva and A.A. Firsov, Science **306**, 666 (2004).
9. H. Terrones and M. Terrones, Carbon **36**, 725 (1998).
10. B. An, S. Fukuyama, K. Yokogawa, M. Yoshimura, M. Egashira, Y. Korai and I. Mochida, Appl. Phys. Lett. **78**, 3696 (2001).
11. J.-C. Charlier and G.-M. Rignanese, Phys. Rev. Lett. **86**, 5970 (2001).
12. P. Kim, T.W. Odom, J.-L. Huang and C.M. Lieber, Phys. Rev. Lett. **82**, 1225 (1999).
13. S. Berber, Y.-K. Kwon and D. Tomanek, Phys. Rev. **B62**, R2291 (2000).
14. P.R. Wallace, Phys. Rev. **71**, 622 (1947).
15. J.C. Slonczewski and P.R. Weiss, Phys. Rev. **109**, 272 (1958).
16. D.P. DiVincenzo and E.J. Mele, Phys. Rev. **B29**, 1685 (1984).
17. D.V. Kolesnikov and V.A. Osipov, Eur. Phys. J. **B49**, 465 (2006).
18. V.A. Osipov, Phys. Lett. **A164**, 327 (1992).
19. E.A. Kochetov and V.A. Osipov, J. Phys. A: Math. Gen. **32**, 1961 (1999).
20. J. González, F. Guinea and M.A.H. Vozmediano, Nucl. Phys. **B406**, 771 (1993).
21. P.E. Lammert and V.H. Crespi, Phys. Rev. **B69**, 035406 (2004).
22. A. Krishnan, E. Dujardin, M.M.J. Treacy, J. Hugdahl, S. Lynum and T.W. Ebbesen, Nature **388**, 451 (1997).
23. P.E. Lammert and V.H. Crespi, Phys. Rev. Lett. **85**, 5190 (2000).
24. V.A. Osipov and E.A. Kochetov, JETP Lett. **73**, 631 (2001).
25. D.R. Nelson and L. Peliti, J. Phys. (Paris) **48**, 1085 (1987).
26. D.V. Kolesnikov and V.A. Osipov, Romanian Journal of Physics **50**, 435 (2005).
27. D.V. Kolesnikov and V.A. Osipov, JETP Letters **79**, 532 (2004).
28. V.A. Osipov, E.A. Kochetov and M. Pudlak, JETP **96**, 140 (2003).
29. J.-M. Bonard, R. Gaál, S. Garaj, L. Thien-Nga, L. Forró, K. Takahashi, F. Kokai, M. Yudasaka and S. Iijima, J. Appl. Phys. **91**, 10107 (2002).
30. S. Iijima, P.M. Ayayan and T. Ichihashi, Phys. Rev. Lett. **69**, 3100 (1992).
31. D.V. Kolesnikov and V.A. Osipov, Europhysics Letters (2007) in press.
32. C.L. Kane and E.J. Mele, Phys. Rev. Lett. **78**, 1932 (1997).
33. R. Jackiw and C. Rebbi, Phys. Rev. **D13**, 3398 (1976).
34. A.A. Abrikosov jr., Int. Journ. of Mod. Phys. **A17**, 885 (2002).
35. E. Manousakis, Phys. Rev. **B44**, 10991 (1991).
36. Y.-L. Lin and F. Nori, Phys. Rev. **B49**, 5020 (1994).

37. A.C. Tang, F.Q. Huang and R.Z. Liu, Phys. Rev. **B53**, 7442 (1996).
38. A. Pérez-Garrido, F. Alhama and D.J. Katada, Chem. Phys. **278**, 77 (2002).
39. M. Pudlak, R. Pincak and V.A. Osipov, Phys. Rev. **B74**, 235435 (2006).
40. M. Pudlak, R. Pincak and V.A. Osipov, Phys. Rev. **A75**, 025201 (2007).
41. M. Pudlak, R. Pincak and V.A. Osipov, Phys. Rev. **A75**, 065201 (2007).

QUANTUM MOLECULAR DYNAMICS SIMULATIONS FOR WARM DENSE MATTER AND APPLICATIONS IN ASTROPHYSICS

R. REDMER, N. NETTELMANN, B. HOLST, A. KIETZMANN and M. FRENCH

University Rostock, Institute of Physics, D–18051 Rostock, Germany

We give an introduction into the method of quantum molecular dynamics simulations which combines density functional theory with classical molecular dynamics. The method is appropriate to determine thermophysical properties of matter under extreme conditions. We give exemplary results for hydrogen and helium. Finally, we describe the standard interior structure model for giant planets and apply this equation of state data in models for Jupiter.

1. Introduction

The physical properties of warm dense matter (WDM) have attracted increasing interest over the last years. Located between cold condensed matter and hot plasmas in the density-temperature plane (see Fig. 1), WDM is characterized by high densities typical for condensed matter and temperatures of several eV. Partial ionization, strong correlations and quantum effects are important under these conditions so that the WDM region is perfectly suited to test concepts of many-particle theory. On the other hand, progress in shock-wave experimental technique (e.g. high-power lasers, gas guns, Z pinches, chemical explosions) and the availability of intense x-ray sources (e.g. at the Laboratory for Laser Energetics in Rochester/USA or at the free electron laser facility FLASH at DESY Hamburg) has allowed to probe and diagnose WDM states.[1]

These activities are mainly driven by inertial confinement fusion (ICF) research and astrophysics. For instance, the path for the compression of deuterium-tritium pellets up to the ignition of fusion processes at ultra-high densities just goes through the WDM region which is, therefore, of great importance for ICF research.[4] On the other hand, most of the interior of giant planets such as Jupiter and Saturn is in WDM states.[5] The detection of a great number of Jupiter-like extrasolar planets since 1995[6]

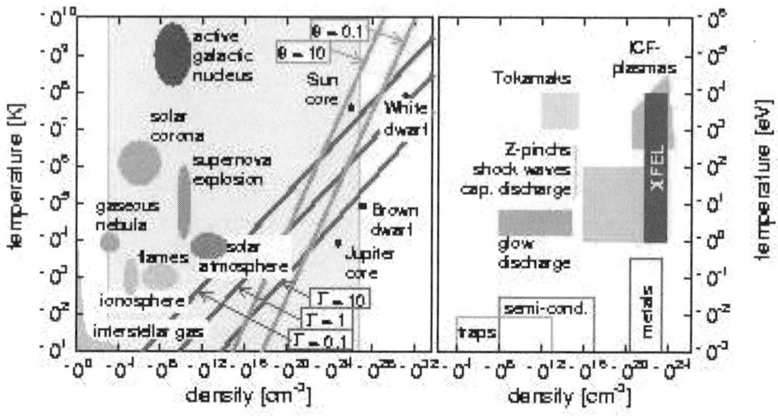

Fig. 1. Density-temperature plane with typical astrophysical and laboratory plasmas.[2] The coupling parameter $\Gamma = e^2/(4\pi\varepsilon_0 k_B T)(4\pi n_e/3)^{1/3}$ and the degeneracy parameter $\Theta = 2m_e k_B T/\hbar^2(3\pi^2 n_e)^{-2/3}$ are defined as usual.[3]

has boosted the interest in the phase diagram of planetary materials such as hydrogen, helium, water and their mixtures at extreme conditions of pressure and temperature. While the limiting cases of low and high densities are well understood within chemical and plasma models, the intermediate WDM region is much more complex. For instance, a nonmetal-to-metal transition occurs in hydrogen[7] and helium[8] at pressures of several megabar and temperatures of few eV which implies a strong state-dependence of the interparticle interactions and, thus, also of the thermodynamic variables. Water has a rich phase diagram at high pressures. A new exotic superionic phase has been predicted under these conditions[9] where mobile protons diffuse almost freely through an oxygen lattice.

In order to treat the strong correlation and quantum effects in the WDM domain adequately, *ab initio* simulation techniques have been developed. In Path Integral Monte Carlo (PIMC) simulations[10] for instance, the key elements are the representation of the density matrix via a path-integral and the evaluation of the respective integrals via Monte Carlo methods. Wave-Packet Molecular Dynamics (WPMD) simulations treat the N-particle wave function as Slater determinant composed of Gaussian wave packets and calculate the dynamics of the electron (and ion) coordinates by solving the classical equations of motion.[11] Alternatively, finite-temperature density functional theory (FT-DFT) calculations can be performed for the electron system at every time step of a classical molecular dynamics simulation for the ions. Such quantum molecular dynamics (QMD) simulations have been applied to warm dense hydrogen,[12,13] helium,[14,15] and water.[9,16] We give here a brief introduction into the method of QMD simulations and present exemplary results for these most abundant planetary materials covering equation of state (EOS) data as well as the electrical conductivity.

2. Quantum molecular dynamics simulations

2.1. *Description of the method*

DFT has become an extremely useful and predictive method for *ab initio* electronic structure calculations for a broad field of applications.[17] A successful combination of DFT with classical MD simulations for the heavy particles (ions) is possible within the Born-Oppenheimer approximation. A full electronic structure calculation is performed for a given ion configuration at every time step of the MD simulation. Utilizing parallelized DFT codes on state-of-the-art compute server, about 100 atoms with respective numbers of nuclei and electrons can be considered in the simulation box. Periodic boundary conditions are applied as usual. The forces on the ions are derived according to the Hellmann-Feynman theorem. Convergence of the method is achieved after several 100 or 1000 time steps dependent on the location of the state in the density-temperature plane.

QMD simulations yield the structural, thermodynamic, and optical properties of WDM.[9,12,13,18–20] We employ a finite temperature Fermi occupation of the electronic states using Mermins approach (FT-DFT)[21] which is implemented in the plane wave density functional code VASP (Vienna Ab Initio Simulation Package).[22] Usually, we consider 32 to 64 atoms in the simulation box and periodic boundary conditions. The electron wave functions are calculated using the projector augmented wave potentials[23] supplied with VASP which yield more accurate conductivity results compared with other pseudopotentials. The exchange-correlation functional is calculated within generalized gradient approximation (GGA) using the parameterization of PBE.[24]

The convergence of the thermodynamic quantities in QMD simulations is an important issue.[17] The plane wave energy cutoff has to be chosen such that the pressure is converged within few percent accuracy, i.e. for H at 1200 eV and for He at 700 eV. The convergence with respect to a systematic enlargement of the **k**-point set in the representation of the Brillouin zone has to be checked as well. Higher-order **k** points modify the EOS data only within few percent and the conductivity results up to a maximum of 10% relative to a one-point result. Therefore, we have restricted our calculations in all cases to the Γ point for the EOS and the mean value point $(1/4, 1/4, 1/4)$ for the electrical conductivity.

The simulations were performed for a canonical ensemble where the temperature, the volume of the simulation box, and the particle number in the box are conserved quantities. The ion temperature is controled by a Nosé-Hoover thermostat and the electronic temperature is fixed by Fermi weighting the occupation of bands.[22] After about hundred time steps the system is equilibrated and the subsequent 400 to 1000 time steps are taken to calculate the EOS data as running averages.

2.2. *Equation of state*

In this Section we present QMD results for the EOS of warm dense hydrogen and helium in order to illustrate the capacity of this *ab initio* approach. Figs. 2 and 3 show pressure isotherms as function of the density in comparison with chemical models.[25,26] The QMD isotherms behave very systematically with temperature and density and show no indications of an instability such as the plasma phase transition (PPT) at lower temperatures, contrary to results derived within some chemical models.[25–28] These are based on a free energy minimization schema for a mixture of atoms and molecules (N_0), ions (N_i), and free electrons (N_e),

$$F(T, V, N_0, N_i, N_e) = F_0(T, V, N_0, N_i, N_e) + F_\pm(T, V, N_0, N_i, N_e) + F_{\text{pol}}(T, V, N_0, N_i, N_e)\,. \tag{1}$$

Correlations are taken into account based on effective two-particle potentials as, e.g., the Debye potential for charged particles and the Morse or exponential-6 potential for neutral particles. The neutral component F_0 can be treated using standard methods of liquid state theory such as fluid variational or fluid perturbation theory. For the plasma component $F_\pm$ one has to apply perturbation theory up to higher-orders[29] or self-consistent integral equation schemas for the correlations[3] which go well beyond the simple Debye-Hückel theory. The interaction between charged and neutral particles is of special importance in the partially ionized domain and can be parameterized in terms of a polarization potential.[30]

For H the chemical models SCvH-i[5] and FVT$^+$[25] agree with the QMD results[32] up to about 0.2 g/cm^3 with respect to the pressure and for ultrahigh densities above 2 g/cm^3. For He the Winisdoerffer-Chabrier (WC) chemical model agrees with the QMD results[15] up to about 1 g/cm^3 and for ultra-high densities above 50 g/cm^3. The chemical model shows a systematic trend towards lower pressures in the intermediate, strongly coupled region where the QMD results approach an almost temperature-independent behavior as characteristic of a degenerate electron gas. These results for H and He underline the significance of *ab initio* calculations for WDM states and have a strong impact on calculations of planetary interiors,[31] see Section 3. On the other hand, we can identify the region where efficient chemical models are applicable in favor of time-consuming *ab initio* calculations.

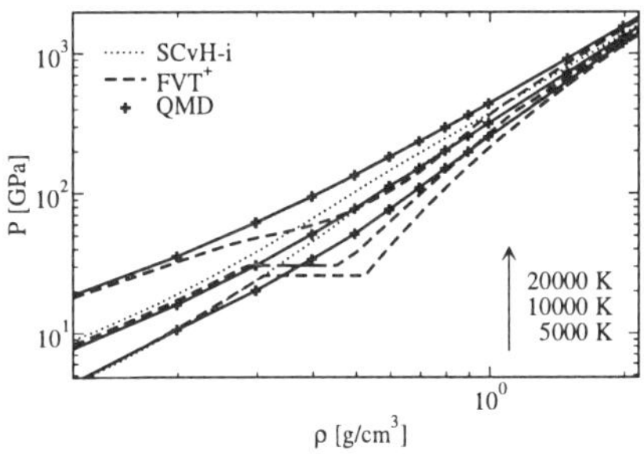

Fig. 2. Pressure isotherms for H: QMD results[32] in comparison with the chemical models SCvH-i[5] and FVT+.[25]

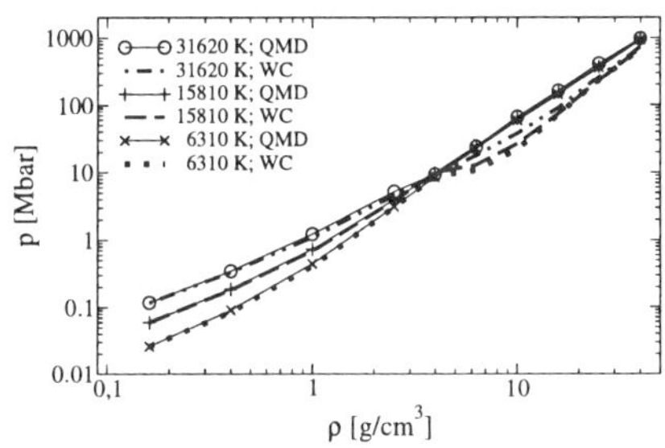

Fig. 3. Pressure isotherms for He: QMD results[15] in comparison with the WC chemical model.[26]

The Hugoniot curve is the location of all (E, P, V) points which can be reached within a single-shock experiment starting from the initial point (E_0, P_0, V_0). It can be calculated using the EOS data via the Hugoniot relation

$$(E - E_0) = \frac{1}{2}(P + P_0)(V_0 - V)\,. \tag{2}$$

A detailed comparison of shock-wave experimental data with theoretical predictions can be found elsewhere.[33,34] QMD simulations yield a very good agreement with experimental data for hydrogen[13] and helium.[15] Key issues are in this context the maximum compression ratio and the location of the metallization transition.

2.3. *Dynamic conductivity*

The dynamic conductivity $\sigma(\omega)$ is a key quantity to determine optical properties via the dielectric function. The index of refraction, the absorption and transmission coefficient as well as the reflectivity can be calculated via standard relations,[12]

$$\varepsilon_1(\omega) = 1 - \frac{1}{\varepsilon_0 \omega}\sigma_2(\omega) \tag{3}$$

$$\varepsilon_2(\omega) = \frac{1}{\varepsilon_0 \omega}\sigma_1(\omega) \tag{4}$$

are the real and imaginary part of the dielectric function and can be obtained from the real and imaginary part of the conductivity. The imaginary part of the conductivity is calculated via the Kramers-Kronig relation:

$$\sigma_2(\omega) = -\frac{2}{\pi}\mathrm{P}\int \frac{\sigma_1(\nu)\omega}{(\nu^2 - \omega^2)} d\nu. \tag{5}$$

The real and imaginary part of the index of refraction follows as:

$$n(\omega) = \sqrt{\frac{1}{2}(|\varepsilon(\omega)| + |\varepsilon_1(\omega)|)} \tag{6}$$

$$k(\omega) = \sqrt{\frac{1}{2}(|\varepsilon(\omega)| - |\varepsilon_1(\omega)|)}\,. \tag{7}$$

Starting point is the Kubo-Greenwood formula[35] for the dynamic conductivity

$$\sigma(\omega) = \frac{2\pi e^2 \hbar^2}{3 m_e^2 \omega \Omega} \sum_{\mathbf{k}} W(\mathbf{k}) \sum_{j=1}^{N} \sum_{i=1}^{N} \sum_{\alpha=1}^{3} [F(\epsilon_{i,\mathbf{k}}) - F(\epsilon_{j,\mathbf{k}})] \times |\langle \Psi_{j,\mathbf{k}} | \nabla_\alpha | \Psi_{i,\mathbf{k}} \rangle|^2 \delta(\epsilon_{j,\mathbf{k}} - \epsilon_{i,\mathbf{k}} - \hbar\omega)\,, \tag{8}$$

where e is the electron charge and m_e its mass. The summations over i and j run over N descrete bands considered in the electronic structure calculation for the cubic supercell volume Ω. The three spatial directions are averaged by the α sum. $F(\epsilon_{i,\mathbf{k}})$ describes the occupation of the ith band corresponding to the energy $\epsilon_{i,\mathbf{k}}$ and the wavefunction $\Psi_{i,\mathbf{k}}$ at $\mathbf{k}$. The δ-function has to be broadened because a discrete energy spectrum results from the finite simulation volume.[18] Integration over the Brillouin zone is performed by sampling special $\mathbf{k}$ points,[36] where $W(\mathbf{k})$ is the respective weighting factor. Restricting the evaluations to the Baldereschi mean value point[37] yields converged results within 10% accuracy.

The dc conductivity follows from Eq. (8) in the limit $\omega \to 0$. In order to illustrate the capacity of the QMD approach described above, we compare the dc conductivity with isentropic compression experiments of Ternovoi *et al.*[8] performed for He in the range $(15 - 25) \times 10^3$ K and predictions of the chemical model COMPTRA04[38,39] in Fig. 4.

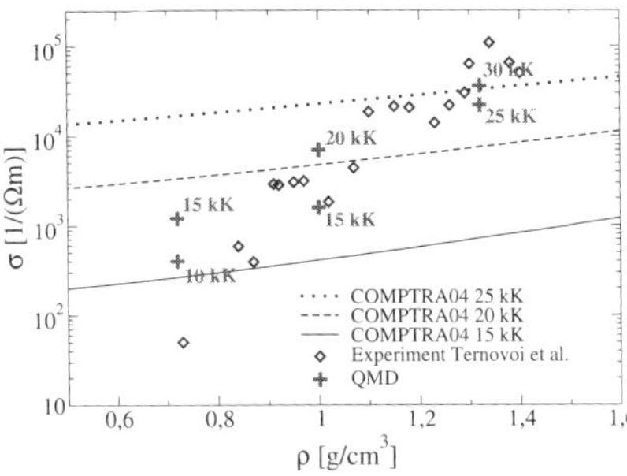

Fig. 4. QMD results for the dc conductivity in helium[15] are compared with shock-wave experiments[8] between $(15 - 25) \times 10^3$ K and isotherms of the COMPTRA04 model;[38,39] temperatures are indicated.

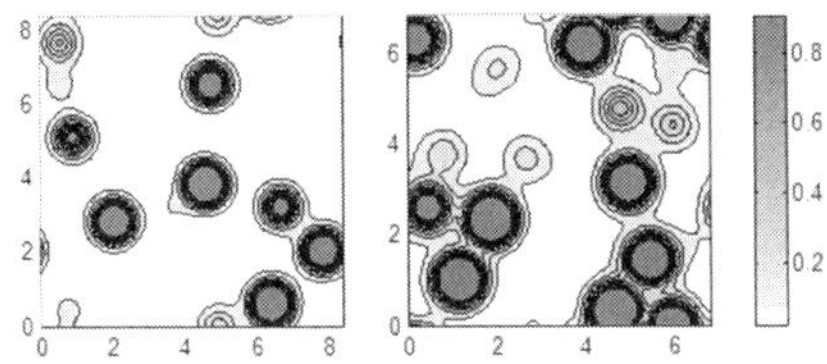

Fig. 5. Contour plot of the charge density[15] in units of $e/\text{Å}^3$ for 0.72 g/cm^3 at 10.000 K (left) and 1.32 g/cm^3 at 30.000 K (right) along a slice through the simulation box. The axis are given in units of Å.

The experimental points show a strong increase between 0.7 and 1.4 g/cm^3 that indicates a nonmetal-to-metal transition. Using the Mott criterion for the minimum metallic conductivity also for finite temperatures,[40] this transition can be located at about 20000/Ωm and 1.3 g/cm^3. The QMD results reproduce the strong increase found experimentally very well except for the lowest density of 0.72 g/cm^3 where the experimental value is substantially lower than the QMD result. This discrepancy stems probably from the band gap problem of DFT. In order to solve this problem, DFT calculations beyond the GGA have to be performed by using, e.g., exact exchange formalisms[41] or quasi-particle calculations.[42] Although the computational requirements of such extended electronic structure calculations in QMD is well beyond the capacity of present resources, their implementation will be an important subject of future work in this area.

Inspection of the density of states (DOS) shows[15] that the gap is slightly reduced with increasing density. The nonmetal-to-metal transition is mainly caused by an increasing number of electronic states which fill up the region of the Fermi energy with increasing temperature so that a higher, metal-like conductivity follows.[12]

The COMPTRA04 code[38,39] calculates the ionization degree and, simultaneously, the electrical conductivity accounting for all scattering processes of free electrons in a partially ionized plasma. This chemical model describes the general trends of the electrical conductivity with the density and temperature as found experimentally.[43] The isotherms displayed in Fig. 4 cover the range of the experimentals and agree with the QMD data so that the nonmetal-to-metal transiton is reproduced qualitatively.

The nonmetal-to-metal transition in helium is also illustrated by means of the charge density with increasing mass density shown in Fig. 5. The electrons are still localized for the lower density representing an atomic fluid (left panel), while transient filaments and clusters occur for the higher density which indicates a metal-like behavior (right panel). A similar behavior has been found for the expanded fluid alkali metals Rb and Cs.[44]

A similar nonmetal-to metal transition can be identified in warm dense hydrogen[32] at about 1 Mbar from the behavior of the electrical conductivity and the reflectivity.

3. Planetary models and EOS of WDM

The interior of giant planets is mainly in WDM states.[45] Besides shock-wave experiments, models for giant planets can be used to check the EOS for planetary materials. Furthermore, modelling giant planets plays a significant role in our understanding of the evolution and formation of planetary systems which has become a great challenge since the detection of extrasolar giant planets in 1995.[46,47] In this section we shortly describe a method applicable to solar giant planets and present new results for Jupiter.

3.1. *General structure of giant planets*

In the simplest model that is capable of matching the observational constraints, a giant planet consists of three layers: two fluid envelopes and a solid core. For Jupiter-like planets, the envelopes are mainly composed of a hydrogen-helium mixture with small amounts of heavier elements, called metals. In the outer envelope, the elements form neutral molecules, e.g. H_2, H_2O, NH_3, CH_4 etc. With increasing pressure up to several Mbar further inside, molecules are pressure ionized, forming a metallic fluid – the 'metallic layer'. With temperatures up to 20000 K, the metallic layer covers the region of WDM in the phase diagram. Both of the envelopes are supposed to be fully convective with adiabatic temperature profiles. They may differ in their content of helium and metals. The core is assumed to consist of isothermal rocks and ices of water, ammonia and methane. The precise values of the parameters describing the planet as, e.g., the size of the core of the fractions of metals are either constraint by observation or inherent results of the planetary model or they are free parameters.

Observational parameters are (values in parenthesis given for Jupiter) the total mass of the planet M (318 $M_{\rm Earth}$), its equatorial radius R_{eq} (11.2 $\rm R_{Earth}$), the temperature T (170 K) at a pressure of 1 bar, the frequency of rotation Ω ($2\pi/9h55$) at the equator, the mass abundance of helium in the molecular layer, the average helium content which is required to equal the helium content of the protosolar cloud (27.5%). From orbits and fly-bys of spacecrafts the values of the first gravitational moments J_{2n} have been derived. The choice of the abundances of metals Z_{mol}, Z_{met} in the envelopes are used to adjust J_2 and J_4. Unfortunately, the large error bars of higher moments do not allow to constrain a model of Jupiter, Saturn, Uranus or Neptune any further.

Such a 3-layer structure model is closely related to major questions concernig the EOS of warm dense hydrogen and helium. A discontinuity in the abundances of helium and metals can result from the existence of a first order plasma phase transition or from demixing and subsequent sedimentation of helium. Both effects are predicted to occur at several megabar and temperatures below 20000 K. As long as the accurate location of these transitions is not known, the transition pressure separating the molecular from the metallic layer is considered as a free parameter.

The resulting values of the free parameters reproducing the observational constraints depend mainly on the compressibility of the EOS for the planetary materials, i.e. for Jupiter-like planets on the EOS of hydrogen, see Section 2.2. We apply here the following hydrogen EOS for the interior model of Jupiter: (i) the standard Sesame data tables[48] which has no PPT, (ii) the chemical model of Saumon, Chabrier, and Van Horn[5] with a PPT (SCVH-ppt) and, alternatively, (iii) an interpolated version without (SCVH-i). Furthermore, we apply the QMD EOS data shown in Section 2.2 (iv).

If not otherwise mentioned, the EOS of metals is represented by the EOS of helium scaled in density by a factor of 4, denoted by He4. We apply linear mixing to combine EOS of different materials. In combination with the Sesame-EOS for hydrogen we use also the Sesame-EOS for helium. To be consistent within chemical models, the SCVH-ppt or SCVH-i EOS for H is combined with the respective He EOS described there.[5] QMD EOS data have been calculated for densities higher than 0.2 g/cm^3. For lower densities, chemical models are applicable so that we interpolate to EOS data calculated within fluid variational theory.[33]

3.2. *Numerical modelling of giant planets*

The interior structure of a giant planet is calculated by numerical integration of the conservation equations for the mass

$$\nabla m = 4\pi r^2 \rho(\vec{r}) \tag{9}$$

and the hydrostatic equation of motion,

$$\frac{1}{\rho}\nabla P = \nabla\left[G\int_{V_0} d^3r' \,\frac{\rho(r')}{|\vec{r'} - \vec{r}|}\right] - \vec{\Omega}\times(\vec{\Omega}\times\vec{r})\,, \tag{10}$$

where the term in brackets is the gravitational potential and the last term the centrifugal force accounting for rotation. In contrast to extremely fast rotating compact objects such as neutron stars where relativity effects predominate, rotation leads to a significant oblateness of giant planets. The deviation of the gravitational potential from spherical symmetry can be expressed by a multipole expansion into Legendre polynomials,

$$V(r,\theta) = \frac{GM}{r}\left(1 - \sum_{i=1}^{\infty}\left(\frac{R_{eq}}{r}\right)^{2i} J_{2i}P_{2i}(\cos\theta)\right)\,, \tag{11}$$

where the expansion coefficients are just the gravitational moments

$$J_{2i} = -\frac{1}{MR_{eq}^{2i}}\int d^3r' r'^{2i}\rho(r')P_{2i}(\cos\theta')\,. \tag{12}$$

They are integrals of the density weighted by some power of the radius and, therefore, are sensitive measures of the mass distribution in the planet. Eq. (12) is not appropriate for the calculation of the graviational moments, because the integrals run over the still unknown volume of the planet. The problem of calculating the shape and the gravitational potential of a body simultaneously can be solved within the *theory of figures* developed by Zharkov and Trubitsyn.[49] For Jupiter at least a third-order approximation is required to ensure that the computational error is less than the measured accuracy for J_4.

The numerical procedure to solve the equations for a given EOS is outlined in detail elsewhere.[50] A certain choice of values for the free parameters defines a certain planetary model. If this model satisfies the observational constraints within their error bars, it is accepted. Properties of different EOS are, thus, mapped onto different ranges of accepted solutions. For some EOS it is not possible to find any acceptable solution which indicates that the respective compressibility is too large or small in the WDM region.

4. Results for Jupiter

Fig. 6 shows the resulting core mass and the total amount of heavy elements of accepted Jupiter models for different EOS. At pressures smaller/higher than the transition pressure a first-order PPT lowers/enhances the compressibility compared with a continuous transition. Therefore more/less fractions of metals are needed to arrange the density profile in order to reproduce J_2 and J_4, see Eq. 12. Due to the large extension of the metallic envelope in Jupiter, a reduced Z_{met} always reduces the total heavy element abundance Z. In the deep interior, a reduced Z requires a larger core mass to conserve the total mass. This opposite case is true for the Sesame EOS. The QMD EOS shows a relatively small compressibility at high pressures that allows for both a large M_c and a large M_Z. Obviously, representing metals by H_2O instead of He4 decreases the compressibility.

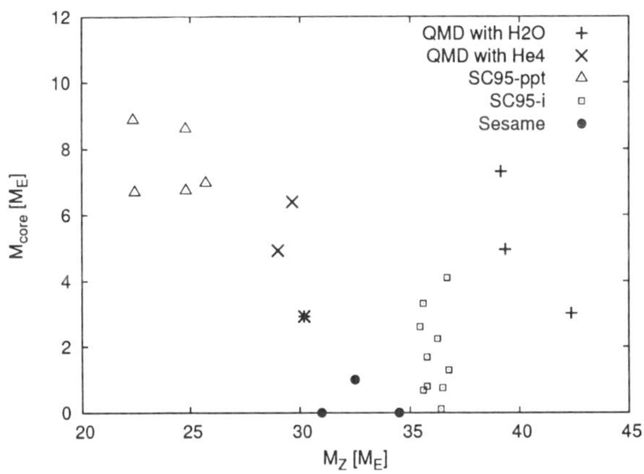

Fig. 6. Core mass and abundance of heavy elements M_Z in units of M_E derived from accepted models for Jupiter using different EOS. Cross, Star: QMD-EOS with metals represented by H_2O, Plus: QMD-EOS with metals represented by He4,[15] Triangle: SCVH-ppt,[5] Box: SCVH-i,[5] Filled Circle: Sesame.[48]

All models that use the more accurate QMD-EOS data in the Mbar range are largely enriched with metals in the inner envelope compared with the outer one. For the model highlighted by a star this is illustrated in Fig. 7 via the mass abundance of the species along the radius.

The species considered are H_2-molecules, neutral or ionized H-atoms, neutral or ionized helium, metals and a core of rocks. A full arc corresponds to 100% in mass. The onset of dissociation begins far out in the molecular

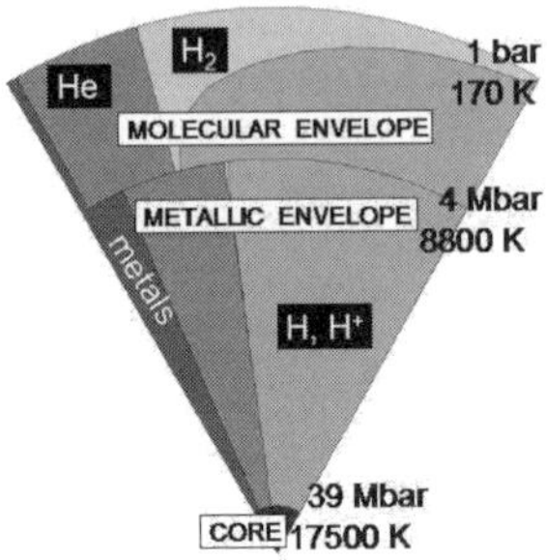

Fig. 7. Mass abundances of various species inside Jupiter using the QMD EOS for hydrogen[32] and helium[15] and a core of rocks.

envelope and is almost completed with the onset of the metallic envelope. The transition pressure has been chosen high enough (4 Mbar) to allow for at least solar abundance of metals in the outer envelope. Observations of single elements indicate an even higher abundance. In the framework of the 3-layer structure model of Jupiter based on QMD data it can be concluded that a discontinuity of metals and helium arises above 4 Mbar which is caused either by H-He demixing at those high pressures or by complete ionization of hydrogen and helium.

5. Conclusions

QMD simulations are capable of treating WDM states where correlations and quantum effects are important. A broad spectrum of physical properties can be determined within an integrated *ab initio* approach. New accurate results for the thermodynamic, structural, and transport properties give insight into the behavior of matter under extreme conditions and allow to solve fundamental problems such as the phase diagram of matter at high pressure. This is of major importance for the understanding of the interior structure of giant planets. We have concentrated here on Jupiter which is a laboratory for WDM physics. The observational constraints impose strong conditions for the H-EOS applied, especially for the region near the molecular-metallic transition. Due to uncertainties in the H-EOS and the large error bars of the higher-order gravitational moments, the exact size of the core and the amount of metals in Jupiter and Saturn are still not known. On the other hand, only few published H-EOS are compatible with a realistic model for Jupiter. Higher compressibility at Mbar pressures results in lower abundances of metals and a larger core mass. Thus limits for the allowed compressibility of the H-EOS are given by the condition of non-vanishing content of metals and non-vanishing core mass. The model introduced in this paper can be applied also to other giant planets such as Saturn, Uranus, and Neptune as well as to extrasolar planets.

Acknowledgments

We thank T.R. Mattsson, M.P. Desjarlais, and D. Blaschke for stimulating discussions. This work was supported by the Deutsche Forschungsgemeinschaft (DFG) within the Graduiertenkolleg GRK 567 and the Sonderforschungsbereich SFB 652.

References

1. R.W. Lee, D. Kalantar and J. Molitoris (editors), *Warm Dense Matter: An Overview*, LLNL Report UCRL-TR-203844 (2004).
2. R. Redmer, Phys. Rep. **282**, 35 (1997).
3. S. Ichimaru, *Plasma Physics*, (Addison-Wesley, Redwood City/CA, 1988).
4. J.D. Lindl, *Inertial Confinement Fusion* (Springer, New York, 1998).
5. D. Saumon, G. Chabrier and H.M. Van Horn, Astrophys. J. Suppl. Ser. **99**, 713 (1995).
6. M.A.C. Perryman, Rep. Prog. Phys. **63**, 1209 (2000). See also http://exoplanet.eu for up to date information.
7. S.T. Weir, A.C. Mitchell and W.J. Nellis, Phys. Rev. Lett. **76**, 1860 (1996).
8. V.Ya. Ternovoi *et al.*, *Shock Compression of Condensed Matter 2001*, Eds. M.D. Furnish, N.N. Thadhani and Y. Horie, AIP Conf. Proc. **620**, 107 (2002).
9. T.R. Mattsson and M.P. Desjarlais, Phys. Rev. Lett. **97**, 017801 (2006).
10. D.M. Ceperley, Rev. Mod. Phys. **67**, 279 (1995).
11. D. Klakow, C. Toepffer and P.-G. Reinhard, J. Chem. Phys. **101**, 10766 (1994); M. Knaup, G. Zwicknagel, P.-G. Reinhard and C. Toepffer, J. Phys. A: Math. Gen. **36**, 6165 (2003).
12. L.A. Collins *et al.*, Phys. Rev. **B63**, 184110 (2001).
13. M.P. Desjarlais, Phys. Rev. **B68**, 064204 (2003).
14. B. Militzer, Phys. Rev. Lett. **97**, 175501 (2006).
15. A. Kietzmann, B. Holst, R. Redmer, M.P. Desjarlais and T.R. Mattsson, Phys. Rev. Lett. **98**, 190602 (2007).
16. C. Cavazzoni *et al.*, Science **283**, 44 (1999).
17. A.E. Mattsson *et al.*, Model. Simul. Mater. Sci. Eng. **13**, R1 (2005).
18. M.P. Desjarlais, J.D. Kress and L.A. Collins, Phys. Rev. **E66**, 025401 (2002).
19. Y. Laudernet, J. Clérouin and S. Mazevet, Phys. Rev. **B70**, 165108 (2004).
20. J. Clérouin *et al.*, Phys. Rev. **B71**, 064203 (2005).
21. N.D. Mermin, Phys. Rev. **137**, A1441 (1965).

22. G. Kresse and J. Hafner, Phys. Rev. **B47**, 558 (1993); **49**, 14251 (1994); G. Kresse and J. Furthmüller, Phys. Rev. **B54**, 11169 (1996).
23. P.E. Blöchl, Phys. Rev. **B50**, 17953 (1994); G. Kresse and D. Joubert, Phys. Rev. **B59**, 1758 (1999).
24. J.P. Perdew, K. Burke and M. Ernzerhof, Phys. Rev. Lett. **77**, 3865 (1999).
25. B. Holst, N. Nettelmann and R. Redmer, Contrib. Plasma Phys. **47**, 368 (2007).
26. C. Winisdoerffer and G. Chabrier, Phys. Rev. **E71**, 026402 (2005).
27. A. Förster, T. Kahlbaum and W. Ebeling, Laser Part. Beams **10**, 253 (1992); T. Kahlbaum and A. Förster, Fluid Phase Equil. **76**, 71 (1992).
28. M. Schlanges, M. Bonitz and A. Tschttschjan, Contrib. Plasma Phys. **35**, 109 (1995).
29. W.-D. Kraeft, D. Kremp, W. Ebeling and G. Röpke, *Quantum Statistics of Charged Particle Systems*, (Akademie-Verlag, Berlin, 1986).
30. R. Redmer, G. Röpke and R. Zimmermann, J. Phys. B: At. Mol. Phys. **20**, 4069 (1987).
31. D. Saumon and T. Guillot, Astrophys. J. **609**, 1170 (2004).
32. B. Holst, R. Redmer and M.P. Desjarlais, unpublished.
33. H. Juranek *et al.* Contrib. Plasma Phys. **45**, 432 (2005).
34. W.J. Nellis, Rep. Prog. Phys. **69**, 1479 (2006).
35. R. Kubo, J. Phys. Soc. Jpn. **12**, 570 (1957); D.A. Greenwood, Proc. Phys. Soc. London **71**, 585 (1958).
36. H.J. Monkhorst and J.D. Pack, Phys. Rev. **B13**, 5188 (1976).
37. A. Baldereschi, Phys. Rev. **B7**, 5212 (1973).
38. S. Kuhlbrodt, B. Holst and R. Redmer, Contrib. Plasma Phys. **45**, 73 (2005).
39. S. Kuhlbrodt *et al.*, Contrib. Plasma Phys. **45**, 61 (2005).
40. P.P. Edwards *et al.*, Phil. Trans. R. Soc. Lond. **A356**, 5 (1998).
41. R.P. Muller and M.P. Desjarlais, J. Chem. Phys. **125**, 054101 (2006).
42. P. Rinke *et al.*, New Journal of Physics **7**, 126 (2005); S.V. Faleev *et al.*, Phys. Rev. **B74**, 033101 (2006).
43. V.E. Fortov *et al.*, J. Exp. Theor. Phys. **97**, 259 (2003).
44. A. Kietzmann *et al.*, J. Phys.: Condens. Matter **18**, 5597 (2006).
45. R. Redmer, Plasmas in Planetary Interiors, in: Lecture Notes in Physics, Vol. 670, Eds. A. Dinklage, T. Klinger, G. Marx and L. Schweikhard (Springer, Berlin, 2005), p. 331-348.
46. T. Guillot, Ann. Rev. Earth & Plan. Sci. **33**, 493 (2005).
47. W.B. Hubbard, J.J. Fortney and J.I. Lunine, Astrophys. J. **560**, 413 (2001).
48. G.I. Kerley, Los Alamos Scientific Laboratory Report LA-4776 (Los Alamos, 1972).
49. V.N. Zharkov and V.P. Trubitsyn, *Physics of Planetary Interiors*, in AASeries, edited by W.B. Hubbard (Parchart, Tucson/AZ, 1978).
50. N. Nettelmann, R. Redmer and D. Blaschke, Phys. Part. Nucl. Lett. (2007).

CORRELATED SYSTEMS ON GEOMETRICALLY FRUSTRATED LATTICES: FROM MAGNONS TO ELECTRONS

J. RICHTER[a] and O. DERZHKO[b]

[a] *Institut für Theoretische Physik, Universität Magdeburg*
P.O. Box 4120, 39016 Magdeburg, Germany
[b] *Institute for Condensed Matter Physics*
of the National Academy of Sciences of Ukraine
1 Svientsitskii Street, L'viv-11, 79011, Ukraine
E-mail: Johannes.Richter@Physik.Uni-Magdeburg.DE

In this paper we review recent studies on localized eigenstates of correlated systems on various one-, two-, and three-dimensional frustrated lattices and discuss their influence on the low-temperature physics. We focus on frustrated quantum Heisenberg antiferromagnets where these localized magnon eigenstates become ground states in high magnetic fields. In addition, we consider localized electron states for the Hubbard model on a specific frustrated lattice (the sawtooth chain) and discuss their relation to the localized magnon states.

1. Introduction

The concept of frustration plays an important role in the search for novel quantum states of condensed matter. Though the term frustration probably was introduced in physics only in the 1970ies by Toulouse[1] in the context of spin glasses[2] this concept dates back to the 1950ies, when Wannier[3] and Houtappel[4] considered the Ising antiferromagnet on the triangular lattice. They already found that the triangular arrangement of antiferromagnetic interactions (which is called frustration in present time) leads to an extensive ground-state entropy. Later on the studies on spin glasses have demonstrated that frustration may have an enormous influence on ground-state and thermodynamic properties[2] of spin systems.

In the 1970ies Anderson and Fazekas[5] first considered the quantum spin-1/2 Heisenberg antiferromagnet on a geometrically frustrated lattice, namely the triangular lattice. They proposed a liquid-like ground state without magnetic long-range order. Though it was found later that the spin-

1/2 Heisenberg antiferromagnet on the triangular lattice possesses semiclassical three-sublattice Néel order Anderson's suggestion was the starting point to search for exotic quantum ground states in frustrated spin systems.

Stimulated by the recent progress in synthesizing frustrated magnetic materials with strong quantum fluctuations[6] and by the rich behavior of such magnetic systems there is presently an enormous interest in frustrated quantum spin systems, see e.g., Refs. [7–11]. There are many compounds which correspond to quantum antiferromagnetic models with frustrated spin interactions. We mention as examples: (i) in one dimension the frustrated $J_1 - J_2$ chains ($LiCuVO_4$, Li_2CuO_2, $NaCu_2O_2$),[12] and the frustrated diamond chain ($Cu_3(CO_3)_2(OH)_2$),[13,14] (ii) in two dimensions the frustrated $J_1 - J_2$ square lattice (Li_2VOSiO_4, Li_2VOGeO_4),[15,16] the triangular lattice (Cs_2CuCl_4),[17] the kagomé lattice ($ZnCu_3(OH)_6Cl_2$),[18] the depleted $J_1 - J_2$ square lattice (CaV_4O_9),[19] and the Shastry-Sutherland lattice ($SrCu_2(BO_3)_2$),[20–22] (iii) in three dimensions the pyrochlore lattice ($MgTi_2O_4$).[23] There are also compounds which correspond to electronic (Hubbard, $t - J$, periodic Anderson) models on geometrically frustrated lattices. We mention as examples cobaltates,[24] $CeRh_3B_2$,[25] as well as a possible application of the electronic models for artificial crystals from quantum dots and for atomic quantum wires.[26]

In this paper we will focus on a special property of antiferromagnetic Heisenberg models and some other strongly correlated models on certain geometrically frustrated lattices, namely the existence of localized eigenstates (on perfect lattices) and their relevance for the low-temperature physics of those correlated systems. In general, for perfect lattices an elementary excitation as a noninteracting quasiparticle is spread all over the whole quantum system. For example, a magnon wave function for a simple hypercubic lattice is extended over all the lattice. The same is true for an electron wave function for a simple electronic Hamiltonian on such a lattice. The reason for this is a hopping term in the Hamiltonian which leads to a wave function having finite amplitudes at all sites of the lattice. However, for some lattice geometries owing to destructive quantum interference a wave function of an elementary excitation in a quantum system may have amplitudes which are different from zero in a restricted area, only. It is natural to call such excitations localized excitations (for example, localized magnons or localized electron states). Because of the local character of these excitations one immediately realizes that several independent (i.e. not interacting) localized excitations are again exact many-particle eigenstates of the Hamiltonian even in the presence of interaction. It is also clear that the number of localized excitations cannot exceed a certain maximal value which should depend on the specific lattice in question, and should be proportional to the system size. In the case when one localized excitation is the lowest-energy eigenstate of the Hamiltonian in the one-particle subspace one may expect (provided there is no attractive interaction) that a state with n independent (isolated) localized excitations is the lowest-energy eigenstate of the Hamiltonian in the corresponding n-particle subspace.[27–29] Under certain conditions these states may become the ground states of the model and therefore may substantially contribute to or even completely determine the low-temperature thermodynamic properties of the correlated system. In the present paper we review the effect of localized elementary excitations on the low-temperature thermodynamics focusing mainly on the quantum Heisenberg antiferromagnet and briefly touching the Hubbard model. Although all material presented below can be found in original papers[27–48] (for electronic models see Refs. [49–51] and also Ref. [52]) and in several review papers,[11,30,41,42,47] the present survey differs from the previous reviews in being more pedagogical in character.

In this paper we deal mainly with the quantum Heisenberg antiferromagnet of N spins with quantum number s (in most cases $s = 1/2$) with the Hamiltonian

$$H = \sum_{\langle i,j \rangle} J_{ij} \vec{s}_i \cdot \vec{s}_j - hS^z. \tag{1}$$

Here the sum runs over all neighboring sites on the lattice under consideration, $J_{ij} > 0$ is the antiferromagnetic isotropic Heisenberg exchange interaction between the sites i and j, h is an external magnetic field, and $S^z = \sum_i s_i^z$ is the z-component of the total spin. Note that our consideration is also valid for the anisotropic XXZ exchange interaction,[27–29] i.e. $\vec{s}_i \cdot \vec{s}_j \to (s_i^+ s_j^- + s_i^- s_j^+)/2 + \Delta s_i^z s_j^z$ in Eq. (1). However, in the present paper we focus on the isotropic case $\Delta = 1$.

In addition, we also discuss in this paper the Hubbard model

$$H = \sum_{\langle i,j \rangle, \sigma=\uparrow,\downarrow} t_{ij} \left(c_{i,\sigma}^\dagger c_{j,\sigma} + c_{j,\sigma}^\dagger c_{i,\sigma} \right) + U \sum_i n_{i,\uparrow} n_{i,\downarrow} + \mu \sum_{i,\sigma=\uparrow,\downarrow} n_{i,\sigma}. \tag{2}$$

Here $t_{i,j} > 0$ is the hopping matrix element between the nearest-neighbor sites i and j, $U > 0$ is the on-site Coulomb repulsion, μ is the chemical potential of electrons, and $n_{i,\sigma} = c_{i,\sigma}^\dagger c_{i,\sigma}$.

We consider various one-, two- and three-dimensional geometrically frustrated lattices, such as the sawtooth chain, the diamond chain, the kagomé-like chains, the dimer-plaquette chain, the frustrated ladder, the frustrated bilayer, the kagomé lattice, the square-kagomé lattice, the star lattice, the checkerboard lattice, the pyrochlore lattice etc. Some of the considered lattices are shown in Figs. 1, 2.

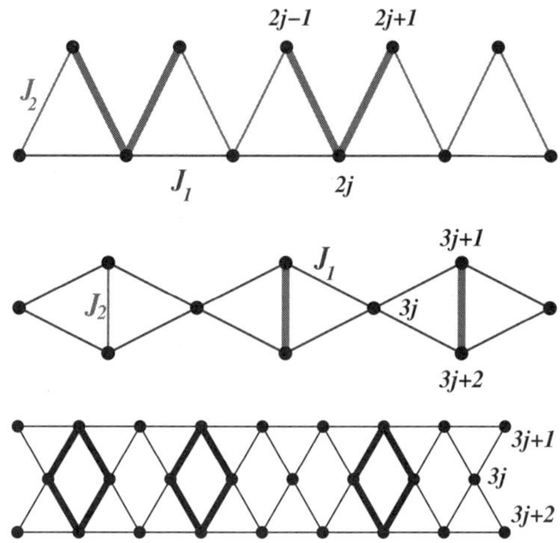

Fig. 1. Some of the lattices considered in this paper. From top to bottom: the sawtooth chain, the diamond chain, the kagomé-like chain. The circles indicate the lattice sites, the lines indicate the exchange bonds. Localized magnons are indicated by bold lines. Note that for the sawtooth and the diamond chain one has two kinds of bonds of different strength, whereas for the kagomé-like chain all bonds have the same strength.

As we have mentioned above, because of frustration the Heisenberg antiferromagnet on such lattices exhibits rich physics. New features appear when we apply an external magnetic field on a frustrated antiferromagnet. The Zeeman interaction of the spins with the magnetic field is an additional source of competition between different interactions in the system. As a result, frustrated quantum antiferromagnets show new features such as plateaus and jumps in the magnetization vs magnetic field curve (see the magnetization curves for the azurite[13,14] (Fig. 3, top) and for $SrCu_2(BO_3)_2$[20–22] (Fig. 3, bottom)). There are also other peculiarities in the low-temperature strong-field regime of frustrated quantum antiferromagnet that attract much attention nowadays (e.g., the temperature behavior of the specific heat near the saturation field[53]). For electronic models the chemical potential μ plays the role of the magnetic field h. Here μ controls the average number of electrons in the system.

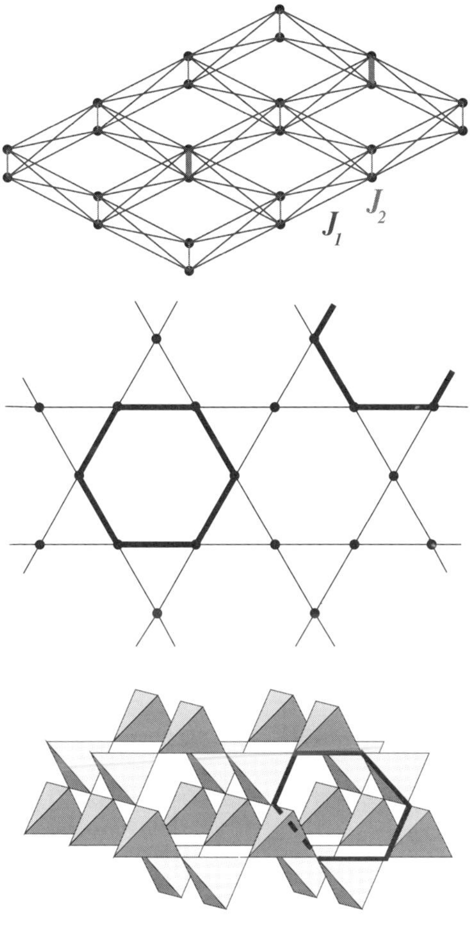

Fig. 2. Some of the lattices considered in this paper. From top to bottom: the frustrated bilayer, the kagomé lattice, the pyrochlore lattice. The circles indicate the lattice sites (for the pyrochlore lattice the lattice sites are located at the corners of the tetrahedra), the lines indicate the exchange bonds. Localized magnons are indicated by bold lines. Note that for the bilayer system one has two kinds of bonds of different strength, whereas for the kagomé and the pyrochlore lattices all bonds have the same strength.

In what follows we discuss some generic properties of frustrated quantum Heisenberg antiferromagnets which are caused by the localized magnon states. In particular, we consider the magnetization process, a field-induced spin-Peierls transition, and the huge ground-state degeneracy of the localized eigenstates leading to a finite residual entropy. We also discuss the low-

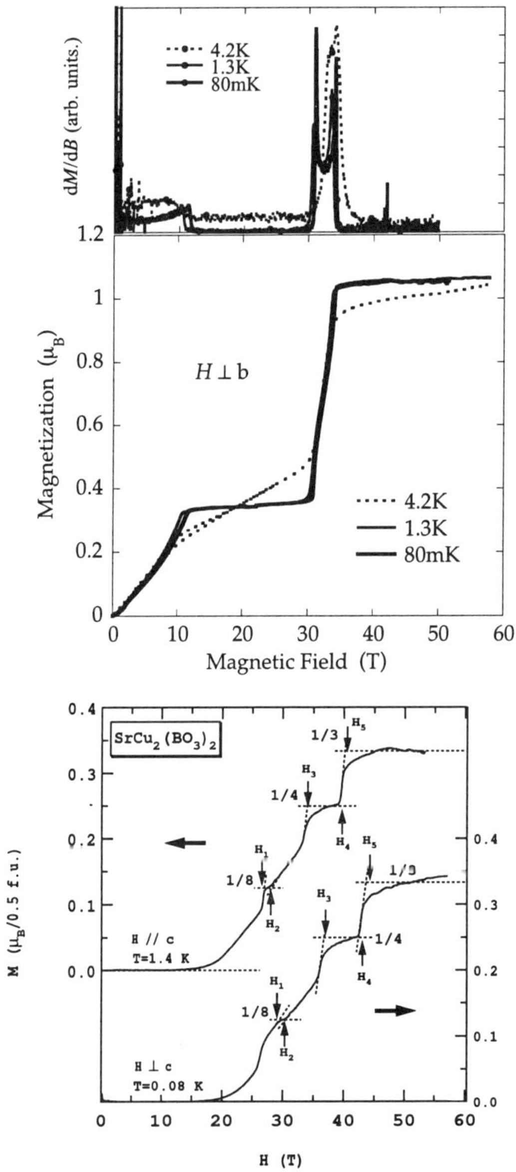

Fig. 3. Magnetization curves for the azurite (top) (this panel corresponds to Fig. 2 from Ref. [14]) and for $SrCu_2(BO_3)_2$ (bottom) (this panel corresponds to Fig. 2 from Ref. [22]). We thank H. Kikuchi and H. Kageyama for providing these figures.

temperature thermodynamics emphasizing a universal behavior of different systems in the vicinity of the saturation field (Sec. 2). A brief illustration of the application of the concepts elaborated for spin systems on electronic (Hubbard) systems is given in Sec. 3. A short summary of our discussion is presented in Sec. 4.

2. Localized magnon states in antiferromagnets on frustrated lattices

2.1. *Flat bands and localized eigenstates*

Let us demonstrate how the localized magnon states emerge for the frustrated quantum Heisenberg antiferromagnet (1). Since S^z commutes with the Hamiltonian (1) we can discuss the eigenstates of the spin model separately in each subspace with different values of $S^z = sN, sN-1, \ldots$. In the subspace with $S^z = sN$ the only eigenstate is the fully polarized ferromagnetic state, $|\mathrm{FM}\rangle = |s, s, s, \ldots\rangle$, which plays the role of the vacuum state for the magnon excitations to be considered below.

In the subspace with $S^z = sN - 1$ (one-magnon subspace) we can use the Holstein-Primakoff transformation to Bose operators, $s^+ \to \sqrt{2s}b$, $s^- \to \sqrt{2s}b^\dagger$, $s^z \to s - b^\dagger b$ taking into account that in the one-magnon subspace the interaction between magnons is irrelevant. Omitting some constant terms which are not essential for further diagonalization the bosonic Hamiltonian then reads

$$H = \sum_{\langle i,j \rangle} sJ_{ij} \left(b_i^\dagger b_j + b_i b_j^\dagger - \Delta \left(b_i^\dagger b_i + b_j^\dagger b_j \right) \right) + h \sum_i b_i^\dagger b_i \,. \tag{3}$$

We can find the eigenstates of the Hamiltonian after bringing H into a diagonal form. Note that on the one-magnon level the energy obtained from (3) is exact (see Eqs. (7), (12), (13), (14) given below).

To be specific, consider, for example, the sawtooth chain with two values of the exchange integrals, J_1 along the base line and J_2 along the zig-zag path (see Fig. 1). Eq. (3) now has the form

$$\begin{aligned} H = \sum_{j=0}^{\frac{N}{2}-1} \Big(& sJ_1 \left(b_{2j}^\dagger b_{2j+2} + b_{2j+2}^\dagger b_{2j} \right) \\ & + sJ_2 \left(b_{2j}^\dagger b_{2j+1} + b_{2j+1}^\dagger b_{2j} + b_{2j+1}^\dagger b_{2j+2} + b_{2j+2}^\dagger b_{2j+1} \right) \\ & + \left(h - s\Delta \left(J_1 + J_2 \right) \right) b_{2j}^\dagger b_{2j} + \left(h - 2s\Delta J_2 \right) b_{2j+1}^\dagger b_{2j+1} \\ & - s\Delta \left(J_1 + J_2 \right) b_{2j+2}^\dagger b_{2j+2} \Big) \,. \end{aligned} \tag{4}$$

After performing the Fourier transformation,

$$A_\kappa = \sqrt{\frac{2}{N}} \sum_{j=0}^{\frac{N}{2}-1} \exp(\mathrm{i}\kappa j)\, b_{2j}, \; B_\kappa = \sqrt{\frac{2}{N}} \sum_{j=0}^{\frac{N}{2}-1} \exp(\mathrm{i}\kappa j)\, b_{2j+1},$$

$$b_{2j} = \sqrt{\frac{2}{N}} \sum_{\kappa} \exp(-\mathrm{i}\kappa j)\, A_\kappa, \; b_{2j+1} = \sqrt{\frac{2}{N}} \sum_{\kappa} \exp(-\mathrm{i}\kappa j)\, B_\kappa, \quad (5)$$

where $\kappa = 4\pi m/N$, $m = 0, 1, \ldots, N/2-1$ (i.e. $\kappa \in [0, 2\pi[$ if $N \to \infty$), in matrix notations we have

$$H = \sum_\kappa \begin{pmatrix} A_\kappa^\dagger & B_\kappa^\dagger \end{pmatrix} \begin{pmatrix} H_{11} & H_{12} \\ H_{21} & H_{22} \end{pmatrix} \begin{pmatrix} A_\kappa \\ B_\kappa \end{pmatrix},$$
$$H_{11} = h - 2s\Delta\,(J_1 + J_2) + 2sJ_1 \cos\kappa, \; H_{12} = sJ_2\,(1 + \exp(\mathrm{i}\kappa)),$$
$$H_{21} = sJ_2\,(1 + \exp(-\mathrm{i}\kappa)), \; H_{22} = h - 2s\Delta J_2. \quad (6)$$

Next we consider a unitary transformation which diagonalizes the 2×2 matrix in Eq. (6)

$$\mathbf{U} \begin{pmatrix} H_{11} & H_{12} \\ H_{21} & H_{22} \end{pmatrix} \mathbf{U}^\dagger = \begin{pmatrix} \varepsilon_-(\kappa) & 0 \\ 0 & \varepsilon_+(\kappa) \end{pmatrix},$$
$$\varepsilon_\pm(\kappa) = h - s\Delta\,(J_1 + 2J_2) + sJ_1 \cos\kappa$$
$$\pm s\sqrt{J_1^2\,(-\Delta + \cos\kappa)^2 + 2J_2^2\,(1 + \cos\kappa)}. \quad (7)$$

After introducing the new operators

$$\begin{pmatrix} \alpha_\kappa \\ \beta_\kappa \end{pmatrix} = \mathbf{U} \begin{pmatrix} A_\kappa \\ B_\kappa \end{pmatrix} \quad (8)$$

the Hamiltonian (6) becomes

$$H = \sum_\kappa \left(\varepsilon_-(\kappa) \alpha_\kappa^\dagger \alpha_\kappa + \varepsilon_+(\kappa) \beta_\kappa^\dagger \beta_\kappa \right). \quad (9)$$

As can be seen from the expression for the magnon energy (7) for a special relation between J_1 and J_2, $J_2 = \sqrt{2(1+\Delta)}J_1$, the lower magnon band becomes completely flat, i.e. $\varepsilon_-(\kappa) = \varepsilon_- = h - 2s\left(\Delta\left(1 + \sqrt{2(1+\Delta)}\right) + 1\right) J_1$, whereas $\varepsilon_+(\kappa) = h + 2s\left(-\Delta\sqrt{2(1+\Delta)} + \cos\kappa + 1\right) J_1$. For the unitary transformation we find

$$\mathbf{U} = \sqrt{\frac{1+\Delta}{2+\Delta+\cos\kappa}} \begin{pmatrix} 1 & -\frac{1+\exp(\mathrm{i}\kappa)}{\sqrt{2(1+\Delta)}} \\ \frac{1+\exp(-\mathrm{i}\kappa)}{\sqrt{2(1+\Delta)}} & 1 \end{pmatrix}. \quad (10)$$

That leads to the following expressions for the creation/annihilation operators of a magnon from the lowest-energy flat band

$$\begin{aligned}
\alpha_\kappa^\dagger &= -\frac{1}{\sqrt{2sN}}\frac{1}{\sqrt{2+\Delta+\cos\kappa}}\sum_{j=0}^{\frac{N}{2}-1}\exp\left(-\mathrm{i}\kappa j\right)l_{2j}^\dagger,\\
l_{2j}^\dagger &= s_{2j-1}^- - \sqrt{2(1+\Delta)}s_{2j}^- + s_{2j+1}^-;\\
\alpha_\kappa &= -\frac{1}{\sqrt{2sN}}\frac{1}{\sqrt{2+\Delta+\cos\kappa}}\sum_{j=0}^{\frac{N}{2}-1}\exp\left(\mathrm{i}\kappa j\right)l_{2j},\\
l_{2j} &= s_{2j-1}^+ - \sqrt{2(1+\Delta)}s_{2j}^+ + s_{2j+1}^+.
\end{aligned} \tag{11}$$

In the usual magnon description the one-particle states belonging to the lowest (flat) band have the form $\alpha_\kappa^\dagger|\mathrm{FM}\rangle$ for $N/2$ values of κ. All these $N/2$ states are degenerate and have the energy ε_-. Due to this degeneracy an arbitrary linear combination of the states $\alpha_\kappa^\dagger|\mathrm{FM}\rangle$ is again a ground state (which is, in general, not characterized by a wave vector κ). In particular, $|1\mathrm{lm}\rangle_{2j} = l_{2j}^\dagger|\mathrm{FM}\rangle$ is also a ground state with a magnon (spin flip) located on a 'valley' of three contiguous sites $2j-1$, $2j$ and $2j+1$, cf. Fig. 1. The set of such states with $j = 0, 1, \ldots, N/2-1$ represents also the complete manifold of ground states in the one-magnon subspace. These states are called the localized magnon states.

We can repeat the calculations explained above for other lattices. For the diamond chain (see Fig. 1) we find

$$\begin{aligned}
H &= \sum_\kappa \left(\varepsilon_1(\kappa)\alpha_\kappa^\dagger\alpha_\kappa + \varepsilon_2(\kappa)\beta_\kappa^\dagger\beta_\kappa + \varepsilon_3(\kappa)\gamma_\kappa^\dagger\gamma_\kappa\right);\\
\varepsilon_1(\kappa) &= h - 2s\Delta J_1 - s\left(\Delta+1\right)J_2,\\
\varepsilon_{2,3}(\kappa) &= h - 3s\Delta J_1 - \frac{1}{2}s\left(\Delta-1\right)J_2\\
&\quad \pm s\sqrt{\left(\Delta J_1 - \frac{1}{2}\left(\Delta-1\right)J_2\right)^2 + 4J_1^2\left(1+\cos\kappa\right)}\ .
\end{aligned} \tag{12}$$

The flat band $\varepsilon_1(\kappa) = \varepsilon_1$ becomes the lowest one when $J_2 > 2J_1$. The localized magnon creation operators are given by $l_{3j}^\dagger = s_{3j+2}^- - s_{3j+1}^-$, $j = 0, 1, \ldots, N/3-1$. For the kagomé-like chain with three sites in the elementary cell (see Fig. 1) we have

$$\begin{aligned} H &= \sum_{\kappa} \left(\varepsilon_1(\kappa)\alpha_\kappa^\dagger \alpha_\kappa + \varepsilon_2(\kappa)\beta_\kappa^\dagger \beta_\kappa + \varepsilon_3(\kappa)\gamma_\kappa^\dagger \gamma_\kappa\right); \\ \varepsilon_1(\kappa) &= h - 2sJ\,(2\Delta + 1)\,, \\ \varepsilon_2(\kappa) &= h - 2sJ\,(2\Delta - \cos\kappa)\,, \\ \varepsilon_3(\kappa) &= h - 2sJ\,(2\Delta - 1 - \cos\kappa)\,. \end{aligned} \tag{13}$$

Again the lowest band $\varepsilon_1(\kappa) = \varepsilon_1$ is flat. The localized magnon creation operators are given by $l_{3j}^\dagger = s_{3j}^- - s_{3j+2}^- + s_{3j+3}^- - s_{3j+1}^-$, $j = 0, 1, \ldots, N/3-1$. From Eq. (13) we see that one propagating state $\kappa = \pi$ with the energy $\varepsilon_2(\pi)$ has the same energy as the localized states ε_1 (for finite N this is true only if N is even).

We can perform similar calculations in the two-dimensional case. For example, for the frustrated bilayer (see Fig. 2)

$$\begin{aligned} H &= \sum_{\mathbf{k}} \left(\varepsilon_1(\mathbf{k})\alpha_\mathbf{k}^\dagger \alpha_\mathbf{k} + \varepsilon_2(\mathbf{k})\beta_\mathbf{k}^\dagger \beta_\mathbf{k}\right); \\ \varepsilon_1(\mathbf{k}) &= h - s\,(J_2 + \Delta(8J_1 + J_2))\,, \\ \varepsilon_2(\mathbf{k}) &= h + s\,(4J_1\,(\cos k_x + \cos k_y) + J_2 - \Delta(8J_1 + J_2))\,. \end{aligned} \tag{14}$$

Here $\mathbf{k} = (k_x, k_y)$, $k_\alpha = 2\pi n_\alpha / M_\alpha$, $n_\alpha = 0, 1, 2, \ldots, M_\alpha - 1$, $\alpha = x, y$, $M_x M_y = N/2$. The flat band $\varepsilon_1(\mathbf{k}) = \varepsilon_1$ becomes the lowest one when $J_2 > 4J_1$. The localized magnon states can be generated by applying the operators $l_{\vec{j}}^\dagger = s_{\vec{j},d}^- - s_{\vec{j},u}^-$ on the vacuum state $|\mathrm{FM}\rangle$, where $\vec{j} = (j_x, j_y)$ enumerates the sites of the underlying square lattice and the indices d and u refer to the lower and the upper planes, respectively.

An alternative approach to explain how a magnon excitation can be localized due to a lattice topology is based on the following arguments. Consider for concreteness the two-dimensional kagomé lattice (see Fig. 2). Let us split the Hamiltonian (1) (with $h = 0$) into the three parts, $H = H_L + H_{L-R} + H_R$, where H_L refers to the bonds connecting six sites which form a hexagon (i.e. the area where the magnon is localized), H_{L-R} refers to the twelve bonds connecting the hexagon with the rest of the lattice, and H_R refers to all other bonds on the lattice. Now, using the well-known relations $s^\pm|s, s^z\rangle = \sqrt{s(s+1) - s^z(s^z \pm 1)}|s, s^z \pm 1\rangle$ and $s^z|s, s^z\rangle = s^z|s, s^z\rangle$ (in the l.h.s. of the latter relation s^z stands for the operator, whereas in the r.h.s. for its eigenvalue) we can check that the one-magnon state

$$\begin{aligned} |1\mathrm{lm}\rangle \propto (&|s-1_1, s_2, s_3, s_4, s_5, s_6\rangle - |s_1, s-1_2, s_3, s_4, s_5, s_6\rangle \\ &+ \ldots - |s_1, s_2, s_3, s_4, s_5, s-1_6\rangle)\,|s, s, \ldots\rangle_e \end{aligned} \tag{15}$$

is an eigenstate of the spin Hamiltonian (1). In Eq. (15) the indices $1, 2, \ldots, 6$ belong to the sites of the hexagon where the spin excitation is localized and $|s, s, \ldots\rangle_e$ stands for the spins on the (fully polarized) rest of the lattice. A straightforward calculation yields

$$\begin{aligned} H_L|1\mathrm{lm}\rangle &= (-2sJ + 2s(3s-1)\Delta J)\,|1\mathrm{lm}\rangle, \\ H_{L-R}|1\mathrm{lm}\rangle &= 2s(6s-1)\Delta J|1\mathrm{lm}\rangle, \\ H_R|1\mathrm{lm}\rangle &= (2N-18)s^2\Delta J|1\mathrm{lm}\rangle, \\ (H_L + H_{L-R} + H_R)\,|1\mathrm{lm}\rangle &= (E_{\mathrm{FM}} - \varepsilon_1)\,|1\mathrm{lm}\rangle, \\ E_{\mathrm{FM}} &= 2Ns^2\Delta J, \quad \varepsilon_1 = 2s(1+2\Delta)J. \end{aligned} \tag{16}$$

In this derivation one can follow the role of destructive quantum interference that occurs due to the topology of the considered lattice. Equilateral triangles (which for the antiferromagnetic sign of exchange couplings lead to geometrical frustration) attached to a polygon (hexagon) are important for the localization of the excitation. The antiferromagnetic sign of the exchange couplings is required to have the lowest energy state.

Let us consider the two-magnon subspace with $S^z = sN - 2$. In this subspace the construction of the eigenstates of the Heisenberg model is, generally, a more difficult problem, see e.g. Ref. [54]. However, for the lattices which support localized magnon states one immediately concludes that a state consisting of two independent (i.e. isolated) localized magnons is an exact eigenstate of the Hamiltonian (1). Using the $l^\dagger$-operators introduced above these states can be written as $l_i^\dagger l_j^\dagger|\mathrm{FM}\rangle$, where i and j are lattice sites separated from each other sufficiently. In general, the construction of n-magnon states is not possible in a rigorous way. However, for lattices having localized one-magnon eigenstates one observes that a state built by n independent localized magnons is an exact eigenstate of the Hamiltonian (1) in the n-magnon subspace with $S^z = sN - n$ for n up to a maximum number localized magnons $n_{\max}$, where $n_{\max}$ is proportional to the number of spins N and depends on the specific lattice geometry. For example, we have $n_{\max} = N/4$ for the sawtooth chain, $n_{\max} = N/3$ for the diamond chain, $n_{\max} = N/9$ for the kagomé lattice etc. The energy of the n-particle state $|n\mathrm{lm}\rangle$ built by n independent localized magnons is $E_{\mathrm{FM}} - n\varepsilon_1$ (for $h = 0$). Since the one-magnon states are the lowest-energy states in the one-magnon subspace we may expect that the states with n independent localized magnons are the lowest-energy states in the n-magnon subspace. Rigorous analysis confirms this expectation.[27,29] Let us denote by $E_{\min}(n)$ the minimal energy within the subspace with $S^z = sN - n$. Then for the spin-s Heisenberg systems with sufficiently general coupling schemes

$$\begin{aligned} E_{\min}(n) \geq (1-n)E_{\mathrm{FM}} + nE_{\min}(1) &= E_{\mathrm{FM}} - n\left(E_{\mathrm{FM}} - E_{\min}(1)\right) \\ &= E_{\mathrm{FM}} - n\varepsilon_1 \end{aligned} \tag{17}$$

for all $n = 0, 1, 2, \ldots 2sN$. Thus, the localized magnons are the lowest-energy eigenstates of the Hamiltonian (1) on the considered geometrically frustrated lattices in the subspaces with $S^z = sN-1, sN-2, \ldots, sN-n_{\max}$.

Some other properties of the localized magnon states are of importance for further discussion. We may focus our consideration on the smallest-area localized magnons, since a one-magnon state with a larger area of localization can be presented as a linear combination of some smallest-area localized magnons. Naturally the following question arises: Are the smallest-area localized magnons linearly independent? This problem has been discussed recently in some detail.[44] It is convenient to divide the known lattices supporting localized magnons into several classes: the orthogonal type (diamond chain, frustrated ladder, frustrated bilayer etc.), the isolated type (sawtooth and kagomé-like chains etc.), codimension 1 type (kagomé lattice, checkerboard lattice etc.), and higher codimension type (pyrochlore lattice). The answer to the question about linear independence is 1) yes, for all $n = 1, 2 \ldots, n_{\max}$ for the lattices of the orthogonal type and of the isolated type; 2) no, for $n = 1$ and yes, for $n = 2, 3, \ldots, n_{\max}$ for the lattices of the codimension 1 type and 3) no, for $n = 1$ and larger values of n for the lattices of the higher codimension type. Note, however, that also in the latter case of higher codimension numerical data for finite pyrochlore lattices indicate a huge degeneracy of the ground state at $h = h_1$ which increases exponentially with N.[44]

Another problem which has to be discussed is the completeness of the independent localized magnon states. In other words, are the smallest-area localized magnons the only ground states in the corresponding sectors of S^z? Again we do not have a general answer to this question. The answer is yes for the diamond chain or the frustrated ladder with $J_2 > 2J_1$, the sawtooth chain ($J_2 = \sqrt{2(1+\Delta)}J_1$), the square-kagomé lattice with $J' > J$ (J' corresponds to the exchange bonds which form squares and J corresponds to all other bonds), the frustrated bilayer with $J_2 > 4J_1$ etc. Moreover, in these cases the ground state is separated by an energy gap from the higher-energy states. In contrast, the answer is no for the diamond chain or the frustrated ladder with $J_2 = 2J_1$, the kagomé-like chains etc. for which extra states in the sector with $S^z = Ns-1$ are present. Even more extra states are present for the kagomé lattice, the square-kagomé lattice with $J' = J$, the checkerboard lattice, the pyrochlore lattice etc. Some of these extra states may be viewed as nested objects, some of them may arise due to periodic boundary conditions which may imply nontrivial winding.[34,42–44,47,48]

At the end of this section we collect some properties of the localized magnon states for different lattices in Table 1. For more detailed information, e.g. about the extra ground states not described by smallest-area localized magnons, we refer to the literature.

Table 1. Some properties of the localized magnons on various lattices.

Lattice	Relation between J_{ij}	ε_1	$n_{\max}$
sawtooth chain	$J_2=\sqrt{2(1+\Delta)}J_1$	$2s\left(\Delta\left(\sqrt{2(1+\Delta)}+1\right)+1\right)J_1$	$N/4$
diamond chain	$J_2 \geq 2J_1$	$2s\Delta J_1 + s(\Delta+1)J_2$	$N/3$
kagomé chain	J for all bonds	$2s(2\Delta+1)J$	$N/6$
frustrated bilayer	$J_2 \geq 4J_1$	$s(J_2+\Delta(8J_1+J_2))$	$N/4$
kagomé lattice	J for all bonds	$2s(1+2\Delta)J$	$N/9$
pyrochlore lattice	J for all bonds	$2s(1+3\Delta)J$	$N/12$

2.2. *Exotic magnetization curves*

In this section we start to illustrate the effect of localized magnon states on physical properties. Let us briefly recall some general aspects of the study of the ground-state magnetization processes. We are interested in the magnetization $m(h) = M/M_{\max}$, $M = S^z$, $M_{\max} = sN$. As a rule, we can find (numerically or analytically) the low-lying energy levels $E(M)$ in the subspaces with different $M = sN, sN-1, \ldots$ for $h = 0$. The energy in the presence of an external magnetic field h is given by

$$E(M,h) = E(M) - hM, \tag{18}$$

where the magnetization M should acquire a value which minimizes $E(M,h)$ (18), i.e. the magnetization can be determined from the equation

$$\frac{\mathrm{d}E(M)}{\mathrm{d}M} = h. \tag{19}$$

Eq. (19) gives the dependence $h(M) = h$ and by inverting we obtain the magnetization curve $M = M(h)$. For a classical Heisenberg antiferromagnet one finds typically a parabolic relation $E(M) \propto M^2$ resulting in straight-line behavior $M \propto h$. In quantum Heisenberg antiferromagnets on bipartite (non-frustrated) lattices such as the square lattice there are small deviations from a parabolic $E-M$ relation (a linear $M-h$ relation), see e.g. Ref. [55]. More interesting $E-M$ $(M-h)$ relations can be observed for frustrated quantum antiferromagnets, cf. Fig. 3. If the $E - M$ curve consists of two parts meeting at M_0 with different slopes h_A (for $M_0 - 0$) and $h_B > h_A$ (for $M_0 + 0$) then $M(h)$ shows a magnetization plateau at M_0 between the two values of the field h_A and $h_B > h_A$. If there is a linear $E - M$ relation between two values of the magnetization M_A and M_B with a slope h_0 then $M(h)$ shows a jump between the two values of the magnetization M_A and M_B at the field h_0. Frustrated quantum spin models provide many examples of intriguing magnetization curves (for example, the 1/3 plateau for the kagomé lattice[56] or the 1/3 and 1/4 plateaus for the Shastry-Sutherland lattice[57]).

Let us now discuss the ground-state magnetization process for the Heisenberg antiferromagnet on a lattice which supports localized magnon states. We have given above the ground-state energy $E(S^z, h = 0) = E_{\rm FM} - sN\varepsilon_1 + \varepsilon_1 M$ for $M = S^z = sN, sN-1, \ldots, sN - n_{\max}$, i.e. there is a linear $E - M$ relation in the M interval between $sN - n_{\max}$ and sN with the slope ε_1. As a result one finds in the $M(h)$ curve a jump between two values of magnetization $sN - n_{\max}$ and sN at the saturation field $h_1 = \varepsilon_1$. In general, the slope of the $E(M)$ curve as M approaches $Ns - n_{\max}$ from below, h_2, is different from $h_1 = \varepsilon_1$[21,58] and therefore the magnetization curve exhibits a plateau between $h_2 < h_1$ and h_1 which precedes the jump at h_1. In Fig. 4 we display the ground-state magnetization curves for two

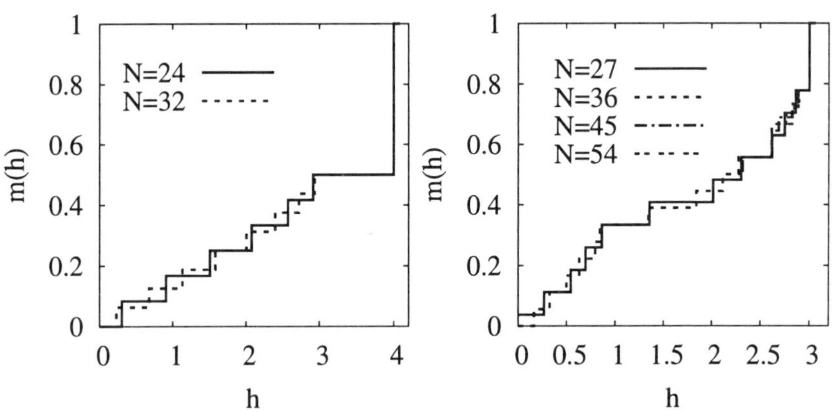

Fig. 4. Ground-state magnetization curves $m(h) = M(h)/M_{\max}$ for the spin-1/2 Heisenberg antiferromagnet on the sawtooth chain (left panel) and on the kagomé lattice (right panel) obtained by exact diagonalization for finite systems.

frustrated lattices which support localized magnons, namely, for the sawtooth chain and the kagomé lattice. The jump to saturation is obvious and has the height $\delta m = 1/2$ ($\delta m = 2/9$) for the sawtooth chain (kagomé lattice). For the sawtooth chain there is a wide plateau at $m = 1/2$ preceding the jump. For the kagomé lattice the plateau at $m = 7/9$ is less obvious, but a finite-size analysis yields a finite plateau width of $\Delta h \approx 0.07J$.[31]

Magnetization curves with a jump to saturation for other lattices can be found in Refs. [27,28,30,36,41,46,55].

We mention finally that the jump is macroscopic, and that there is no finite-size effect, see Fig. 4. Furthermore the height of the jump decreases with $1/s$, i.e. the jump is a true quantum effect and disappears in the classical limit $s \to \infty$.

2.3. *Field-induced spin-Peierls transition*

Inspecting the energy of the independent localized magnon states one may observe, that this energy is a sum of terms referring to separate localized magnons and the energy of a localized magnon is proportional to the exchange coupling along the trapping cell (see, e.g., Eq. (16) for the kagomé lattice). Thus one can expect that an appropriate lattice deformation may decrease the magnetic energy of a localized magnon eigenstate. To find a favorable deformation one needs optimal gain in magnetic energy. For that we use the circumstance that due to the localized nature of the magnons we have an inhomogeneous distribution of nearest-neighbor spin-spin correlations $\langle \vec{s}_i \cdot \vec{s}_j \rangle$.[30] In case that one magnon is distributed uniformly over the lattice the deviation of the nearest-neighbor correlation from the ferromagnetic value, $\langle \vec{s}_i \cdot \vec{s}_j \rangle - 1/4$, is of the order $1/N$. On the other hand, e.g. for a localized magnon on the kagomé lattice we have along the hexagon hosting the localized magnon actually negative nearest-neighbor spin-spin correlations $\langle \vec{s}_i \cdot \vec{s}_j \rangle = -1/12$, but all other spin-spin correlations are positive. Having in mind that the exchange integrals are positive (antiferromagnetic) a lattice deformation which implies a contraction of the trapping cell (i.e. a hexagon for the kagomé lattice) increases the strength of exchange integrals along the trapping cell and therefore decreases the ground-state magnetic energy. This gain of magnetic energy is linear with respect to a displacement δ which, obviously, should be balanced by an increase of the elastic energy which is quadratic with respect to δ. As a result one finds a lattice instability in the subspaces with $S^z = sN - 1, \ldots, sN - n_{\max}$ (note that the undistorted lattice is favorable in the fully polarized ferromagnetic state with $S^z = sN$). In the subspaces with $S^z = sN - n_{\max} - 1, sN - n_{\max} - 2, \ldots$

we can calculate the ground-state magnetic energy for the assumed lattice deformation pattern only numerically for finite systems to check whether it decreases as δ^p with $p < 2$ (in this case the deformation is favorable) or with $p \geq 2$ (in this case the deformation is unfavorable). In the latter case the lattice deformation appears only in strong magnetic fields around the saturation field h_1 and disappears if the field becomes weaker. In the former case the distorted lattice remains favorable with respect to the undistorted one even in weaker fields. Appropriate lattice deformations were constructed and discussed in some detail for the kagomé and square-kagomé lattices[31,40] and for the checkerboard and pyrochlore lattices.[39]

2.4. *Finite ground-state residual entropy*

It has been shown above that the energy of the n-magnon state in a magnetic field is $E_{\rm FM} - n\varepsilon_1 - h(sN - n)$, cf. Eqs. (17) and (18). This energy level is highly degenerate, since there are many ways to place n independent localized magnons on a lattice. The degeneracy further increases at the saturation field $h_1 = \varepsilon_1$, since the energies of the states with different numbers of localized magnons $n = 0, 1, \ldots, n_{\rm max}$ become equal, namely $E_{\rm FM} - h_1 sN$. We denote this degeneracy at $h = h_1$ by $\mathcal{W}$. Let us recall that the localized magnon states are linearly independent in the corresponding subspaces of S^z (except for the three-dimensional pyrochlore lattice).[44] It can be shown that the number of ways to place $n = 0, 1, \ldots, n_{\rm max}$ localized magnons on a lattice $\mathcal{W}$ grows exponentially with the system size N, i.e. $\ln \mathcal{W} \propto N$.[32–35,42,43,45,48]

To demonstrate this we introduce an auxiliary lattice with classical hard-core objects. We put hard-core objects on the auxiliary lattice in accordance with the rules for independent localized magnons. Depending on a specific spin model we can faced (i) with the hard-monomer restriction (double occupancy of sites is forbidden) as for the diamond chain, the square-kagomé lattice etc., (ii) with the hard-dimer restriction on a chain (occupation of neighboring sites is forbidden) as for the sawtooth chain, the kagomé-like chains, the frustrated ladder etc., (iii) with hard squares on a square lattice (occupation of neighboring sites is forbidden) as for the frustrated bilayer, (iv) with large hard squares on a square lattice (occupation of neighboring and second neighboring sites is forbidden) as for the checkerboard lattice, or (v) with hard hexagons on a triangular lattice (occupation of neighboring sites is forbidden) as for the kagomé lattice and the star lattice.

It is easy to realize that the number of ways to put n localized magnons on the spin lattice $g_N(n)$ (that is the degeneracy of the localized n-magnon states) is equal to the number of spatial configurations of n hard-core objects on the auxiliary lattice. The latter quantity is simply the canonical partition function of the hard-core object system $Z(n,\mathcal{N})$ (here $\mathcal{N}$ is number of sites of the auxiliary lattice, e.g., $\mathcal{N}=N/3$ for the diamond chain, $\mathcal{N}=N/2$ for the sawtooth chain, $\mathcal{N}=N/2$ for the frustrated bilayer, $\mathcal{N}=N/3$ for the kagomé lattice etc.). Moreover, the degeneracy at the saturation $\mathcal{W}=\sum_{n=0}^{n_{\max}} g_N(n)$ is simply the grand-canonical partition function of the hard-core object system at the zero value of the chemical potential (i.e. $z=\exp(\mu/k_{\rm B}T)$ is equal to 1)

$$\mathcal{W}=\sum_{n=0}^{n_{\max}} Z(n,\mathcal{N})=\Xi(z=1,\mathcal{N})\,. \tag{20}$$

The grand-canonical partition function for some hard-core object lattice-gas models can be calculated analytically (hard monomers, one-dimensional hard dimers,[59] hard hexagons[60,61]) or it is known from precise numerical computations (hard squares,[62] large hard squares[48,62]). Therefore we can give a theoretical prediction for the residual ground-state entropy $S/N \geq k_{\rm B}\ln\mathcal{W}/N$ which arises owing to the localized magnon states at the saturation field. For example, $S/k_{\rm B}N=(1/2)\ln\left((1+\sqrt{5})/2\right)\approx 0.2406$ for the sawtooth chain,[33,35] $S/k_{\rm B}N\approx 0.1111$ for the kagomé lattice[34,35] etc. Note, however, that for models with extra ground states not described by smallest-area localized magnons (e.g. the kagomé lattice) this prediction gives a lower bound for the residual entropy.

At the end of this section we illustrate the general discussion of the mapping for one example, namely the sawtooth chain, in more detail, see Fig. 5. The auxiliary chain (Fig. 5, bottom) consists of $\mathcal{N}=N/2$ sites

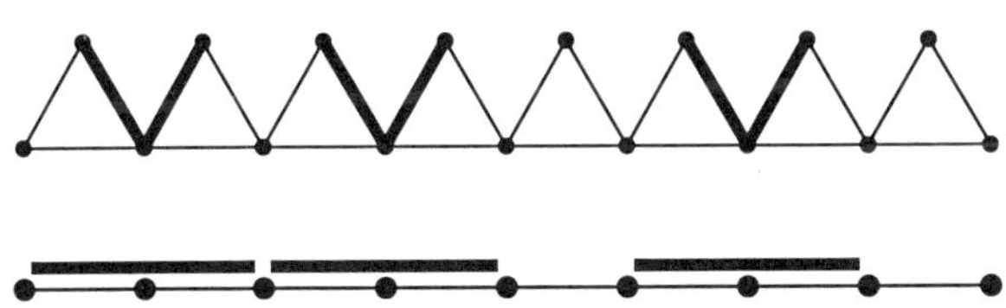

Fig. 5. Sawtooth chain hosting three localized magnons (thick lines) and the corresponding auxiliary lattice-gas model (three hard dimers on a linear chain).

which may be occupied by hard dimers, see also Refs. [33–35,43]. Taking the number of hard-dimer distributions from the literature[59] we can use this mapping to find the ground-state degeneracy at the saturation field $\mathcal{W}$. For $\mathcal{N} \to \infty$ one finds $\mathcal{W} = \left((1+\sqrt{5})/2\right)^{\mathcal{N}} \approx \exp(0.4812\mathcal{N})$ leading to the finite residual entropy given above.

2.5. *Universal low-temperature thermodynamics*

We may exploit the correspondence between the localized magnon states and the spatial configurations of hard-core objects introduced above to calculate the contribution of the localized magnon states to the thermodynamic quantities. Under certain conditions this contribution may dominate the low-temperature thermodynamics and therefore we may find predictions for the low-temperature behavior of the magnetic quantities for frustrated quantum antiferromagnets in the vicinity of the saturation field h_1.

To be specific, the contribution of the localized states to the partition function of the spin model can be written as follows

$$\begin{aligned} Z_{\rm lm}(T,h,N) &= \sum_{n=0}^{n_{\max}} g_N(n) \exp\left(-\frac{E_{\rm FM} - hsN - n(\varepsilon_1 - h)}{k_{\rm B}T}\right) \\ &= \exp\left(-\frac{E_{\rm FM} - hsN}{k_{\rm B}T}\right) \sum_{n=0}^{n_{\max}} g_N(n) \exp\left(\frac{h_1 - h}{k_{\rm B}T} n\right) \\ &= \exp\left(-\frac{E_{\rm FM} - hsN}{k_{\rm B}T}\right) \Xi(T,\mu,\mathcal{N}) . \end{aligned} \tag{21}$$

Here $g_N(n)$ is the canonical partition function of a certain classical lattice-gas model of hard-core objects $Z(n,\mathcal{N})$, $h_1 - h = \mu$ is the chemical potential of hard-core objects, and $\Xi(T,\mu,\mathcal{N})$ (or $\Xi(z,\mathcal{N})$, $z = \exp(\mu/k_{\rm B}T)$) is the grand-canonical partition function of the hard-core object lattice gas. This formula describes the low-temperature thermodynamics near the saturation field accurately, i.e. $Z(T,h,N) \approx Z_{\rm lm}(T,h,N)$, provided that (i) there are no other ground states (apart from the considered localized-magnon states) in the corresponding sectors of S^z or that the contribution of such extra states is vanishingly small as $N \to \infty$, and that (ii) excited states in these sectors are separated by a finite energy gap from the ground states. Indeed this seems to be valid for several lattices, cf. Refs. [34,42,43,45,47,48].

Now the contribution of the localized magnon states to the Helmholtz free energy F of the spin model is given by

$$\frac{F_{\rm lm}(T,h,N)}{N} = \frac{E_{\rm FM}}{N} - hs - k_{\rm B}T \frac{\ln \Xi(z,\mathcal{N})}{\mathcal{N}} \frac{\mathcal{N}}{N} . \tag{22}$$

The entropy S, the specific heat C, the magnetization M and the susceptibility χ follow from Eq. (22) according to usual relations

$$\begin{aligned}
S_{\rm lm}(T,h,N) &= -\frac{\partial F_{\rm lm}(T,h,N)}{\partial T}\,,\\
C_{\rm lm}(T,h,N) &= T\frac{\partial S_{\rm lm}(T,h,N)}{\partial T}\,,\\
M_{\rm lm}(T,h,N) &= sN - \langle n\rangle = sN - k_{\rm B}T\frac{\partial \ln \Xi(T,\mu,\mathcal{N})}{\partial \mu}\,,\\
\chi_{\rm lm}(T,h,N) &= \frac{\partial M_{\rm lm}(T,h,N)}{\partial h}\,. \qquad (23)
\end{aligned}$$

For all spin models with localized magnon states which can be mapped on the same hard-object system the contribution of the localized magnon states to thermodynamic quantities is identical (up to a factor $\mathcal{N}/N$), i.e. they belong to the same universality class.

In what follows we present numerical results for low-temperature thermodynamic quantities (obtained by exact diagonalization of finite systems of up to $N = 20$ spins) for several spin models corresponding to different hard-object universality classes. We compare the results for the full spin model with data for the corresponding hard-object model this way checking to what extent the localized magnon states cover the main features of the low-temperature thermodynamics.

We start with the simplest case of hard monomers, which describes the low-energy physics around the saturation field of, e.g., the diamond chain (with $N = 3\mathcal{N}$), the dimer-plaquette chain (with $N = 4\mathcal{N}$) or the square-kagomé lattice (with $N = 6\mathcal{N}$). The relevant quantities for the hard-monomer universality class read

$$\begin{aligned}
\Xi(T,\mu,\mathcal{N}) &= (1+\exp x)^{\mathcal{N}}\,,\\
\frac{S_{\rm lm}(T,h,N)}{k_{\rm B}N} &= \frac{\mathcal{N}}{N}\left(\ln(1+\exp x) - \frac{x\exp x}{1+\exp x}\right),\\
\frac{C_{\rm lm}(T,h,N)}{k_{\rm B}N} &= \frac{\mathcal{N}}{N}\frac{\left(\frac{x}{2}\right)^2}{\cosh^2\frac{x}{2}}\,,\\
\frac{M_{\rm lm}(T,h,N)}{sN} &= 1-\frac{1}{s}\frac{\mathcal{N}}{N}\frac{\exp x}{1+\exp x}\,,\\
\frac{k_{\rm B}T\chi_{\rm lm}(T,h,N)}{N} &= \frac{\mathcal{N}}{N}\frac{1}{4\cosh^2\frac{x}{2}}\,;\\
x &= \frac{\mu}{k_{\rm B}T} = \frac{h_1-h}{k_{\rm B}T}\,. \qquad (24)
\end{aligned}$$

Note that the results for S, C, M, χ in the r.h.s. of Eq. (24) do not depend on N and thus they are valid for finite system sizes N but also for $N \to \infty$. Note further that the thermodynamic quantities depend on T and h via the parameter $x = (h_1 - h)/k_{\rm B}T$, only.

Next we consider the one-dimensional case of hard dimers which corresponds, e.g., to the sawtooth chain ($N = 2\mathcal{N}$), the frustrated ladder ($N = 2\mathcal{N}$) or the kagomé chains ($N = 3\mathcal{N}$ or $N = 5\mathcal{N}$). The thermodynamic quantities for the one-dimensional hard-dimer universality class read

$$
\begin{aligned}
\Xi(T,\mu,\mathcal{N}) &= \lambda_1^{\mathcal{N}} + \lambda_2^{\mathcal{N}}\,, \quad \lambda_{1,2} = \frac{1}{2} \pm \sqrt{\frac{1}{4} + \exp x}\,,\\
\frac{S_{\rm lm}(T,h,N)}{k_{\rm B}N} &= \frac{\mathcal{N}}{N}\left(\ln\left(\frac{1}{2} + \sqrt{\frac{1}{4} + \exp x}\right)\right.\\
&\qquad \left. - x\left(\frac{1}{2} - \frac{1}{4\sqrt{\frac{1}{4} + \exp x}}\right)\right),\\
\frac{C_{\rm lm}(T,h,N)}{k_{\rm B}N} &= \frac{\mathcal{N}}{N}\frac{x^2 \exp x}{8\left(\frac{1}{4} + \exp x\right)^{\frac{3}{2}}}\,,\\
\frac{M_{\rm lm}(T,h,N)}{sN} &= 1 - \frac{1}{s}\frac{\mathcal{N}}{N}\left(\frac{1}{2} - \frac{1}{4\sqrt{\frac{1}{4} + \exp x}}\right),\\
\frac{k_{\rm B}T\chi_{\rm lm}(T,h,N)}{N} &= \frac{\mathcal{N}}{N}\frac{\exp x}{8\left(\frac{1}{4} + \exp x\right)^{\frac{3}{2}}}\,;\\
x &= \frac{\mu}{k_{\rm B}T} = \frac{h_1 - h}{k_{\rm B}T}\,.
\end{aligned}
\tag{25}
$$

Again the thermodynamic quantities depend on T and h via the parameter $x = (h_1 - h)/k_{\rm B}T$, only. Note that the formulas for S, C, M, χ in the r.h.s. of Eq. (25) are valid in the thermodynamic limit $N \to \infty$, only. For finite systems corresponding formulas can be found using $\Xi(T,\mu,\mathcal{N})$ from the first line in Eq. (25) in combination with Eqs. (22) and (23).

The typical behavior of some thermodynamic quantities for two frustrated quantum antiferromagnets at low temperatures around the saturation field is shown in Fig. 6. We emphasize here some prominent features: a low-temperature peak in the dependence C vs T for fields slightly below or slightly above h_1 and an enhanced entropy at h_1 at low temperatures

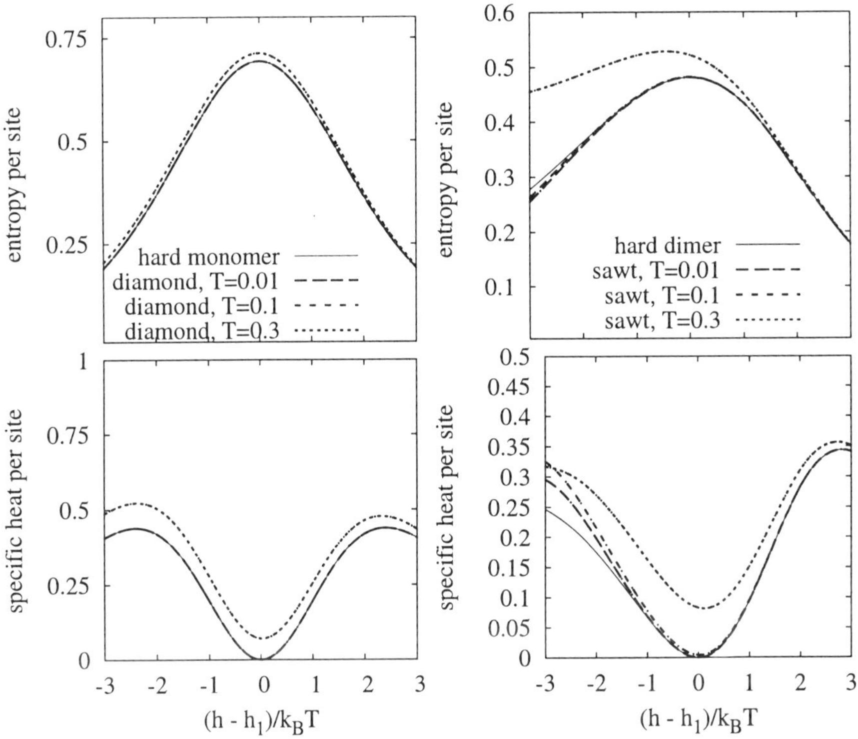

Fig. 6. The universal dependences of the entropy (top) and the specific heat (bottom). Left panel: Spin-1/2 Heisenberg antiferromagnet on the diamond chain with $N = 18$, $J_1 = 1$ and $J_2 = 3$ and the hard-monomer gas. (Note that the curves for the spin model with $k_BT = 0.01$ and $k_BT = 0.1$ and the hard-monomer gas almost coincide.) Right panel: Spin-1/2 Heisenberg antiferromagnet on the sawtooth chain with $N = 20$, $J_1 = 1$ and $J_2 = 2$ and the one-dimensional hard-dimer gas.

(see above, the section on residual ground-state entropy). Furthermore from Fig. 6 it becomes evident the hard-object description works excellent for temperatures up to 10% of the exchange coupling and reproduces qualitatively the characteristic features of the spin model for higher temperatures.

The case of two-dimensional systems such as the kagomé lattice, the star lattice, the checkerboard lattice, or the frustrated bilayer are even more intriguing. The simple reason for that is the order-disorder phase transition inherent in the corresponding two-dimensional hard-core object lattice-gas models. We will discuss this question in the next section in more detail.

2.6. *Finite-temperature phase transitions*

The degeneracy of localized magnon states and the contribution of these states to the partition function for two-dimensional lattices such as the kagomé lattice or the star lattice can be calculated by mapping onto a hard-hexagon model on a triangular lattice whereas for the frustrated bilayer (the checkerboard lattice) the hard-square model (large-hard-square model) on the square lattice is relevant. An important feature of such two-dimensional hard-object lattice gases is an order-disorder phase transitions of pure geometrical origin.[61] While for small densities of hard-squares (hard-hexagons) the two (three) sublattices of the square (triangular) lattice are equally occupied, with increasing of the density of hard-squares (hard-hexagons) at a critical value the hard-squares (hard-hexagons) preferentially occupy mainly one of the two (three) sublattices.[61] The order-disorder phase transition belongs to the two-dimensional Ising (3-state Potts) model universality class for the hard-square (hard-hexagon) model and manifests itself through corresponding singularities in thermodynamic quantities in the vicinity of the phase transition point. Thus, the ordering of localized magnons on these two-dimensional lattices at low-temperatures just below the saturation field (as their number exceeds a critical value) should also produce singularities in some thermodynamic quantities.

In order to make the differences to the hard-core-object models discussed in the last section more evident we will now discuss the hard-hexagon model (i.e. the contribution of the corresponding localized magnon states to the thermodynamic quantities) in more detail. Due to Baxter[60,61] we know the grand-canonical partition function $\Xi(T,\mu,\mathcal{N})$ of the hard-hexagon model in the thermodynamic limit $\mathcal{N}\to\infty$. It has the form[60,61]

$$\begin{aligned}(\Xi(T,\mu,\mathcal{N}))^{\frac{1}{\mathcal{N}}} &= \frac{H^3(x)Q^2(x^5)}{G^2(x)}\prod_{n=1}^{\infty}\frac{\left(1-x^{6n-4}\right)\left(1-x^{6n-3}\right)^2\left(1-x^{6n-2}\right)}{\left(1-x^{6n-5}\right)\left(1-x^{6n-1}\right)\left(1-x^{6n}\right)^2},\\ z &= -x\frac{H^5(x)}{G^5(x)}\end{aligned} \tag{26}$$

for $0<z=\exp(\mu/kT)<z_c$ $(-1<x<0)$ and

$$\begin{aligned}(\Xi(T,\mu,\mathcal{N}))^{\frac{1}{\mathcal{N}}} &= \frac{x^{-\frac{1}{3}}G^3(x)Q^2(x^5)}{H^2(x)}\prod_{n=1}^{\infty}\frac{\left(1-x^{3n-2}\right)\left(1-x^{3n-1}\right)}{\left(1-x^{3n}\right)^2},\\ z &= \frac{1}{x}\frac{G^5(x)}{H^5(x)}\end{aligned} \tag{27}$$

for $z_c<z$ $(0<x<1)$. The critical activity z_c is $z_c=(11+5\sqrt{5})/2\approx 11.0902$ $(x=\pm 1)$. The functions H,G,Q in Eqs. (26) and (27) are given

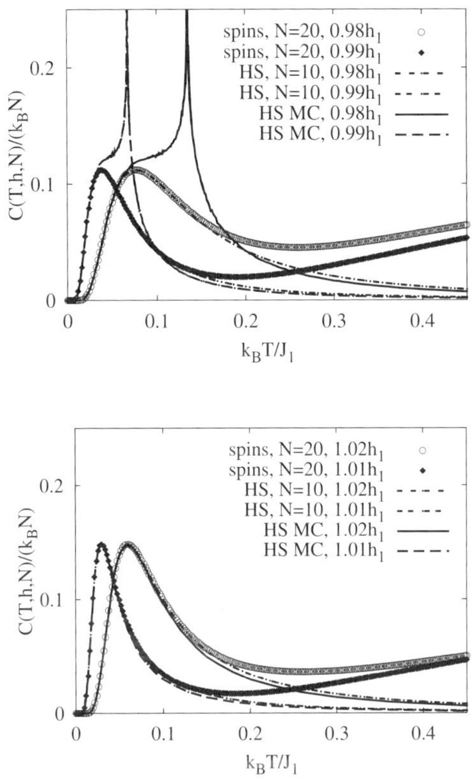

Fig. 7. The specific heat vs temperature for the spin-1/2 Heisenberg antiferromagnet on the frustrated bilayer ($s = 1/2$, $J_1 = 1$, $J_2 = 5$, $\Delta = 1$, $h_1 = 9$) for $h = 0.98h_1$ and $h = 0.99h_1$ (upper panel) and $h = 1.02h_1$ and $h = 1.01h_1$ (lower panel). Exact diagonalization data ($N = 20$) versus hard-square data ($\mathcal{N} = 10$ and Monte Carlo simulations for up to 800×800 sites).

by

$$
\begin{aligned}
H(x) &= \prod_{n=1}^{\infty} \left(\left(1 - x^{5n-4}\right) \left(1 - x^{5n-1}\right) \right)^{-1} , \\
G(x) &= \prod_{n=1}^{\infty} \left(\left(1 - x^{5n-3}\right) \left(1 - x^{5n-2}\right) \right)^{-1} , \\
Q(x) &= \prod_{n=1}^{\infty} \left(1 - x^{n}\right) . \qquad (28)
\end{aligned}
$$

Similarly to hard-monomer and one-dimensional hard-dimer systems, cf. Eqs. (24) and (25), the chemical potential μ and the temperature T enter only via the combination $\mu/k_BT = \ln z$, but in difference to (24) and (25) for the hard-hexagon problem the auxiliary real parameter x in Eqs. (26) and (27) cannot be eliminated to obtain explicit dependence of $\Xi^{1/\mathcal{N}}$ on μ/k_BT. Furthermore, the equations for the hard-hexagon model are much more complex and no explicit formulas for the thermodynamic quantities such as entropy or specific heat can be given.

The order-disorder phase transition in the hard-hexagon model belongs to the two-dimensional 3-state Potts model universality class and implies, in particular, a divergency of the specific heat with exponent $\alpha = 1/3$. This model was discussed in the context of localized magnon effect on low-temperature magnetothermodynamics of the kagomé quantum Heisenberg antiferromagnet by Zhitomirsky and Tsunetsugu.[34,42]

Although the hard-hexagon model for the kagomé lattice correctly describes the contribution of the (smallest-area) localized magnons on hexagons to thermodynamics, both general arguments and numerical data show that there are other states which also contribute to thermodynamics.[34,42,47] Their contribution is not accounted by hard hexagons and it remains unclear whether the contribution of these extra states remains finite in the thermodynamic limit.[47] Therefore, the simple scenario described above may be not sufficient for explaining the low-temperature strong-field behavior of the quantum Heisenberg antiferromagnet on the kagomé lattice. Similar arguments can be given for the checkerboard lattice discussed recently in Ref. [48].

A more promising candidate is the Heisenberg antiferromagnet on a frustrated bilayer,[45] see Fig. 2. The low-temperature strong-field thermodynamics for this model is described exclusively by the hard-square model. This statement has been confirmed by considering finite systems up to $N = 64$ and comparing the degeneracy of the ground state in the high-S^z sectors $g_N(n)$ and the canonical partition function of hard squares on a square lattice $Z(n, \mathcal{N})$ $(N = 2\mathcal{N})$.[45,47] The hard-square model shows an order-disorder phase transition at $z_c = 3.7962\ldots$ which belongs to the two-dimensional Ising model universality class and implies, in particular, a logarithmic divergency of the specific heat ($\alpha = 0$) (for recent studies of the hard-square model, see Ref. [62] and references therein). In Fig. 7 we show the temperature behavior of the specific heat just below and above the saturation field. For smaller systems the comparison of the results for the spin model and hard-square model shows that the data almost coincide up to temperatures of about 20% of the exchange constant J_1. For large systems the logarithmic singularity in the specific heat for $h \lesssim h_1$ is obvious.

Finally, we would like to emphasize that such a phase transition in a low-dimensional Heisenberg spin system does not contradict the Mermin–Wagner theorem,[63] since only a discrete symmetry is broken spontaneously. This demonstration of a finite-temperature phase transition in a two-dimensional interacting many-body spin model is an interesting example for the impact of the localized-magnon states on the physical properties of a frustrated magnet.

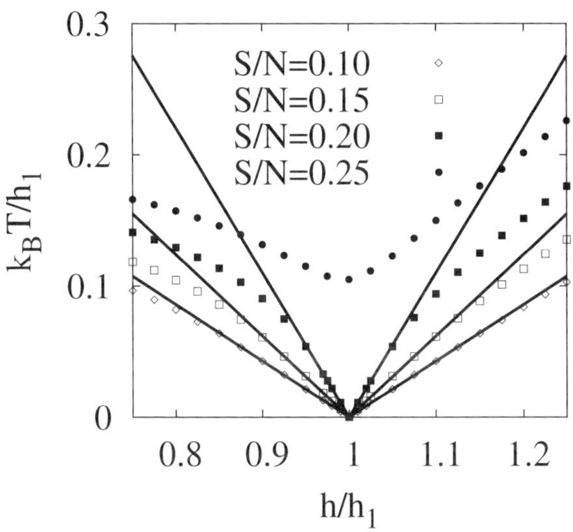

Fig. 8. Curves of constant entropy as a function of magnetic field and temperature ($S(T, h, N)/(k_B N)$ = const, adiabatic (de)magnetization curves) for the spin-1/2 Heisenberg antiferromagnet on the diamond chain with $N = 18$, $J_1 = 1$ and $J_2 = 3$. Solid lines correspond to analytical results which follow from Eq. (24), symbols correspond to exact diagonalization data for the finite spin system.

2.7. *Magnetic cooling*

Let us very briefly discuss an aspect of the localized magnon scenario which might have some relevance for a possible application of highly frustrated magnets.

Due to the huge degeneracy of the localized magnon states and the resulting residual entropy at $h = h_1$ there is a well pronounced low-temperature peak in the entropy S versus field h curve, see Fig. 6. As it has been pointed out recently by Zhitomirsky[64] considering the classical kagomé Heisenberg antiferromagnet such a degeneracy leads to an enhanced magnetocaloric effect. Later on this point has been discussed for quantum spin systems, e.g. in Refs. [33,42,43,65].

The magnetocaloric effect consists in cooling or heating of a magnetic system in a varying magnetic field. The adiabatic (i.e. isentropic, $S = \text{const}$) temperature change as function of temperature and applied magnetic field is given by

$$\left(\frac{\partial T}{\partial h}\right)_S = -\frac{T}{C(T,h)}\left(\frac{\partial S}{\partial h}\right)_T, \tag{29}$$

where $C(T,h)$ is the temperature- and field-dependent specific heat of the system. This rate is also called cooling rate. It is obvious that the cooling rate is large if $(\partial S/\partial h)_T$ is large, i.e. at the trailing and rising edges of a peak in $S(h)$.

The universal dependence which emerges when localized magnons dominate thermodynamics implies that the entropy depends on the magnetic field h and the temperature T only through the combination $x = (h_1 - h)/k_{\rm B}T$. Therefore, in the vicinity of the point $T = 0$, $h = h_1$ the constant entropy curves are straight lines just like in the case of an ideal paramagnet. As we move away from this point deviations from the paramagnetic type behavior become visible. In Fig. 8 we illustrate the constant entropy curves at low temperatures around the saturation field.

3. Correlated Hubbard electrons on geometrically frustrated lattices

In the last part of this paper we discuss briefly how and to what extent the localized-magnon scenario can be applied to other strongly correlated models (for the Hubbard model see Refs. [49,51], for the $t-J$ model see Ref. [50]). For concreteness we focus on the Hubbard model (2) on a sawtooth chain with the Hamiltonian

$$\begin{aligned} H = \sum_{j=0}^{\frac{N}{2}-1}\sum_{\sigma=\uparrow,\downarrow}\Big\{ & t c^\dagger_{2j,\sigma}c_{2j+2,\sigma} + t'\left(c^\dagger_{2j,\sigma}c_{2j+1,\sigma} + c^\dagger_{2j+1,\sigma}c_{2j+2,\sigma}\right) + \text{h.c.} \\ & + \mu\left(c^\dagger_{2j,\sigma}c_{2j,\sigma} + c^\dagger_{2j+1,\sigma}c_{2j+1,\sigma}\right)\Big\} \\ & + U\sum_{j=0}^{\frac{N}{2}-1}\left(c^\dagger_{2j,\uparrow}c_{2j,\uparrow}c^\dagger_{2j,\downarrow}c_{2j,\downarrow} + c^\dagger_{2j+1,\uparrow}c_{2j+1,\uparrow}c^\dagger_{2j+1,\downarrow}c_{2j+1,\downarrow}\right). \end{aligned} \tag{30}$$

Here $t > 0$ and $t' > 0$ are the hopping integrals along the base line and the zig-zag path, respectively, and $U > 0$ is the on-site Coulomb repulsion. The sawtooth-chain Hubbard model attracts much attention since the 1990ies and was discussed earlier within different approaches.[66]

On the one-particle level (or when $U = 0$) the description of the electron system is the same as of the $XX0$ Heisenberg spin system (compare Eq. (30) and Eq. (4) with $\Delta = 0$). Thus, if $t' = \sqrt{2}t$ the lowest single electron energy is completely flat. The N one-electron localized states $l^{\dagger}_{2j,\sigma}|0\rangle$, $l^{\dagger}_{2j,\sigma} = c^{\dagger}_{2j-1,\sigma} - \sqrt{2}c^{\dagger}_{2j,\sigma} + c^{\dagger}_{2j+1,\sigma}$ (i.e. localized in any of $N/2$ valleys labelled by the index $2j$ and having either spin up or spin down) are the ground states with the energy $\varepsilon_- = -2t + \mu$. Note that the indices at the $l^{\dagger}$ and $c^{\dagger}$ operators correspond to the lattice sites as illustrated in Fig. 1.

Next we pass to the two-electron subspace where the Hubbard repulsion becomes relevant. Obviously a two-particle ground state can be constructed by two independent localized electrons with arbitrary spin trapped on two valleys which do not touch each other. However, in difference to the Heisenberg model two electrons can also be trapped on two neighboring valleys, e.g. with indices $2j$ and $2j+2$, i.e. $l^{\dagger}_{2j,\uparrow}l^{\dagger}_{2j+2,\uparrow}|0\rangle$ and $l^{\dagger}_{2j,\downarrow}l^{\dagger}_{2j+2,\downarrow}|0\rangle$ are also eigenstates. It is obvious that these states do not feel U, since both electrons have the same spin and hence the Pauli principle forbids the simultaneous occupation of the site $2j+1$ contained in both valleys. More interesting, for two electrons having different spin the linear combination

$$l^{\dagger}_{2j,\uparrow}l^{\dagger}_{2j+2,\downarrow}|0\rangle + l^{\dagger}_{2j,\downarrow}l^{\dagger}_{2j+2,\uparrow}|0\rangle\,, \tag{31}$$

is also a ground state in the two-electron subspace. By simple calculations one can convince oneself that (31) does not feel the Coulomb repulsion U at the common site $2j+1$. This is not an accident. Due to the SU(2) symmetry of the Hubbard Hamiltonian the state (31) together with the states $l^{\dagger}_{2j,\uparrow}l^{\dagger}_{2j+2,\uparrow}|0\rangle$ and $l^{\dagger}_{2j,\downarrow}l^{\dagger}_{2j+2,\downarrow}|0\rangle$ form a triplet, i.e. the state (31) can be obtained e.g. by acting of the spin lowering operator of the total spin $S^- = \sum_i c^{\dagger}_{i,\downarrow}c_{i,\uparrow}$ on the state $l^{\dagger}_{2j,\uparrow}l^{\dagger}_{2j+2,\uparrow}|0\rangle$. Of course, all states belonging to one triplet have the same energy.

We can generalize this procedure to construct the ground states in the subspaces with $n = 3, \ldots, N/2$ electrons. Similarly to the case of the Heisenberg antiferromagnet these ground states are constructed from the localized electron states, however, their degeneracies $g_N(n)$ do not coincide with the ones for the Heisenberg sawtooth chain (which were equal to the canonical partition functions of n hard dimers on a chain of $\mathcal{N} = N/2$ sites, see above). Nevertheless, $g_N(n)$, $n = 0, 1, 2, \ldots, N/2$ for the Hubbard sawtooth chain is still related to the one-dimensional hard dimers, however, in a more intricate way. Namely, $g_N(n) = Z(n,N)$ for $n = 0,1,\ldots,N/2-1$ and $g_N(N/2) = N/2+1 = Z(N/2,N)+N/2-1$. These formulas follow from a mapping[51] between the localized electron states and the spatial configurations of hard dimers on a chain of N sites (but not $N/2$ sites, the increase of the number of sites with respect to the spin model by 2 corresponds to the 2 spin polarizations of the electron). As for spin systems using this mapping we can calculate the contribution of localized electron states to the partition function. Hence, again we can give the formulas for the low-temperature thermodynamic quantities for a non-trivial quantum many-body problem. In particular, the grand-canonical partition function of the electron system for a chemical potential μ in the vicinity of $\mu_0 = 2t$ has the form

$$\begin{aligned}\Xi(T,\mu,N) &= \lambda_1^N + \lambda_2^N + \lambda_3^N\,,\\ \lambda_{1,2} &= \frac{1}{2} \pm \sqrt{\frac{1}{4} + \exp x}\,, \quad \lambda_3 = \left(\frac{N}{2}-1\right)^{\frac{1}{N}} \exp\frac{x}{2}\,,\\ x &= \frac{2t-\mu}{k_{\mathrm{B}}T}\,. \end{aligned}\tag{32}$$

In the thermodynamic limit $N \to \infty$ only the largest eigenvalue λ_1 of the transfer matrix survives and we arrive at formulas reflecting the thermodynamics of one-dimensional hard dimers

$$\begin{aligned}\frac{\Omega(T,\mu,N)}{N} &= -k_{\mathrm{B}}T \ln\left(\frac{1}{2} + \sqrt{\frac{1}{4} + \exp x}\right)\,,\\ \frac{S(T,\mu,N)}{k_{\mathrm{B}}N} &= \ln\left(\frac{1}{2} + \sqrt{\frac{1}{4} + \exp x}\right) - x\left(\frac{1}{2} - \frac{1}{4\sqrt{\frac{1}{4}+\exp x}}\right)\,,\\ \frac{C(T,\mu,N)}{k_{\mathrm{B}}N} &= \frac{x^2 \exp x}{8\left(\frac{1}{4}+\exp x\right)^{\frac{3}{2}}}\,,\\ \frac{\langle n\rangle}{N} &= \frac{1}{2} - \frac{1}{4\sqrt{\frac{1}{4}+\exp x}}\end{aligned}\tag{33}$$

(compare Eq. (33) with Eq. (25)). Some results for the low-temperature specific heat for three values of the chemical potential $\mu = 0.98\mu_0, \mu_0, 1.02\mu_0$ are shown in Fig. 9. Similar as for the spin systems we see (i) that the hard-object model, Eqs. (32) and (33), yields a good description of the electronic model at low temperatures and (ii) that there is an extra low-temperature

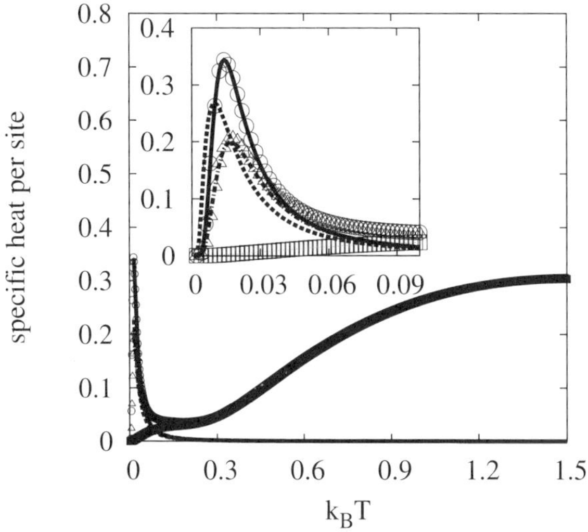

Fig. 9. The specific heat $C(T, \mu, N)/(k_B N)$ vs temperature for the sawtooth Hubbard chain of $N = 10$ sites and with $U = 10$ (symbols) compared with the analytical formula for $N = 10$ which follows from Eq. (32) (lines). $\mu = 0.98\mu_0$: triangles and dashed-dotted lines, $\mu = \mu_0$: squares and the line $C(T, \mu, N) = 0$, $\mu = 1.02\mu_0$: circles and solid lines. The dashed line corresponds to the hard-dimer prediction for $\mu = 0.98\mu_0$ and $N \to \infty$.

maximum in the specific heat due to the manifold of localized electron ground states.

At the end of this section we would like to illustrate the relation of our considerations to the so-called flat-band ferromagnetism in the Hubbard model found by Mielke and Tasaki in the early 1990ies.[52] The Hubbard sawtooth chain is a particular case of the so-called Tasaki's model for flat-band ferromagnetism. Note that other lattices discussed in the preceding sections in connection with spin models such as the kagomé lattice also exhibit flat-band ferromagnetism.[52] Note further that in Mielke's and Tasaki's papers one can also find a discussion of the localized nature of the eigenstates but no discussion of a possible relation to hard-core object models and the corresponding thermodynamics. For the sawtooth chain it was found that the ground state at quarter filling (i.e. for the number of electrons $n = N/2$) is ferromagnetic with the degeneracy $2S + 1$, $S = n/2 = N/4$. We can illustrate this rigorous result as follows. Consider for $n = N/2$ electrons a ground state in which all valleys are occupied by spin-up electrons. This state is fully polarized, i.e. its total spin is $\vec{S}^2 = (n/2)(n/2 + 1)$, and has the energy $N\varepsilon_-/2$. Because of the SU(2) invariance of the Hubbard model (2) we can construct $N/2$ more fully polarized states (i.e. with maximum total spin $\vec{S}^2 = (n/2)(n/2+1)$) with the same energy by acting of $(S^-)^m$, $m = 1, 2, \ldots, N/2$. No further states which do not not feel U can be constructed, since all valleys are already occupied with precisely one electron. As a result only the $N/2+1$ states belonging to the spin-$N/4$ SU(2)-multiplet remain the ground states all exhibiting full polarization (saturated ferromagnetism). Similar reasoning can be applied for the case of $n = N/2 - 1$ electrons, but not for the case of $n = N/2 - 2$ electrons. In the latter case one can construct states with energy $(N/2-2)\varepsilon_-$ where two separated clusters of connected valleys carry full polarization. But for each cluster the spin flip operator S^-_{cluster} can be applied separately resulting in a not fully polarized ground state, i.e. a state with $\vec{S}^2 < (n/2)(n/2+1)$.

For further details of flat-band ferromagnetism in the sawtooth-chain Hubbard model the interested reader is referred to original papers of Tasaki[52] but also to Ref. [51].

4. Summary

To summarize, we have reviewed some basic concepts of localized eigenstates in correlated systems on highly frustrated lattices and their effect on low-temperature thermodynamics. Usually noninteracting electrons or magnons on a lattice are delocalized, i.e. are described by a propagating wave function. These excitations may become localized due to randomness or after switching on interactions. In difference to that frustrating lattice topology may provide another mechanism for localization. Localized states may survive in the presence of interaction and under certain conditions they can control the properties of the system.

Strong frustration may lead to localization of magnons, i.e. the quantum XXZ Heisenberg antiferromagnetic Hamiltonian on certain lattices, e.g. the kagomé lattice, exhibits eigenstates built by a product of excitations localized on restricted areas of the lattice. These states are relevant at low temperatures T and for magnetic fields h in the vicinity of the saturation field h_1 and they produce a universal and characteristic thermodynamic behavior, for instance an extra low-temperature peak in the specific heat and an enhanced magnetocaloric effect. We have distinguished several universality classes: hard monomers, one-dimensional hard dimers, hard squares, hard hexagons, large hard squares. The antiferromagnets which belong to the hard-square, large-hard-square or hard-hexagon universality classes should exhibit an order-disorder phase transition just below the saturation field which has a pure geometrical origin (ordering of localized magnons at sufficiently high densities).

The elaborated concepts can be applied to other strongly correlated systems, for example, to Hubbard electrons. Considering as an example the Hubbard model on a sawtooth chain we have shown that the localized electron states dominate the low-temperature thermodynamics around the chemical potential $\mu = \mu_0 = 2t$ corresponding to quarter filling.

There are still many open questions in this field. One problem should be illustrated here as an example. The localized excitations exist only under certain restrictions on the lattice geometry. Only if these restrictions are fulfilled the lowest energy excitations are completely localized (or the lowest-energy single-particle band is completely flat). Of course, it would be desirable to examine the case of small deviations from the ideal geometry, when localized excitations acquire some dispersion. Some preliminary results give evidence that the localized magnon effects survive in this case.[35] The systems with small deviations from the ideal geometry are also interesting from the point of view of possible experiments on the corresponding compounds. Although we do not have an unambiguous experimental proof for the localized state effects at the present time, the recent measurements for azurite[13,14] at high magnetic fields may present a closely related physics.

Acknowledgments

The authors would like to thank A. Honecker, T. Krokhmalskii, R. Moessner, H.-J. Schmidt, J. Schnack, and J. Schulenburg for fruitful collaboration in this field. The most of the results reviewed in this paper were taken from papers published together with these colleagues. We mention that most of the numerical work was preformed using J. Schulenburg's *spinpack*. We thank H. Kikuchi and H. Kageyama for providing us Fig. 3. The authors acknowledge kind hospitality of the Organizers of the 43rd Karpacz Winter School of Theoretical Physics "Condensed Matter Physics in the Prime of XXI Century: Phenomena, Materials, Ideas, Methods" in Lądek Zdrój in February 2007. O. D. is indebted to Magdeburg University for hospitality in the summer of 2007.

References

1. G. Toulouse, Commun. Phys. **2**, 115 (1977).
2. K. Binder and A. P. Young, Rev. Mod. Phys. **58**, 801 (1986).
3. G.H. Wannier, Phys. Rev. **79**, 357 (1950).
4. R.M.F. Houtappel, Physica **16**, 425 (1950).

5. P.W. Anderson, Mater. Res. Bull. **8**, 153 (1973);
P.W. Anderson and P. Fazekas, Phil. Mag. **30**, 423 (1974).
6. P. Lemmens and P. Millet, in: *Quantum Magnetism*, U. Schollwöck, J. Richter, D.J.J. Farnell, R.F. Bishop, Eds. (Lecture Notes in Physics, **645**) (Springer, Berlin, 2004), pp. 433-477.
7. P. Schiffer, Nature **413**, 48 (2001).
8. R. Moessner, Can. J. Phys. **79**, 1283 (2001).
9. R. Moessner and A.P. Ramirez, *Physics Today*, February 2006, p. 24.
10. *Frustrated Spin Systems*, H.T. Diep, Ed. (World Scientific, Singapore, 2004).
11. *Quantum Magnetism*, U. Schollwöck, J. Richter, D.J.J. Farnell and R.F. Bishop, Eds. (Lecture Notes in Physics, **645**) (Springer, Berlin, 2004).
12. T. Masuda, A. Zheludev, A. Bush, M. Markina and A. Vasiliev, Phys. Rev. Lett. **94**, 039706 (2005);
M. Enderle, C. Mukherjee, B. Fak, R.K. Kremer, J.-M. Broto, H. Rosner, S.-L. Drechsler, J. Richter, J. Malek, A. Prokofiev, W. Assmus, S. Pujol, J.-L. Raggazzoni, H. Rakoto, M. Rheinstädter and H.M. Rønnow, Europhys. Lett. **70**, 237 (2005);
S.-L. Drechsler, O. Volkova, A. N. Vasiliev, N. Tristan, J. Richter, M. Schmitt, H. Rosner, J. Málek, R. Klingeler, A.A. Zvyagin and B. Büchner, Phys. Rev. Lett. **98**, 077202 (2007).
13. H. Kikuchi, Y. Fujii, M. Chiba, S. Mitsudo, T. Idehara, T. Tonegawa, K. Okamoto, T. Sakai, T. Kuwai and H. Ohta, Phys. Rev. Lett. **94**, 227201 (2005);
B. Gu and G. Su, Phys. Rev. Lett. **97**, 089701 (2006);
H. Kikuchi, Y. Fujii, M. Chiba, S. Mitsudo, T. Idehara, T. Tonegawa, K. Okamoto, T. Sakai, T. Kuwai and H. Ohta, Phys. Rev. Lett. **97**, 089702 (2006).
14. H. Kikuchi, Y. Fujii, M. Chiba, S. Mitsudo, T. Idehara, T. Tonegawa, K. Okamoto, T. Sakai, T. Kuwai, K. Kindo, A. Matsuo, W. Higemoto, K. Nishiyama, M. Horvatić and C. Bertheir, Prog. Theor. Phys. Suppl. **159**, 1 (2005).
15. R. Melzi, P. Carretta, A. Lascialfari, M. Mambrini, M. Troyer, P. Millet and F. Mila, Phys. Rev. Lett. **85**, 1318 (2000).
16. H. Rosner, R.R.P. Singh, W.H. Zheng, J. Oitmaa, S.-L. Drechsler and W.E. Picket, Phys. Rev. Lett. **88**, 186405 (2002).
17. R. Coldea, D.A. Tennant, A.M. Tsvelik and Z. Tylczynski, Phys. Rev. Lett. **86**, 1335 (2001).
18. P. Mendels, F. Bert, M.A. de Vries, A. Olariu, A. Harrison, F. Duc, J.C. Trombe, J.S. Lord, A. Amato and C. Baines, Phys. Rev. Lett. **98**, 077204 (2007).
19. S. Taniguchi, T. Nishikawa, Y. Yasui, Y. Kobayashi, M. Sato, T. Nishioka, M. Kontani and K. Sano, J. Phys. Soc. Jpn. **64**, 2758 (1995).
20. H. Kageyama, K. Yoshimura, R. Stern, N.V. Mushnikov, K. Onizuka, M. Kato, K. Kosuge, C.P. Slichter, T. Goto and Y. Ueda, Phys. Rev. Lett. **82**, 3168 (1999).
21. T. Momoi and K. Totsuka, Phys. Rev. **B61**, 3231 (2000).

22. K. Onizuka, H. Kageyama, Y. Narumi, K. Kindo, Y. Ueda and T. Goto, J. Phys. Soc. Jpn. **69**, 1016 (2000).
23. M. Isobe and Y. Ueda, J. Phys. Soc. Jpn. **71**, 1848 (2002).
24. K. Takada, H. Sakurai, E. Takayama-Muromachi, F. Izumi, R.A. Dilanian and T. Sasaki, Nature **422**, 53 (2003);
Y. Wang, N.S. Rogado, R.J. Cava and N.P. Ong, Nature **423**, 425 (2003);
M.L. Foo, Y. Wang, S. Watauchi, H.W. Zandbergen, T. He, R.J. Cava and N.P. Ong, Phys. Rev. Lett. **92**, 247001 (2004).
25. H.N. Kono and Y. Kuramoto, J. Phys. Soc. Jpn. **75**, 084706 (2006).
26. H. Tamura, K. Shiraishi, T. Kimura and H. Takayanagi, Phys. Rev. **B65**, 085324 (2002);
R. Arita, K. Kuroki, H. Aoki, A. Yajima, M. Tsukada, S. Watanabe, M. Ichimura, T. Onogi and T. Hashizume, Phys. Rev. **B57**, R6854 (1998).
27. J. Schnack, H.-J. Schmidt, J. Richter and J. Schulenburg, Eur. Phys. J. **B24**, 475 (2001).
28. J. Schulenburg, A. Honecker, J. Schnack, J. Richter and H.-J. Schmidt, Phys. Rev. Lett. **88**, 167207 (2002).
29. H.-J. Schmidt, J. Phys. **A35**, 6545 (2002).
30. J. Richter, J. Schulenburg, A. Honecker, J. Schnack and H.-J. Schmidt, J. Phys.: Condens. Matter **16**, S779 (2004).
31. J. Richter, O. Derzhko and J. Schulenburg, Phys. Rev. Lett. **93**, 107206 (2004).
32. J. Richter, J. Schulenburg and A. Honecker, in: *Quantum Magnetism*, U. Schollwöck, J. Richter, D.J.J. Farnell, R.F. Bishop, Eds. (Lecture Notes in Physics, **645**) (Springer, Berlin, 2004), pp. 85-153.
33. M.E. Zhitomirsky and A. Honecker, J. Stat. Mech.: Theor. Exp., P07012 (2004).
34. M.E. Zhitomirsky and H. Tsunetsugu, Phys. Rev. **B70**, 100403(R) (2004).
35. O. Derzhko and J. Richter, Phys. Rev. **B70**, 104415 (2004).
36. J. Richter, J. Schulenburg, A. Honecker and D. Schmalfuß, Phys. Rev. **B70**, 174454 (2004).
37. J. Richter, J. Schulenburg, P. Tomczak and D. Schmalfuß, arXiv:cond-mat/0411673.
38. R. Schmidt, J. Richter and J. Schnack, J. Magn. Magn. Mater. **295**, 164 (2005).
39. O. Derzhko and J. Richter, Phys. Rev. **B72**, 094437 (2005).
40. O. Derzhko, J. Richter and J. Schulenburg, Phys. Stat. Sol. (b) **242**, 3189 (2005).
41. J. Richter, Fizika Nizkikh Temperatur (Kharkiv) **31**, 918 (2005), [Low Temperature Physics **31**, 695 (2005)].
42. M.E. Zhitomirsky and H. Tsunetsugu, Prog. Theor. Phys. Suppl. **160**, 361 (2005).
43. O. Derzhko and J. Richter, Eur. Phys. J. **B52**, 23 (2006).
44. H.-J. Schmidt, J. Richter and R. Moessner, J. Phys. **A39**, 10673 (2006).
45. J. Richter, O. Derzhko and T. Krokhmalskii, Phys. Rev. **B74**, 144430 (2006).

46. J. Schnack, H.-J. Schmidt, A. Honecker, J. Schulenburg and J. Richter, J. Phys.: Conf. Ser. **51**, 43 (2006).
47. O. Derzhko, J. Richter, A. Honecker and H.-J. Schmidt, Fizika Nizkikh Temperatur (Kharkiv) **33**, 982 (2007) [Low Temperature Physics **33**, 745 (2007)].
48. M.E. Zhitomirsky and H. Tsunetsugu, Phys. Rev. **B75**, 224416 (2007).
49. A. Honecker and J. Richter, Condensed Matter Physics (L'viv) **8**, 813 (2005).
50. A. Honecker and J. Richter, J. Magn. Magn. Mater. **310**, 1331 (2007).
51. O. Derzhko, A. Honecker and J. Richter, Phys. Rev. **B76**, 220402(R) (2007).
52. A. Mielke, J. Phys. **A24**, L73 (1991); A. Mielke, J. Phys. **A24**, 3311 (1991); A. Mielke, J. Phys. **A25**, 4335 (1992); H. Tasaki, Phys. Rev. Lett. **69**, 1608 (1992); A. Mielke and H. Tasaki, Commun. Math. Phys. **158**, 341 (1993); H. Tasaki, Prog. Theor. Phys. **99**, 489 (1998).
53. T. Radu, H. Wilhelm, V. Yushankhai, D. Kovrizhin, R. Coldea, Z. Tylczynski, T. Lühmann and F. Steglich, Phys. Rev. Lett. **95**, 127202 (2005).
54. D.C. Mattis, *The Theory of Magnetism I* (Springer, Berlin, 1988).
55. A. Honecker, J. Schulenburg and J. Richter, J. Phys.: Condens. Matter **16**, S749 (2004).
56. D.C. Cabra, M.D. Grynberg, P.C.W. Holdsworth and P. Pujol, Phys. Rev. **B65**, 094418 (2002);
D.C. Cabra, M.D. Grynberg, P.C.W. Holdsworth, A. Honecker, P. Pujol, J. Richter, D. Schmalfuß and J. Schulenburg, Phys. Rev. **B71**, 144420 (2005).
57. G. Misguich, Th. Jolicoeur and S.M. Girvin, Phys. Rev. Lett. **87**, 097203 (2001).
58. M. Oshikawa, Phys. Rev. Lett. **84**, 1535 (2000).
59. M.E. Fisher, Phys. Rev. **124,** 1664 (1961).
60. R.J. Baxter, J. Phys. **A13**, L61 (1980).
61. R.J. Baxter, *Exactly Solved Models in Statistical Mechanics* (Academic Press, London, 1982).
62. H.C.M. Fernandes, J.J. Arenzon and Y. Levin, J. Chem. Phys. **126**, 114508 (2007).
63. N.D. Mermin and H. Wagner, Phys. Rev. Lett. **17**, 1133 (1966);
N.D. Mermin and H. Wagner, Phys. Rev. Lett. **17**, 1307 (1966).
64. M.E. Zhitomirsky, Phys. Rev. **B67**, 104421 (2003).
65. J. Schnack, R. Schmidt and J. Richter, Phys. Rev. **B76**, 054413 (2007).
66. K. Penc, H. Shiba, F. Mila and T. Tsukagoshi, Phys. Rev. **B54**, 4056 (1996);
H. Sakamoto and K. Kubo, J. Phys. Soc. Jpn. **65**, 3732 (1996);
Y. Watanabe and S. Miyashita, J. Phys. Soc. Jpn. **66**, 2123;
Y. Watanabe and S. Miyashita, J. Phys. Soc. Jpn. **66**, 3981 (1997).

FULL-POTENTIAL LOCAL-ORBITAL APPROACH TO THE ELECTRONIC STRUCTURE OF SOLIDS AND MOLECULES

M. RICHTER, K. KOEPERNIK and H. ESCHRIG

Leibniz Institute for Solid State and Materials Research, IFW Dresden e.V., P.O. Box 270 116, D-01171 Dresden, Germany

The issue of numerical accuracy in modern electronic structure theory is addressed. It is pointed out, that the main difference between high-accuracy electronic structure codes consists in the choice of the basis functions for an algebraic solution of the Kohn-Sham equations. Different strategies to define appropriate basis sets in linear combination of local orbitals methods are reviewed and numerical examples are given. Finally, features of the full-potential local-orbital (FPLO) code are summarized.

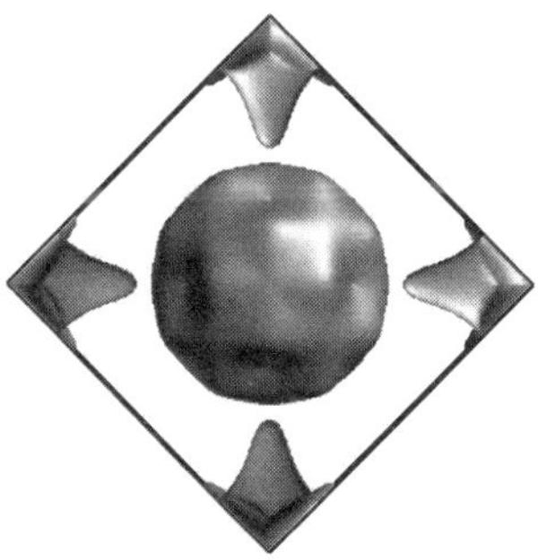

1. Introduction

Each chapter of the well-known textbook *Principles of the Theory of Solids* by J.M. Ziman is headed by a quotation. In particular, the chapter about *Electronic states* is introduced by R. Kipling's words

There are nine and sixty ways of constructing tribal lays,
And-every-single-one-of-them-is-right.

Kipling's wisdom he "learned [...] when the moose and the reindeer roard where Paris roars to-night" is a pray for tolerance of variety, and Ziman makes the point that even in aiming at a result of mathematical rigor there may be all the good reason of the world to justify a variety of ways to reach the goal. Everybody who has been working in the field of electronic structure theory knows, however, that *quantitative* comparisons between recent codes, on a physically relevant scale of accuracy, can still be unsatisfactory. This statement does not refer to the never ending discussion if LSDA, GGA, LSDA+U, EXX, SIC-LSDA or yet another functional is the preferable approximation[a] to density functional theory (DFT) for a wanted answer on a given system. What is meant is the purely numerical implementation of a well-defined task. Take six different band structure codes and let them calculate the lattice constant of fcc thorium in local density approximation (LDA). You will get answers deviating from each other by a much larger amount than the scatter of related experimental data, see Figure 1. Remember, we do not want to discuss the so-called problem of over-binding in LDA![b] ***This problem can only be tackled if we know what the numerically well defined LDA result is.***

At this point it is fair to state that tremendous advances in the numerical techniques have been achieved since John Slater's days. A number of full-potential methods with a high degree of reliability and flexibility have been developed.[c] The price to be paid for accuracy is computing time and main storage: *accurate* linear methods require large basis sets, while non-linear schemes are slow by construction.

These Lecture Notes are devoted to the recently developed full-potential local-orbital (FPLO) code. Already in its original version,[1] this highly accurate code is numerically (almost) as efficient as wide-spread lower-accuracy schemes. This efficiency is based on employing a self-adjusting minimum valence basis. The valence basis is completed by a few upper core states and polarization states. All basis states are nearly exactly orthogonalized to true crystal potential core states computed separately in an on-site problem. In combination with sophisticated numerical techniques to solve the self-consistent Kohn-Sham equations, the basis set ensures a numerical accuracy that competes with other all-electron full-potential schemes. As the description of the complete scheme would by far exceed the frame of these Lecture Notes, we focus on the basis set construction and explain why it is

[a] We do not explain the meaning of these abbreviations, since they are not relevant in the context of these Notes.

[b] Transition metals of the 5d series are exceptions from the LDA over-binding feature. Here, LDA yields equilibrium volumes only a few per mille off the experimental values.

[c] On purpose, we do not quote the individual methods: to forget one of them would be worse than not to mention any...

superior to the traditional linear combination of atomic orbitals (LCAO) method. Notwithstanding, the self-adjusting basis can destabilize the self-consistent iterations and, more severe, does not allow to calculate forces. For this reason, the latest version of the code, FPLO-7, makes use of a fixed, predefined basis set. The construction of the latter set is based on experience with FPLO and ideas developed and implemented in the DMol code[2] by Bernard Delley.

Before we present the different approaches, a few words should be spent on the history of this development. The ancestor of our present method is the linear combination of atomic orbitals method (LCAO), well-known from textbooks and, besides the complementary plane-wave method, most frequently explained to students in solid state physics. LCAO has two major disadvantages. At first, many Slater-Koster integrals have to be calculated since the atomic functions are far-ranging. Second and more severe, the atomic basis is incomplete if only bound states are included and would be over-complete if all scattering states were included. In the mid-seventies, the development of an optimized (O)LCAO method was started being exempt from the mentioned shortcomings. The trick was to smoothly squeeze the atomic orbitals by an attractive potential which pushes them also to higher energies and thus improves basis completeness.[3] The same idea is used in the presented code in refined form. The ins and outs of OLCAO were published in in the year 1988 together with calculated data for all light metals of the periodic table up to Zn.[4]

As time went by, the calculation of band structures and densities of states alone was not sufficient anymore, and total energy calculations with much higher accuracy demands came into focus. The original OLCAO implementation could not compete in this field, since it relied on a representation of radial functions (wave functions, densities, and potentials) in terms of analytic Slater-type orbitals. This choice had been taken on the background of the available main storage (up to at most 1 MByte) of computers accessible to Dresden physicists before 1990. In the early ninetieth, it became clear[5] that the Slater-type representation had to be abandoned. In addition, we were heading for uttermost accuracy. Thus, a mixed basis scheme that achieved this goal at the price of a comparably poor performance was implemented.[6] The accuracy obtained by this method is perhaps not surpassed yet, but it cannot be used for unit cells larger than a few atoms. On the basis of the existing experience, the development of FPLO was launched. Its available versions seem to represent a most satisfactory compromise between high accuracy and reasonable performance.[1]

This development through more than three decades was almost from the beginning paralleled by the implementation of important extensions to the basic codes. Chemical disorder was treated in the OLCAO code within non charge-selfconsistent coherent potential approximation (CPA), implemented by late Reinhard Richter.[7] A much extended and completely charge-selfconsistent CPA version[8,9] was added to FPLO on the basis of the nearly forgotten pseudo-spin approach of Blackman, Esterling, and Berk. Relativistic versions have been developed for the old code[10,11] as well as for FPLO.[12,13] Both implementations are based on a full four component representation of the Bloch states. Finally, LSDA+U has been implemented[14] and a slab version has just been completed.[15]

2. Why yet another DFT solver?

A vast number of numerical methods to solve the Kohn-Sham equations, a set of nonlinear integro-differential equations, has been implemented in past decades. Though these equations are included in every modern solid state theory textbook and, thus, seemingly standard, their accurate solution for the needs of solid state physics problems of steadily growing complexity requires a large arsenal of numerical methods and tricks. The related codes, mostly written in FORTRAN and/or C, usually consist of several hundred-thousand lines. Depending on the number of add-ons and the level of accuracy and sophistication, their development takes up to several 10 person-years.

DFT solvers can be classified according to

(i) the level of accuracy in the construction of potential and charge density: muffin-tin (MT) approximation, atomic-sphere approximation (ASA), or full-potential (FP or F), the latter being the most accurate approach;

(ii) the treatment of the atomic core states: pseudo-potential (PP) or all-electron;

(iii) the choice of the basis set for the Bloch states or molecular orbitals: augmented plane waves (APW), muffin-tin orbitals (MTO), plane waves (PW), atomic orbitals (AO), just to name some.

As a rule of thumb, a better accuracy of the numerical method consumes more computational resources. However, the level of sophistication in the numerical approaches and, in particular, in the choice of the basis states, can essentially influence the relation between accuracy and effort. This is the reason why the following discussion will be focussed on the choice of an

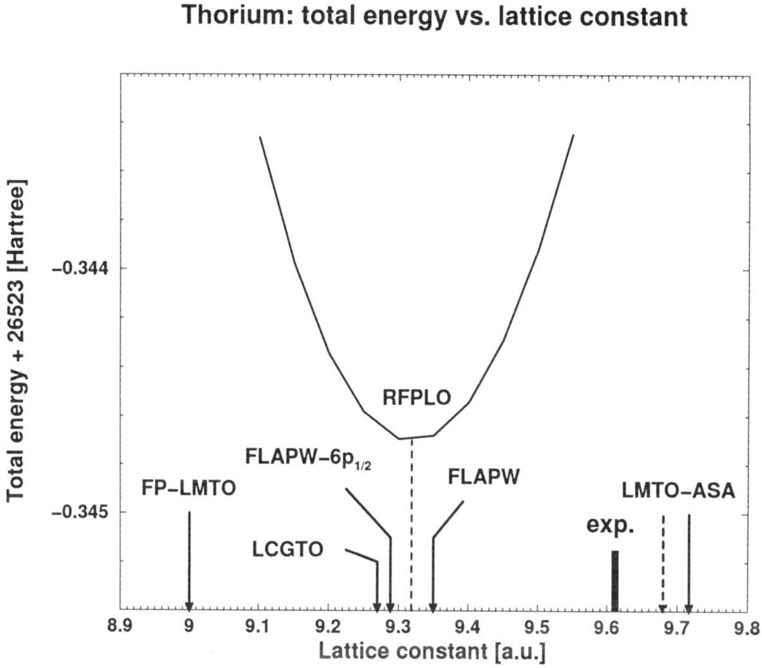

Fig. 1. Total energy of fcc thorium obtained with the fully relativistic FPLO method (RFPLO)[12,13] and comparison of different published values of the LDA lattice constant with the experimental value. The thickness of the line denoted "exp." indicates the scatter of different experimental data.[16] Theoretical data are taken from the following references: FP-LMTO,[17] LCGTO and FLAPW,[18] FLAPW-6$p_{1/2}$ unpublished data (Jan Kuneš, using the method published in Ref. [19]), LMTO-ASA.[20] The LMTO-ASA value published in Ref. [20] refers to room temperature (full line), and the related zero-temperature value (dashed line) was estimated from the linear coefficient of thermal expansion.

appropriate set of basis states. Still, the demands on accuracy, efficiency, and stability may be different for different problems in the vast realm of solid state theory, and, most importantly, since there is less than no hope to get exact analytical solutions to a representative selection of them, the only way to judge numerical accuracy is to compare output numbers of different approaches corresponding to exactly the same input numbers.

Why is accuracy an issue at all? The answer is given in Figure 1, showing the most recent published LDA results for the lattice constant of fcc thorium. While the scatter of the experimental data amounts to about 0.1%, the difference between ASA and FP variants of the same method (LMTO) amounts to more than 7%, i.e. about 20% in volume. Even results obtained with different sophisticated full-potential methods scatter within about 3.5%. Only if this number is reduced to less than 1% it is possible to judge the quality of approximations to DFT, like the local density approximation (LDA) or the generalized gradient approximation (GGA), with respect to the evaluation of lattice geometries and elastic properties.

3. LCLO equations and core-valence transformation

We solely focus on the task of solving a given single-particle Schrödinger-like equation, which is part of one step of iterating the self-consistency explained below,

$$\hat{H}\psi = \epsilon\psi \,. \tag{1}$$

To be specific, $\hat{H}$ should denote the Kohn-Sham operator of a (periodic) arrangement, ψ are Kohn-Sham orbitals, and ϵ are the related energies. More explicitly, the Kohn-Sham equations read

$$(-\frac{\Delta}{2} + V_{\text{ext}} + V_{\text{Hartree}} + V_{\text{xc}})\, \psi_{\mathbf{k}n}(\mathbf{r}) = \psi_{\mathbf{k}n}(\mathbf{r})\, \epsilon_{\mathbf{k}n} \,, \tag{2}$$

$$\rho(\mathbf{r}) = \sum_{\mathbf{k}n}^{\text{occ}} |\psi_{\mathbf{k}n}(\mathbf{r})|^2 \,, \tag{3}$$

$$V_{\text{Hartree}}(\mathbf{r}) = \int d^3r' \rho(\mathbf{r}')/|\mathbf{r} - \mathbf{r}'| \,, \tag{4}$$

$$V_{\text{xc}}(\mathbf{r}) = \delta E_{\text{xc}}[\rho]/\delta\rho \,, \tag{5}$$

with the kinetic energy $-\frac{\Delta}{2}$ which may be replaced in case of need by a relativistic operator, the external potential V_{ext}, mostly produced by the atomic nuclei, the Hartree potential V_{Hartree} produced by the electronic charge density $\rho(\mathbf{r})$, and the exchange-correlation potential V_{xc} which is the functional derivative of the unknown and hence modeled exchange-correlation energy functional $E_{\text{xc}}[\rho]$. The quantum numbers $\mathbf{k}n$ refer to Bloch states but can be replaced by any other appropriate set in a situation lacking three-dimensional periodicity.

As both Hartree and exchange-correlation potential depend on the charge density in a non-trivial way, the Kohn-Sham equations form a set of non-linear integro-differential equations that has to be solved by iteration, intuition, and precise numerics. Once this is done, one can evaluate the total energy and other quantities of interest, e.g. matrix elements, in a straightforward manner. There are two classes of approximations that have to be applied in order to make the task tractable:

(1) The unknown functional $E_{\rm xc}[\rho]$ has to be defined. Particular choices are the (LDA), the (GGA), and a number of others. We will not dwell on this point and presume that an appropriate choice has been taken. Let us only insist, that any density functional application leaves *ab initio* level at this stage and enters model treatment, although in many cases without recourse to empirical parameters.

(2) Numerical approximations are unavoidable and numerous, the most critical being the choice of a reasonably complete basis set for the expansion of the wave function ψ, if an algebraic solution of the single-particle equation is attempted. This choice usually determines the name of the band structure code.

The reader should be reminded, that *the choice of a model functional is an issue independent of the numerical accuracy.* Unfortunately, both issues are frequently not distinguished from each other in the discussion. This distinction may even be impossible if the numerical accuracy cannot be controlled at a sufficient level.

From now on, we will focus on one specific Ansatz for the basis set and discuss three different implementations. The other ingredients for the solution of the Kohn-Sham equations will not be discussed: the calculation of the charge density, the solution of the Poisson equation, the evaluation of the exchange-correlation potential, and the iteration of the self-consistent equations.

It was Felix Bloch who suggested an Ansatz for ψ in terms of localized, e.g. atomic, orbitals: linear combination of atomic orbitals (LCAO).[21]

Consider the following Ansatz for the Bloch states $\psi_{\mathbf{k}n}(\mathbf{r})$ with band index n and wave vector $\mathbf{k}$,

$$\psi_{\mathbf{k}n}(\mathbf{r}) = \frac{1}{\sqrt{N}} \sum_{\mathbf{R}\mathbf{s}L} \phi_{\mathbf{s}L}(\mathbf{r}-\mathbf{R}-\mathbf{s}) C_{L\mathbf{s},\mathbf{k}n} e^{i\mathbf{k}(\mathbf{R}+\mathbf{s})} \; . \tag{6}$$

Here, the basis states $\phi_{\mathbf{s}L} = \langle \hat{\mathbf{r}} | \mathbf{R}\mathbf{s}L \rangle$ are local overlapping orbitals at sites $\mathbf{s}$ in the cell at $\mathbf{R}$. The index L comprises any complete set of quantum numbers, e.g., $L = \{\nu, l, m\}$, including the principle, orbital, and magnetic quantum numbers ν, l, and m, respectively. The variable N denotes the number of unit cells (using Born-von Kármán boundary conditions). Finally, C is the transformation matrix from the basis states ϕ to the Bloch states ψ, up to a phase factor determined by translational symmetry. The prefactor $\sqrt{N}$ makes C independent of N. The columns of C contain the coefficients of the eigenstates of the problem.

This Ansatz is not restricted to lattices with three-dimensional periodicity but can equally well be applied for lattices with two-, one- or zero-dimensional periodicity (slabs, rods, or molecules/clusters, respectively) with the appropriate modification in the dimensionality of the vectors $\mathbf{R}$ and $\mathbf{k}$. For molecules/clusters, $\mathbf{R}$, $\mathbf{k}$, and the phase factor are simply skipped. Further, the Ansatz and the following considerations are not restricted to atomic orbitals ϕ but are also valid for the more general *linear combination of local orbitals (LCLO)*, specified in the next section.

For a non-orthogonal basis, the secular equation reads

$$HC = SC\epsilon\,. \tag{7}$$

$$H_{\mathbf{s}'L',\mathbf{s}L} = \sum_{\mathbf{R}} \langle \mathbf{0}\mathbf{s}'L'|\hat{H}|\mathbf{R}\mathbf{s}L\rangle e^{i\mathbf{k}(\mathbf{R}+\mathbf{s}-\mathbf{s}')}, \tag{8}$$

$$S_{\mathbf{s}'L',\mathbf{s}L} = \sum_{\mathbf{R}} \langle \mathbf{0}\mathbf{s}'L'|\mathbf{R}\mathbf{s}L\rangle e^{i\mathbf{k}(\mathbf{R}+\mathbf{s}-\mathbf{s}')}\,. \tag{9}$$

To reduce the rank of the Hamilton and overlap matrices $H_{\mathbf{s}'L',\mathbf{s}L}$ and $S_{\mathbf{s}'L',\mathbf{s}L}$, the basis states are divided into the valence states $\phi_{\mathbf{s}L_v}$ and the core states $\phi_{\mathbf{s}L_c}$. The latter are defined by the condition

$$\langle \mathbf{R}'\mathbf{s}'L'_c|\mathbf{R}\mathbf{s}L_c\rangle \propto \delta_{\mathbf{R}\mathbf{R}'}\delta_{\mathbf{s}\mathbf{s}'}\,. \tag{10}$$

There are several options on how to treat the core sector. In FPLO-5 the core orbitals are chosen to be eigenstates of the spherically averaged crystal Hamiltonian, thus forming good approximations to the Kohn-Sham core states. In FPLO-6 and later versions, the Kohn-Sham core states are obtained by diagonalizing the core-core sector of the full Hamiltonian matrix separately. Due to the vanishing overlap from different sites, Eq. (10), this diagonalization is purely on-site and hence $\mathbf{k}$-independent and cheap. The core states thus obtained are rather good approximants to the true Kohn-Sham core states; in particular they exhibit the local crystal symmetry.

In both schemes described above the core states can be removed from the $\mathbf{k}$-dependent diagonalization of the full crystal Hamiltonian by orthogonalization of the valence sector to the Kohn-Sham core states, which is the core orbitals in FPLO-5 and a linear combination of the core orbitals in FPLO-6 and later versions. This exact transformation allows to keep all electrons in the calculation, while only the valence state problem needs to be diagonalized. Since matrix eigenvalue problems scale with the third power of the rank, the transformation can reduce the effort by a factor of up to 30 for heavy atoms. It should be noted that Eq. (10) is of course only fulfilled to a certain level of accuracy. Accuracy requirements determine which states can be treated as core states.

4. Three strategies to define the local basis

Ancestor to all local basis schemes is the linear combination of *atomic* orbitals, LCAO. This method has been used in numerous early implementations.[22,23] In order to define the atomic orbitals, one first solves the self-consistent Kohn-Sham equations for a spherical atom or ion and obtains the density:

$$\rho_0(r) = \sum_i n_i |\phi_i(r)|^2 \ , \quad N_e = \sum_i n_i \ , \tag{11}$$

ϕ_i being radial orbital functions, $i = \{\nu, l\}$, and n_i appropriate occupation numbers. If an atom is considered, the number of electrons equals the atomic number, $N_e = Z$. Using this density, one evaluates the spherical atomic Kohn-Sham potential,

$$V[\rho_0] = V_{\text{ext}} + \quad V_{\text{Hartree}} \quad + V_{\text{xc}} \tag{12}$$

$$= -\frac{Z}{r} + \int d^3r' \frac{\rho_0(r')}{|\mathbf{r} - \mathbf{r}'|} + V_{\text{xc}}(\rho_0(r)) \ . \tag{13}$$

The atomic orbitals ϕ_L are finally obtained by solving a non-selfconsistent Schrödinger equation with the potential from Equation (12). Figure 2 schematically shows the resulting energy levels. Lowest in energy, we find the levels of the inner core shells. The related radial wave functions do not extend beyond half nearest neighbor distance and fulfill Equation (10).

At higher energies but still clearly separated from the energies of the valence states we find the levels of the outer core shells. These shells are fully occupied but due to their relatively high energies their radial functions overlap with those of the outer core shells of neighboring atoms. Equation (10) is not fulfilled and these states have to be included in the valence basis. We will refer to these states as semi-core states from now on, in order to distinguish them from those core states that can be removed from the basis by orthogonalization. In practice, Equation (10) has to be replaced by a weaker condition

$$|\langle \mathbf{R}'\mathbf{s}'L'_c|\mathbf{R}\mathbf{s}L_c\rangle - \delta_{cc'}\delta_{\mathbf{R}\mathbf{R}'}\delta_{\mathbf{s}\mathbf{s}'}| < \varepsilon \ , \tag{14}$$

with ε adjusted to the required level of numerical accuracy.

Finally, there is a number of bound states just below potential zero, the valence states. They belong to incompletely filled shells and are responsible for the chemical bonding. Their radial functions usually considerably extend beyond the nearest neighbors which necessitates the computation of a large number of multi-center integrals to evaluate the Hamilton and overlap matrices, Equations (8,9). While this can yield a bad numerical performance, there is another incurable problem inherent to the described basis construction. At the left side of Figure 2, the energy range of the valence and conduction bands (and of the lower lying semi-core bands) is indicated. This range extends considerable into positive energies, but there are no basis states available for $\epsilon > 0$. In other words, the LCAO basis is incomplete by construction for energies above zero.

One possibility to cure the basis incompleteness of LCAO is to add plane waves to the basis, which have to be orthogonalized to the atomic states. So-called orthogonalized plane wave (OPW) schemes can be very accurate, if the core-orthogonalization is carefully done and a very small value of ε in Eq. (14) is chosen. [ε is the symbol on the right hand site of Eq. (14)].[6] However, they are numerically inefficient in particular compared to pseudo-potential methods and did not succeed on the market of electronic structure methods. We now will describe two variants of an alternative local basis construction, that is free of either of the mentioned problems, basis set incompleteness and high number of multi-center integrals.

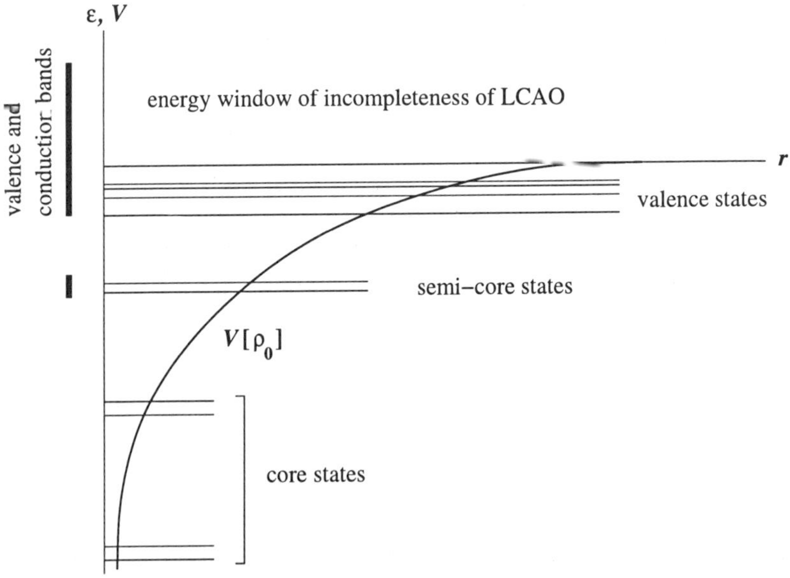

Fig. 2. Bound states in an atomic potential.

The first variant has been applied in the Dresden electronic structure code that was in use from 1975 till about 2000.[d] It is described in detail in the book by H. Eschrig.[4] We will refer to this variant as optimized linear combination of atomic orbitals, OLCAO. It has also been implemented in the early releases of the FPLO code, up to FPLO-5.

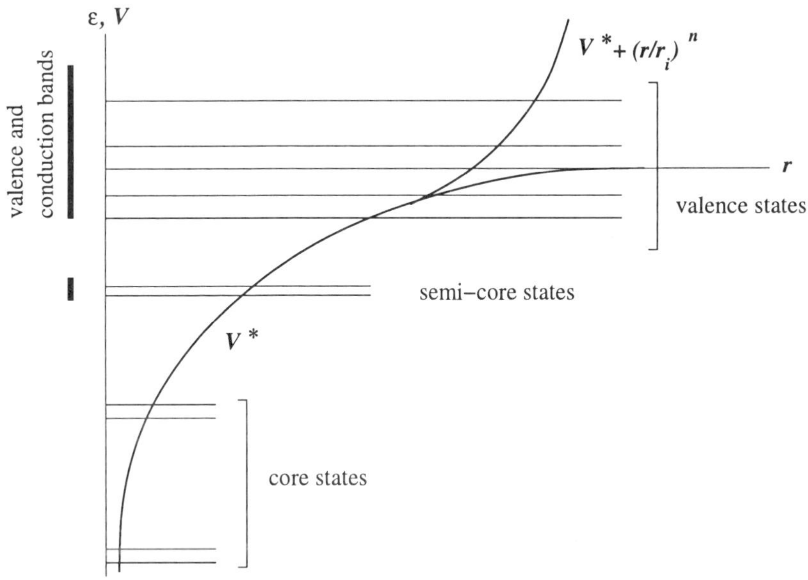

Fig. 3. All states are bound in a potential for basis optimization according to Eq. (15). Thus, sufficient completeness is achieved in the energy window of valence and conduction bands.

The key idea is to add an attractive potential to the atomic potential such, that the atomic potential is only marginally changed in the interior of the atom and that all states are bound. Polynomial potentials can fulfill both requirements, as they can be chosen small for small distances r from the nucleus and approaching $+\infty$ for $r \to \infty$. In OLCAO, the following potential is used to evaluate the local basis states (Figure 3):

$$V_0^{\text{olcao}}(r) = V^* + \left(\frac{r}{r_i}\right)^n . \tag{15}$$

Here, r_i is an optimization parameter, and $n = 4 \ldots 6$. By appropriate

[d]It is also at the roots of the still widely used density functional tight-binding method[24] for large clusters.

choice of the r_i, the energy levels of the valence basis states can be evenly distributed over the energy region of the valence and conduction bands, see Figure 4. This ensures a sufficient completeness of the basis within this region even if only a minimum basis set is used, e.g. 3s, 3p, 3d, 4s, and 4p states for a 3d transition metal or 1s, 1p, 2s, 2p, and 3d states for a first row main group element. Note, that the "minimum" basis includes the semicore-states and also low-lying empty states.

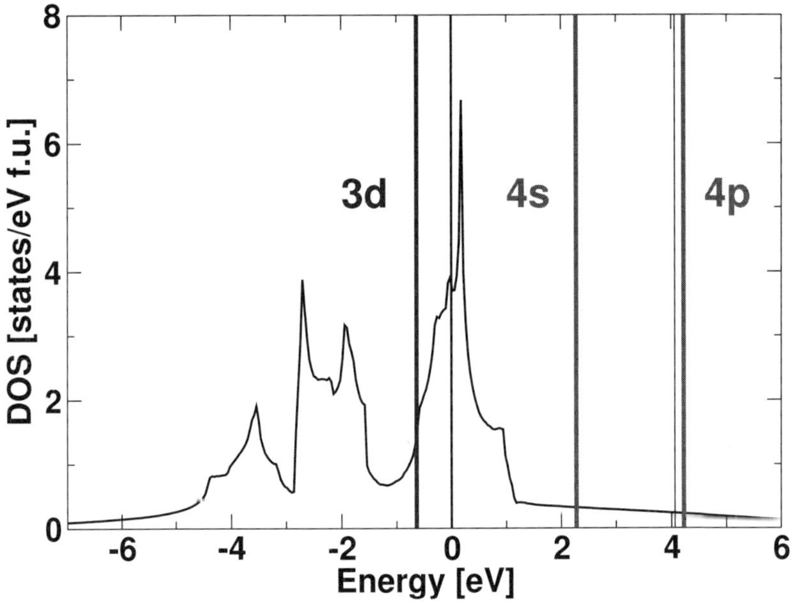

Fig. 4. Calculated density of states of non-magnetic bcc iron (FPLO-5, $a = 5.4$ Bohr radii). The positions of the local state energies (OLCAO scheme with optimized basis) are indicated by vertical lines. While the 3d states are located close to the barycenter of the 3d band, the 4s and 4p states are distributed over the lower part of the unoccupied bands.

The challenge of the minimum basis is that the basis orbitals must be well adjusted to the self-consistent potential. Therefore, in a calculation for a solid the basis must be recalculated in every self-consistency cycle. Thus, the potential V^* used in Eq. (15) is a spherical potential similar to $V[\rho_0]$ of Eq. (12) but evaluated by taking a spherical average of the Kohn-Sham potential from the current iteration step. Singularities at the neighbor atom distance are suppressed by a Gaussian smearing.

The chance of this strategy is that an automatic basis optimization by modification of the parameters r_i can be incorporated into the self-consistency cycle. By applying a Hellman-Feynman type approach, implemented into FPLO-5, the total energy is optimized with respect to the r_i. This way, a very high accuracy of total energies and densities is achieved. For the first time ever, *in 2002 two radically independent all-electron methods (FLAPW-WIEN and FPLO) provided the same total energies*, $\Delta E/E \approx 10^{-6} \ldots 10^{-7}$, for a number of close-packed structures.

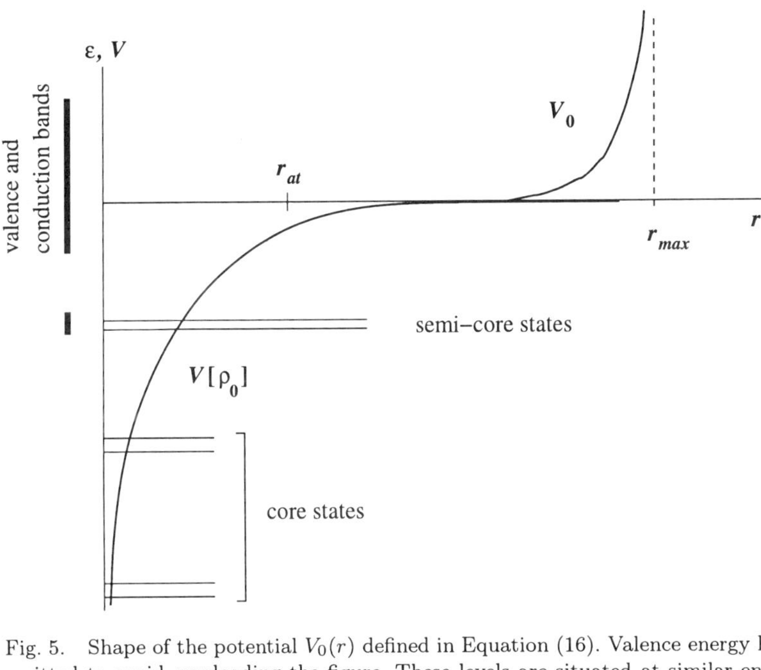

Fig. 5. Shape of the potential $V_0(r)$ defined in Equation (16). Valence energy levels are omitted to avoid overloading the figure. These levels are situated at similar energies as the levels in the optimized atomic potential, Figure 3.

Unfortunately, the optimization introduces highly complicated additional Pulay terms for the calculation of forces via the dependence of r_i on the atom positions. For this reason, another choice of a local basis is more appropriate to ensure full flexibility of a many-purpose electronic structure code: an adjusted fixed basis. Such a basis has been implemented in the DMol code[2] and also in FPLO-6 and later versions. Since the basis construction described below is much more general and much better suited for the description of both Bloch states and molecular orbitals than the original atomic basis of LCAO, we will use a distinct name for the approach: linear combination of local orbitals (LCLO).

Consider the following spherical potential, depicted in Figure 5:

$$V_0(r) = V[\rho_0] + \left(\frac{r}{\alpha_0\, r_{\max}}\right)^{n_0} + \bar{\Theta}(r_{\max} - r) , \tag{16}$$

with fixed parameters $\alpha_0 = 0.7 \ldots 0.8$, $n_0 = 14 \ldots 18$, $r_{\max} = 8 \ldots 12$ Bohr radii, and $\bar{\Theta}(x) = 0$ or $+\infty$ for positive or negative values of x, respectively. Valence states evaluated in the potential $V_0(r)$ also fulfill the two requirements discussed above, a reasonable completeness in the energy region above potential zero and compactness of the radial wave function. There is one major difference between the OLCAO and the LCLO constructions: The OLCAO basis is optimized for each individual chemical composition and each individual geometry during the calculation. At variance, the parameters of the LCLO basis are kept fixed, with related simplifications in the calculation scheme and performance. However, as its variational freedom is more restricted than that of the LCAO basis, the simple LCLO basis is less complete. This can be well compensated by introduction of additional states, at the price of a somewhat larger secular matrix.

The following choice of additional basis states is used in the package DMol. It was brought to our attention by Bernard Delley during the 4th FPLO workshop and adopted in FPLO. Take self-consistent atomic states ϕ_i as in Eq. (11) and construct an *ionic* charge density ρ_1,

$$\rho_1(r) = \sum_i n_i |\phi_i(r)|^2 , \quad Z > N_e = \sum_i n_i , \tag{17}$$

with $N_e = Z - Q$ and $Q = 0 \ldots 6 \leq Z$. Then, an ionic Kohn-Sham potential $V[\rho_1]$ is constructed according to Eq. (12) and used to define a second LCLO potential $V_1(r)$:

$$V_1(r) = V[\rho_1] + \left(\frac{r}{\alpha_1\, r_{\max}}\right)^{n_1} + \bar{\Theta}(r_{\max} - r) . \tag{18}$$

The values for α_1 and n_1 may be chosen slightly smaller than α_0 and n_0. Finally, the basis is adjusted by minimization of dimer and close-package bonding energies with respect to the values of Q. This procedure can also be applied to the valence states.

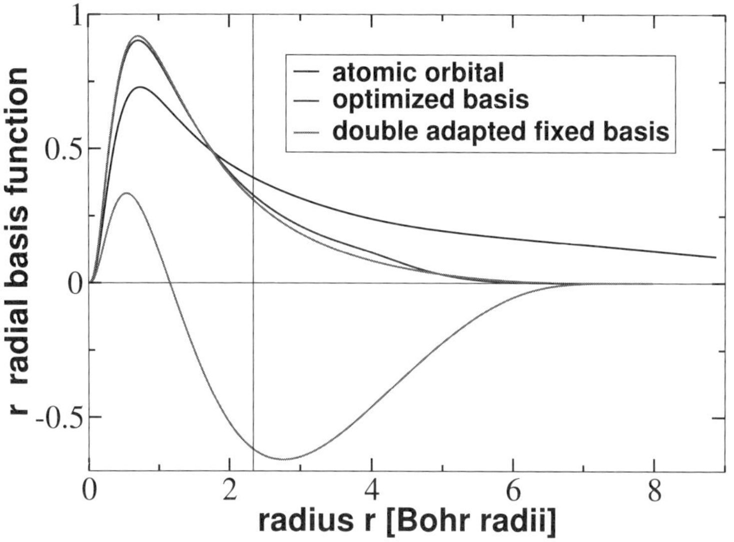

Fig. 6. Comparison of radial d basis functions (multiplied with the radius r) of iron. Black line: atomic 3d radial function (LCAO); blue line: optimized 3d radial function (OLCAO); red lines: fixed basis 3d and 4d radial functions with $r_{\mathrm{max}} = 8$ (LCLO). The vertical line indicates half nearest neighbor distance in bcc Fe.

Before we show examples of LCLO basis states and of the related adjustment, let us summarize the different types of basis states, their definition and the potential used to evaluate their radial functions in the following table:

core states:	occupied, no overlap	from V_0
semi-core states:	occupied, overlap	from V_0
valence states:	partly occupied	from V_0 (V_1)
polarization states:	empty	from V_1

Core, semi-core, and valence states form the primary basis set defined in the potential V_0. The completeness is improved by adding polarization states to the valence basis. These are states which are empty in the free atom (e.g., 3d states of first- or second-row main group elements). The polarization states are evaluated in the ionic potential V_1. As the ionic potential is more attractive than V_0, the polarization states reside in the same energy region as the valence states, improving the variational freedom.

In order to get an idea of the basis set size, consider two typical elements:

	core	semi-core	valence	polarization	rank of matrix
Al:	(1s)	(2s,2p)	(3s,3p)	(3d,4s,4p)	17
Fe:	(1s,2s,2p)	(3s,3p)	(3d,4s)	(4p,4d,5s)	19

One should compare the rank of valence state matrix with related numbers for other high-precision methods. For example, FLAPW implementations need about 100 basis functions per atom, and APW+lo implementations need about 60 basis functions per atom. Accordingly, systems with 5(3) times larger unit cells can be treated in FPLO with comparable numerical effort. On the other hand, it is possible to refine the mesh for k space integrations which is frequently an issue for high-precision calculations[25] by a factor of 5(3) in three dimensions, if the system size is unchanged. (Note, that the computation time scales with the third power of the matrix rank.)

Figure 6 shows a comparison of radial basis functions for iron d states used in the three methods discussed. The atomic radial 3d function used in the traditional LCAO method (black line) extends much beyond the nearest neighbor distance of about 4.7 Bohr radii in bcc Fe and requires the calculation of a large number of overlap and Hamilton integrals. The extension of the optimized atomic radial 3d function used in the OLCAO method (blue line) is restricted to the nearest neighbor shell. Thus, it overlaps with 3d functions of neighbors up to the fourth shell only (34 atoms in a bcc lattice). This is a another aspect of what is called nearsightedness in the density functional theory.[26] Other basis functions like the 4s functions extend somewhat further, but the number of directly interacting neighbor atoms is restricted to a few hundred in any case. The same holds for the fixed LCLO basis, red lines in Figure 6. The main difference between the OLCAO and the LCLO 3d radial functions consists in the small hump present in the OLCAO basis function between half and full nearest neighbor distance. This hump comes about, since the atomic-like potential V^* used to calculate the basis contains attractive contributions from the neighbor potentials at variance with $V[\rho_0]$ and $V[\rho_1]$. It is essential in the minimum basis representation used in OLCAO to describe the enhancement of d-like (in the present case) charge density due to bonding in the interstitial region. In the LCLO basis set, this task is fulfilled by the polarization states,

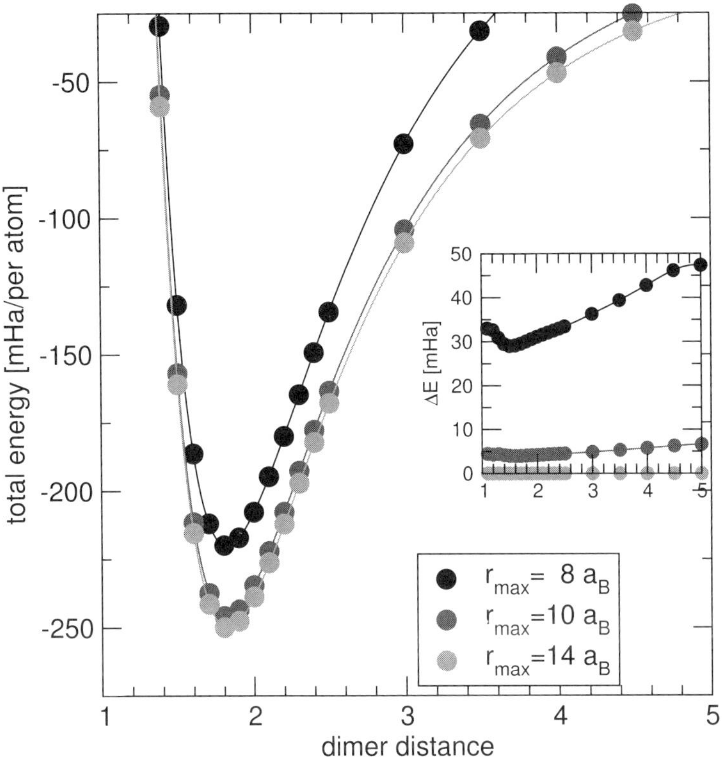

Fig. 7. Total energy of an iron dimer with respect to the energy sum of confined atoms in dependence of the distance and of the finite support radius $r_{\max}$. The inset shows energies normalized with respect to the $r_{\max} = 14$ values. While the data for $r_{\max} = 10$ differ almost only by a constant shift, the data for $r_{\max} = 8$ show different first and second derivatives.

4d in the present case (lower red line). The outer maximum of the adapted 4d radial function is close to the hump in the OLCAO 3d radial function. The completion of the basis by polarization states in LCLO provides additional variational freedom and allows to use a fixed basis with the related numerical advantages.

Next, we consider the dependence of total energies on the finite support radius $r_{\max}$ of the LCLO basis functions, Figure 7. The figure shows bonding energy curves, i.e. the energy difference between the dimer energy and the energy sum of two confined atoms in dependence of the distance between the nuclei. The finite support radius $r_{\max}$ is used as a parameter: a small value of $r_{\max} = 8$ Bohr radii (black curve) is compared with an intermediate value of $r_{\text{max}} = 10$ Bohr radii (red curve) and a converged value of $r_{\text{max}} = 14$ Bohr radii (green curve). One notes that the curves for $r_{\text{max}} = 14$ Bohr radii and $r_{\text{max}} = 10$ Bohr radii mainly differ by a constant energy. Thus, a moderate compression of the atomic states by reducing the compact support radius has almost no influence on calculated geometries (energy minima), forces (first derivatives) and elastic properties (second and higher derivatives). This allows to reduce the numerical effort (number of overlap and Hamilton integrals and sparsity of the eigenvalue problem) with modest reduction of accuracy.

Finally, we consider the dependence of total energies on the optimization charges, Figure 8. The upper panel shows the dependence of bonding energies for an Fe dimer on the optimization charge $Q = Z - N_e$ according to Eq. (17), the lower panel shows related bonding energies for bcc Fe. One notes that the total energies hardly change upon variation of the value of Q that defines the 5s basis states. This can be explained by the vanishingly low occupation of the 5s state. The 4p and 4d states are more occupied and thus have considerably more impact on the basis completeness in the occupied states region. Both are optimum at about the same value of Q, which is close to 3 in the dimer and close to 2 in the bcc bulk case. If a value of $Q = 3$ is taken as a compromise, the total energy is less than 1 mHartree above the lowest value for both the dimer and the bcc bulk geometry. This means, that a common value of Q can be applied for the most open (dimer) and more or less close packed geometries.

5. Summary

These Lecture Notes were devoted to a description of strategies to define appropriate basis sets in linear combination of local orbitals methods. Two of these strategies have been implemented in different FPLO releases.[27] The number of basis functions needed to achieve an absolute accuracy in the order of 0.1 eV/atom amounts to 10 ... 38 local functions per atom, depending on the atomic number. In this way, the code is one of the most efficient and accurate all-electron methods. It comprises a number of features, such as

- a cluster version on the same footing as the periodic version;
- calculation of total energies and forces;
- a 4-component Dirac-Kohn-Sham implementation;[12,13]
- a new and advantageous scalar relativistic method;
- a relative accuracy (numerical noise level) of $10^{-6\,(\ldots-8)}$ eV/atom,

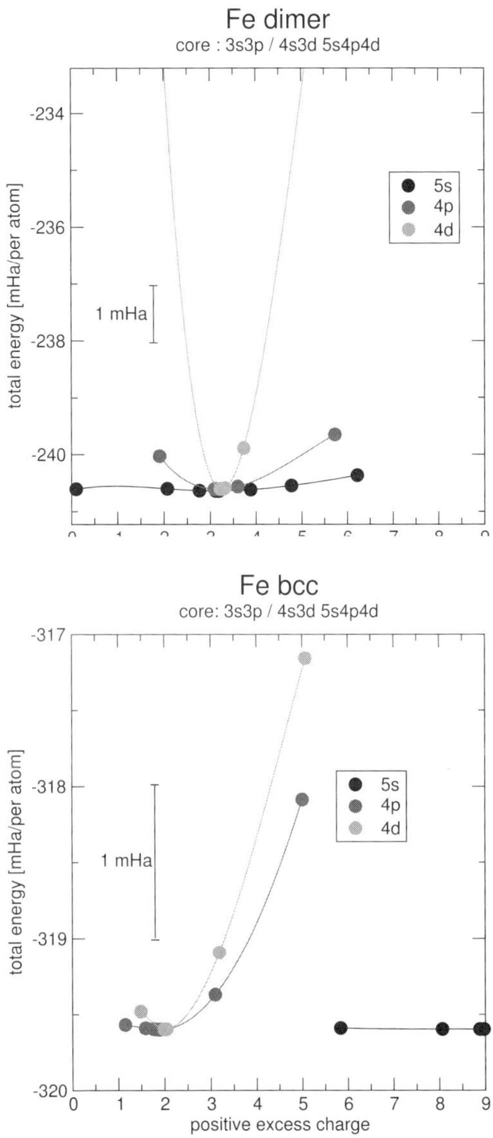

Fig. 8. Dependence of total energies of Fe in dimer and bcc bulk geometries on the ionicities Q defining the polarization states. Each curve is evaluated with Q fixed at its optimum value for the other states. Note that the leading digits of the total energies are omitted.

depending on the choice of numerical parameters;
- LSDA+U[14] in different versions;
- orbital polarization correction[28] schemes;
- band structures and 'fat bands';
- Fermi surfaces and Fermi velocities;
- coherent potential approximation[8,9] and disordered local moments (up to now, only in FPLO-5);
- slab- and rod versions (test stage).[15]

On top of these features, the FPLO package comes with a handy user interface that parses the input and thus restricts the trivial errors to a minimum. The whole package contains several 10^5 lines source code in FORTRAN90 and C, and its development took about 20 person years. The package is being used by about 70 groups worldwide, and the growing number of FPLO publications can be checked at the FPLO homepage.[27]

Acknowledgments

We are grateful for permanent technical assistance by Ulrike Nitzsche, for major contributions by Ingo Opahle (relativistic version), Igor Chaplygin (LSDA+U), and Ferenc Tasnadi (slab and rod versions in preparation) as well as for critical application, testing, completion, and distribution by Helge Rosner. Funding has been provided by DFG through SFB 463 and SPP 1145 and by the EC (psi-k f-electron).

References

1. K. Koepernik and H. Eschrig, Phys. Rev. **B59**, 1743–1757 (1999).
2. http://people.web.psi.ch/delley/dmol3.html/.
3. H. Eschrig and I. Bergert, phys. stat. sol. (b). **90**, 621 (1978).
4. H. Eschrig, *Optimized LCAO Method and the Electronic Structure of Extended Systems*, (Springer-Verlag, Berlin 1989).
5. U. Nitzsche, unpublished.
6. A. Ernst, PhD thesis, TU Dresden (1997).
7. R. Richter, H. Eschrig and B. Velicky, J. Phys. **F17**, 351 (1987).
8. K. Koepernik, B. Velicky, R. Hayn and H. Eschrig, Phys. Rev. **B55**, 5717–5729 (1997).
9. K. Koepernik, B. Velicky, R. Hayn and H. Eschrig, Phys. Rev. **B58**, 6944–6962 (1998).
10. M. Richter, PhD thesis, TU Dresden (1988).
11. M. Richter and H. Eschrig, Solid State Commun. **72**, 263–265 (1989).
12. I. Opahle, PhD thesis, TU Dresden (2001).

13. H. Eschrig, M. Richter and I. Opahle, in: *Relativistic Electronic Structure Theory*, Part II (Ed. P. Schwerdtfeger), pp. 723–776, Elsevier, Amsterdam (2004).
14. H. Eschrig, K. Koepernik and I. Chaplygin, J. Solid State Chem. **176**, 482–495 (2003).
15. F. Tasnadi, J. Comp. Theoret. Nanosci. **4**, 1–12, (2007).
16. P. Villars and L. Calvert, *Pearson's Handbook of Crystallographic Data for Intermetallic Phases, 2nd ed.* (ASM International, Materials Park, 1991).
17. P. Söderlind, O. Erikson, B. Johansson and J.M. Wills, Phys. Rev. **B50**, 7291–7294 (1994).
18. M.D. Jones, J.C. Boettger, R.C. Albers and D.J. Singh, Phys. Rev. **B61**, 4644–4650 (2000).
19. J. Kuneš, P. Novák, R. Schmid, P. Blaha and K. Schwarz, Phys. Rev. **B64**, 153102 (2001).
20. P. Söderlind, L. Nordström, L. Yongming and B. Johansson, Phys. Rev. **B42**, 4544–4552 (1990).
21. F. Bloch, Z. f. Physik. **52**, 555–600 (1928).
22. E. E. Lafon and C. C. Lin, Phys. Rev. **152**, 579–584 (1966).
23. C. S. Wang and J. Callaway, Phys. Rev. **B9**, 4897–4907 (1974).
24. G. Seifert, D. Porezag and T. Frauenheim, Int. J. Quantum Chem. **58**, 185 (1996).
25. G. Steinle-Neumann, L. Stixrude and R. E. Cohen, Phys. Rev. **B63**, 054103 (2001).
26. W. Kohn, Phys. Rev. Lett. **76**, 3168–3171 (1996).
27. http://www.fplo.de/.
28. H. Eschrig, M. Sargolzaei, K. Koepernik and M. Richter, Europhys. Lett. **72**, 611–617 (2005).

THEORY OF DYNAMICAL THERMAL TRANSPORT COEFFICIENTS IN CORRELATED CONDENSED MATTER *

B. SRIRAM SHASTRY

Physics Department, University of California, Santa Cruz, Ca 95064

We present a recently developed formalism for computing certain dynamical transport coefficients for standard models of correlated matter, such as the Hubbard and the $t-J$ model. The case of the Hall constant in correlated matter is used to motivate the method of high frequency. This method is pointed out to be closer to DC results for models describing low energy properties after eliminating high energy degrees of freedom. Successful predictions of this method are also noted. The extension of this method is made to evaluate and estimate the Seebeck coefficient, the Lorentz number L, and the figure of merit ZT, in terms of novel equal time correlation functions. Along the way, we uncover a new sum rule for the dynamical thermal conductivity for many standard models, precisely analogous to the f-sum rule for the electrical conductivity.

The new formalism is tested in simple settings, such as the Sommerfeld model of non interacting electrons within the Boltzmann approach. Further, recent computational results are displayed for testing the frequency dependence of these variables in certain standard models. Finally some new predictions made regarding triangular lattice systems, motivated by the sodium cobaltate $Na_{.68}CoO_2$, are displayed.

Contents

1. Introduction
2. Hall constant
3. Thermoelectric response

    3.1 Luttinger's gravitational field analogy
    3.2 Finite ω thermal response functions
    3.3 Onsager reciprocity at finite frequencies
    3.4 General formulas for $L_{ij}(\omega)$

*Based on lectures given at the 43rd Karpacz Winter School of Theoretical Physics on *Condensed Matter Physics in the Prime of XXI Century: Phenomena, Materials, Ideas, Methods*, Lądek Zdrój, Poland, 5-11 February 2007.

3.5 High frequency behaviour
3.6 Sum rules for electrical and thermal conductivity
3.7 Dispersion relations for thermopower, Lorentz number and figure of merit

4. Thermoelectric phenomena in correlated matter

4.1 Limiting case of free electrons, S^* the Heikes Mott and Mott results
4.2 Kelvin's thermodynamical formula for thermopower
4.3 Applications to sodium cobaltates in the Curie Weiss metallic phase
4.4 High temperature expansion for thermopower
4.5 Lorentz number and figure of merit for the triangular lattice $t-J$ model

1. Introduction

A major problem in condensed matter physics is to understand transport phenomena in correlated matter. Traditional approaches such as the Boltzmann equation have served long and distinguished tenures to explain transport coefficients in terms of a few measurable objects (relaxation times, and effective masses etc). However these methods run into severe problems in the most interesting and important problem of metals, with strong Mott Hubbard correlations. These correlations give a Mott insulating state at commensurate (half) filling, with localized spins interacting with each other, and away from half filling, one has metallic states that carry the distinguishing marks and signatures of the parent Mott insulator.

The High T_c systems provide one outstanding set of materials that have dominated the community for the last 20 years. These are widely believed to be strongly correlated, following Anderson's original and early identification of these as doped Mott insulators.[1] Another important material, sodium cobaltate Na_xCoO_2 has recently been popular in studies of thermoelectricity,[2,3] this is strongly correlated too, but the underlying lattice is triangular rather than square. These two systems have in common the presence of spin half entities, and have both been modeled in (rather extensive) literature by some variants of the $t-J$ model.

The qualitative reason for the difficulties of the Boltzmann equation approach in these correlated models can be understood in several ways. One is to recognize from a variety of experiments that the wave function renormalization, or quasiparticle residue z_k (defined as the jump of the occupation number at the Fermi surface in momentum space) is either zero, or if non zero, it is certainly very small in these systems. Another simple and yet powerful point of view is to ask what is a charge carrier in a Mott Hubbard system near half filling. From the real space point of view, we see that if a correlated electron hops to a nearby position, it must make sure that there is *no particle* at that site. Hence the motion of an electron of either spin is accompanied by the "backflow" of a vacancy. It is therefore clear that carriers best viewed as holes *measured from half filling.* Thus at a filling of electrons $n \equiv N/N_s$ (where N and N_s are the total number of electrons and the number of lattice sites), the carriers are in fact $\delta = |1-n|$, so that near half filling $\delta \to 0$, and one sees that the carriers are frozen out. Thus several transport coefficients can be guessed by inspection: the Hall number must vanish near half filling, as must the inverse thermopower. One can continue and guess the signs of these objects etc. However, it is already clear that the Boltzmann approach cannot easily capture these "obvious results". The latter starts with the band structure derived quasiparticles, and as $n \to 1$, has no knowledge of the impending disaster, also known as the Mott insulating state! One can also view this issue from the point of view of real space versus momentum space definition of holes: the correlated matter clearly requires a real space picture to make physical sense (as opposed to computational ease), whereas the Boltzmann approach takes a purely momentum space approach to particle and holes. In fact the "Boltzmann holes" are vacancies in momentum space measured from a *completely filled band*, and have no resemblance to the Mott Hubbard holes. Of course the above diatribe obscures a crucial point, the Mott Hubbard real space view point is almost impossible to compute with, at least using techniques that exist so far. On the other hand, the momentum space view is seductive because of the ease of computations exploiting a well oiled machine, namely the perturbative many body framework. Hence it seems profitable to explore methods and techniques that implement the Mott Hubbard correlations at the outset, and give qualitatively correct answers. Our formalism, described below, was motivated by these considerations.

Since this is a "school", there will be little effort at literature survey. Our published papers contain more references to other approaches taken in literature. I would however, like to mention that at a "mean field theory" level, the Mott Hubbard correlations *can be built in*, by various slave Boson or slave Fermion approaches, with varying degrees of success. Since well controlled calculations are difficult to perform for the experimentally relevant case of 2-dimensions with electrons having spin $\frac{1}{2}$, we are most often forced into numerical computations. The formalism developed here provides some guidance towards effective computations. We expect that our formalism is to be supplemented by a heavy dose of numerics, either exact diagonalization or some other means. In the final part of this article, some recent numerical results of this type are presented.

2. Hall constant

The basic idea of this approach is well illustrated by the example of the Hall constant for correlated matter R_H. Here the initial paper of Shastry, Shraiman and Singh[4] pointed out that the *dynamical Hall constant* is better suited for computation in correlated systems. Consider the simplest framework, the Drude theory of electrons, where we know that

$$\begin{aligned}\sigma_{xx}(\omega) &= \frac{\sigma_{xx}(0)}{(1+i\omega\tau)}, \\ \sigma_{xy}(\omega) &= \frac{\sigma_{xy}(0)}{(1+i\omega\tau)^2}, \\ \rho_{xy}(\omega) &= \frac{\sigma_{xy}(\omega)}{\sigma_{xx}(\omega)^2} = \frac{1}{nq_e c},\end{aligned} \tag{1}$$

where $q_e = -|e|$ is the electron charge, n the density of electrons and τ the relaxation time. The relaxation time cancels out in computing the Hall resistivity at arbitrary frequencies, and this gives us a clue. We might as well compute the two conductivities $\sigma_{ab}(\omega)$ at *high frequencies*, since here the notorious difficulties inherent in computing these objects vanish. For this calculation, we need to take the Kubo formulae for the conductivities[a], and take the appropriate ratios to get the dynamical resistivity. We note[4] the general form of the dynamical conductivity from linear response theory:

$$\sigma_{\alpha\beta}(\omega_c) = \frac{i}{\hbar N_s v \omega_c}\left[\langle\tau^{\alpha\beta}\rangle + \hbar\sum_{n,m}\frac{p_n - p_m}{\epsilon_n - \epsilon_m + \hbar\omega_c}\langle n|\hat{J}_\alpha|m\rangle\langle m|\hat{J}_\beta|n\rangle\right], \tag{2}$$

where the "stress tensor" is defined by

$$\tau^{\alpha\beta} = \sum_{k,\sigma}\frac{d^2\varepsilon(k)}{dk_\alpha dk_\beta}c^\dagger_\sigma(k)c_\sigma(k), \tag{3}$$

[a]It is frustrating that despite several ambitious claims in literature, especially from the Mori formulation experts, there is no practical and direct way of computing the dynamical *resistivity* that bypasses the intermediate stage of computing the dynamical *conductivities*.

where v is the atomic volume, and the current operator $\hat{J}_\alpha$ is dressed by a suitable Peierls[5] phase factor in the presence of the uniform magnetic field B along the z axis. Moreover, all the operators include the effect of Gutzwiller Hubbard projection, in the case of a $t - J$ model. One is thus working in the non dissipative (reactive) regime. The main article of faith is the claim that $\rho_{xy}(\omega)$ at large frequencies is related in a simple way to the transport variable $\rho_{xy}(0)$. Is this rationalizable? Further, what is meaning of high frequency, or how "high" is "high enough"?

With regard to the "largeness" of the frequency, the key point is to work with a projected Fermi system rather than a bare one. For example in the case of the Hubbard model versus the $t-J$ model, one sees that the energy scale inequality requirement is

$$\hbar\omega \gg \{|t|, U\}_{max}\,, \tag{4}$$

$$\hbar\omega \gg \{|t|, J\}_{max}\,. \tag{5}$$

Thus in case of the $t - J$ model, one can be in the high frequency limit, and yet have a modest value of ω, in contrast to the Hubbard model since usually U is large, $O(ev's)$. In case of the cobaltates, the energy scale that determines the high frequency limit is presumably the Hunds rule or crystal field energy. Thus the "high frequency limit" is expected to be close to the transport values, for models where the high energy scale is projected out to give an effective low energy Hamiltonian with suitably projected operators. The result of this manipulation with Kubo conductivities gives a formal expression for the Hall constant

$$R_H(\omega)|_{\omega\to\infty} \to R_H^* = \frac{-iN_s v}{B\hbar}\frac{\langle[\hat{J}_x, \hat{J}_y]\rangle}{\langle\tau_{xx}\rangle^2}\,. \tag{6}$$

Subsequent studies show that this simple formula is a particularly useful one, we list some of its merits:

- It is exact in the limit of simple dynamics (e.g. few frequencies involved), as in the Boltzmann equation approach.
- It can be computed in various ways (e.g. using exact diagonalization, high T expansion, etc..) for all temperatures.
- We have successfully removed the dissipational aspect of the Hall constant here and retained the (lower Hubbard subband physics) correlations aspect. This is done by going to high frequencies, and using the Gutzwiller projected Fermi operators in defining the currents.

- It captures the crossover from hopping type incoherent transport at high T, to coherent band type transport at low T in a narrow band system.

We emphasize that this provides a very good description of the $t-J$ model, where this asymptotic formula requires ω to be larger than J, but should not be expected to be particularly useful for Hubbard model (where ω needs to be larger than U!).

It is worthwhile recording a dispersion relation for the Hall constant at this point. Since $R_H(\omega)$ is analytic in the upper half of the complex ω plane, and has a finite limit at infinite ω, we may write

$$R_H(\omega) = R_H(\infty) - \int_{-\infty}^{\infty} \frac{d\nu}{\pi} \frac{\Im m R_H(\nu)}{\omega - \nu + i0+}, \tag{7}$$

therefore in the DC limit we get:

$$\Re e R_H(0) = R_H^* + \frac{2}{\pi} \int_0^{\infty} \frac{\Im m R_H(\nu)}{\nu} d\nu. \tag{8}$$

This equation quantifies the difference between the experimentally measured DC-Hall coefficient and the theoretically more accessible infinite frequency limit. The second term on the right is estimated to be quite small numerically, and interestingly is an independently measurable object.[6]

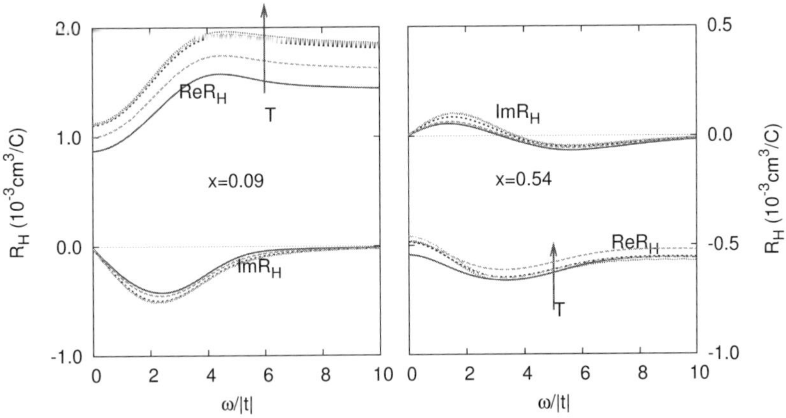

Fig. 1. Frequency-dependence of the Hall coefficient on the simple square lattice for hole doping $\delta = x$ from computation on small clusters of the $t-J$ model Ref. [7]. These are extreme cases of doping, and provide some feel for the range of the frequency dependence.

As an illustration of the above formalism, we note that recent work on triangular lattice system Na_xCoO_2 provides a good example. Theoretically, the "exotic" possibility of the Hall constant behaving linearly with temperature T on a triangular lattice, was recognized in 1993 Ref. [4]. This arises in a simple way from Eq. (6) treated within the high temperature expansion. The numerator is dominated by smallest loops that encircle a flux, and these are of course triangles for the triangular lattice. This leads at high T (or small inverse temperature β) to the numerator $\propto \beta$ whereas the denominator is always $\propto \beta^2$, and hence a T linear Hall constant with a well defined coefficient

$$R_H^* = -\frac{v}{4|q_e|}\frac{k_B T}{t}\frac{1+\delta}{\delta(1-\delta)} + c_1 + \frac{c_2}{T} + \cdots . \quad (9)$$

This result[8] is for the experimentally relevant case of electron filling so that $\delta = \frac{N}{N_s} - 1$, and shows that the Hall constant is a highly non universal quantity, depending upon material parameters such as the magnitude and sign of the hopping! We must stress that this behaviour is obtained provided $k_B T \geq |t|$ and as such is realizable in nature only for very narrow band systems. Interestingly enough, the case of Na_xCoO_2 with $x \sim 0.7$, i.e. the so called Curie Weiss metallic phase, seems to fulfill these conditions of narrow bandwidth, and as Figure (2) shows, the experiments show a clear-cut tendency towards linear T dependence,[9] thereby fulfilling the basic theoretical prediction of Eq. (9). Further recent work[10] attempts to reconcile the coefficient of T with the experimental values. While this is work in progress, we stress the basic point, namely that R_H^* provides a sensible estimate for the physical transport Hall constant for correlated systems. Our task in these notes is to carry this message to the computation of the thermal response functions, and so we terminate our discussion of the interesting problem of the Hall constant.

3. Thermoelectric response

We next address the main topic of these lectures, namely the thermal response functions. In light of the previous discussion of the Hall constant, in the quest for the analog of R_H^*, we expect that the finite (in fact infinite) frequency limit of thermal response functions will be needed. We begin with a quick review of the standard transport theory.[11,12,14–16]

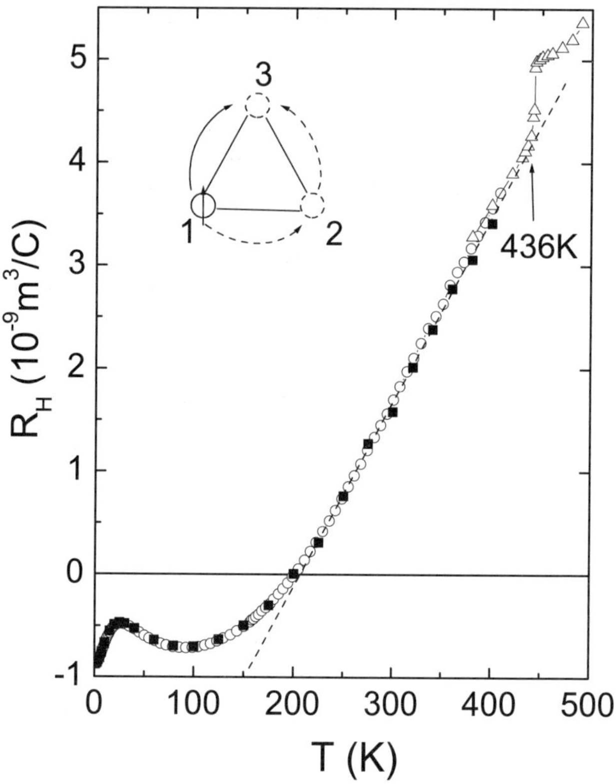

Fig. 2. Experimental temperature dependence of the Hall coefficient of sodium cobaltate Ref. [9]. The sample is in the Curie Weiss metallic phase with $\delta \sim 0.68$. The inset stresses the crucial role of the triangular closed loops in giving rise to the surprising behaviour.

We write down the set of linear response equations following Onsager[11]

$$\frac{1}{\Omega}\langle \hat{J}_x \rangle = L_{11} E_x + L_{12}(-\nabla_x T/T)\,, \tag{10}$$

$$\frac{1}{\Omega}\langle \hat{J}^Q_x \rangle = L_{21} E_x + L_{22}(-\nabla_x T/T)\,, \tag{11}$$

where $(-\nabla_x T/T)$ is regarded as the *external driving thermal force*, and $\hat{J}^Q_x$ is the heat current operator defined more elaborately below in Eq. (29) and

Ω is the total volume of the system[b]. The parameter L_{11} is related to the DC conductivity $\sigma(0) = L_{11}$, the parameter L_{12} is related[c] to the Seebeck coefficient $S = \frac{L_{12}}{TL_{11}}$, also L_{21} is related to the Peltier coefficient $\Pi = L_{21}/L_{11}$. The Onsager constant L_{22} is related to the thermal conductivity $\kappa = \frac{1}{T}L_{22}$ for problems with immobile degrees of freedom (spins, ions, etc). For metallic systems, however, the observed thermal conductivity κ_{zc} requires a small correction (see Eq. (12)). The usually observed thermal conductivity uses the *zero current condition* $\langle \hat{J}_x \rangle = 0$, thus the generated electric field is related by Eq. (10) to the applied thermal force, and using it in Eq. (11) we find

$$\kappa_{zc} = \frac{1}{TL_{11}}(L_{22}L_{11} - L_{12}L_{21}) . \tag{12}$$

We note the celebrated reciprocity relation $L_{12} = L_{21}$, which relates the Peltier and Thomson effects. These are equations in the static limit, and correspond to the most simple non equilibrium states with a steady current flows.

3.1. *Luttinger's gravitational field analogy*

In order to generalize the above transport theory to finite frequencies, we need to borrow a beautiful idea from Luttinger Ref. [12]. In order to derive the so called Kubo formulae Ref. [13], he introduces the mechanical equivalent of the thermal gradient, and we shall use it extensively. The mechanical field is termed as the "gravitational field" $G(\vec{x}) = c^2/g\ \psi(\vec{x},t)$ and can be imagined as being a time dependent gravitational interaction coupling to the effective "mass density" $m_{eff}(\vec{x}) = \frac{1}{c^2}K(\vec{x})$ via

$$K_{tot} = K + \sum_x K(\vec{x})\psi(\vec{x},t) . \tag{13}$$

Here $K = \sum_x K(\vec{x})$, and $K(\vec{x}) = H(\vec{x}) - \mu n(\vec{x})$ is the *grand canonical* Hamiltonian[d], $H(\vec{x}), n(\vec{x}), \mu$ are the local canonical ensemble Hamiltonian, number density and chemical potential. We can compute the standard linear response to a space time dependent $\psi(\vec{x},t)$, and with the help of the ideas

---

[b]Our definition includes the volume factor and this makes L_{11} identical to the usual (intensive) conductivity.

[c]Sometimes in literature [14–16], S is denoted by Q.

[d]The need of introducing the grand canonical Hamiltonian K lies in the construction of the heat current operator $\hat{J}_x^Q$, where the particle current must be subtracted from the energy current Eq. (29).

initiated by Luttinger, deduce the dynamical thermal response functions required in Eq. (32).

Firstly let us note that the local temperature $\delta T(\vec{x},t)$ can be defined in the long wavelength almost static limit through small departures from equilibrium. The local energy fluctuation can be written as $\langle K(\vec{x},t)\rangle = \langle K\rangle_0 + C(T)\ \delta T(\vec{x},t)$, with $C(T)$ as the specific heat at the equilibrium temperature T(at constant volume and μ), provided $\delta T(\vec{X},t) \ll T$. Hence we can invert to define the local temperature through

$$\delta T(\vec{x},t) = \frac{\delta\langle K(\vec{x},t)\rangle}{C(T)} . \tag{14}$$

The connection of $\psi(\vec{x},t)$ with local temperature $\delta T(\vec{x},t)$ emerges from a study of the generalized phenomenological equations proposed by Luttinger.[12] He specializes to long wavelength $\vec{q} \to 0$ and static $\omega \to 0$ limits where equilibrium is rigorously definable, and we will generalize to arbitrary variations. The phenomenological relations are generalizations of the Onsager formulation[11] as in Eq. (10) and Eq. (11), and correspond to adding terms proportional to the gradient of the mechanical term $\propto \psi$ in Eq. (13). Luttinger writes

$$\frac{1}{\Omega}\langle \hat{J}_x\rangle = L_{11}E_x + L_{12}(-\nabla_x T/T) + \hat{L}_{12}(-\nabla_x\psi(\vec{x},t)) , \tag{15}$$

$$\frac{1}{\Omega}\langle \hat{J}^Q_x\rangle = L_{21}E_x + L_{22}(-\nabla_x T/T) + \hat{L}_{22}(-\nabla_x\psi(\vec{x},t)) , \tag{16}$$

where the two new response functions $\hat{L}_{12}, \hat{L}_{22}$ are functions of space and time which can be readily computed from a linear response theory treatment of the mechanical perturbation in Eq. (13). Addition of the ψ term in these equations allows us to take a different perspective[e] of these, as compared to the Eqs. (10,11). In Eqs. (15,16) *we can view the driving term as* ψ, with the temperature fluctuation arising as a consequence of this driving, at least for long wavelengths and slow variations[f].

---

[e]Note that experiments usually employ *open boundary conditions*, and the temperature gradient is externally applied. The usual argument made is that the periodic boundary case and the open boundary case are equivalent, provided we take the wave vector $\vec{q} \to 0$ or the thermodynamic volume $\Omega \to \infty$ limits respectively, while keeping the frequency ω finite and small. This gives a prescription for the DC limit in both cases, namely to take the DC limit at the end of the volume (or wave vector) limits.

[f]This is where Luttinger uses the tactical analogy with the Einstein relation for the relationship between self diffusion and conductivity. In the phenomenological equation $\langle \hat{J}_x\rangle = \sigma E_x + D(-\nabla_x)\langle\rho\rangle$, the driving term is E_x. In Eq. (15) (neglecting the L_{11} term for a moment), the ψ term is analogous to the E_x in the diffusion problem, and the

In these Eqs. (15,16), the idea is to determine the difficult unknowns L_{12}, L_{22} in terms of the easier objects $\hat{L}_{12}, \hat{L}_{22}$. Let us consider one particular example for simplicity, the others follow similarly. Let us focus on Eq. (15), and introduce a single Fourier component $\psi(\vec{x}, t) = \psi_q \exp\{-i(q_x x + \omega t + i0^+ t)\}$, (adiabatic switching implied) and the electric potential $\phi(\vec{x}, t) = \phi_q \exp\{-i(q_x x + \omega t + i0^+ t)\}$ thus write

$$\delta \hat{J}_x = (iq_x) L_{11}(q_x, \omega)\phi_q + (iq_x)\left[L_{12}(q_x, \omega)\frac{\delta T_q}{T} + \hat{L}_{12}(q_x, \omega)\,\psi_q\right], \quad (17)$$

where $\frac{1}{\Omega}\langle \hat{J}_x(q_x)\rangle = \delta J_q$, so that $\delta \hat{J}_x$ is the amplitude of the response, and we have written the arguments of the Onsager-Luttinger functions $L_{ij}, \hat{L}_{ij}$ explicitly. In the limit, $q_x \to 0$ we have a uniform field and hence $\delta T_q \to 0$, i.e. the temperature fluctuation vanishes in the rapid limit. This is most easily seen by inspecting the continuity equation, which can be written, using Eq. (14) as $\delta T_q = \frac{-q_x}{C(T)\ \omega}\langle \hat{J}_x^Q\rangle$. Thus

$$\delta \hat{J}_x = L_{11}(0, \omega) \lim_{q_x \to 0}(iq_x)\phi_q + \hat{L}_{12}(0, \omega) \lim_{q_x \to 0}(iq_x)\psi_q\,. \quad (18)$$

In the opposite limit, with $\omega \to 0$, the system is subject to a (not necessarily slowly) varying potential, and hence this is an equilibrium problem without a net current. Thus $\langle \hat{J}_x(q_x)\rangle = 0$, leading to

$$0 = L_{12}(q, 0)\frac{\delta T_q}{T} + \hat{L}_{12}(q, 0)\psi_q\,. \quad (19)$$

In this equilibrium situation, we can compute the connection between $\frac{\delta T_q}{T}$ and ψ_q readily. Using lowest order thermodynamic perturbation theory Ref. [16,17] we compute the change in energy induced by a small perturbation ψ_q

$$\frac{\delta\langle K(\vec{q})\rangle}{\psi(\vec{q})} = -\sum \frac{p_n - p_m}{\varepsilon_m - \varepsilon_n}|\langle n|K(\vec{q})|m\rangle|^2 + O(\psi^2)\,, \quad (20)$$

---

induced temperature variation is similar to the induced charge fluctuation. For completeness, we summarize Luttinger's argument for this case. For small wave vectors and slow variation of the electric field $E_x = -\nabla\phi(x) = E_0 \exp -i(q_x x + \omega t)$. Upon using the continuity equation $\langle \rho_q\rangle = -\frac{q_x}{\omega}\langle \hat{J}_x\rangle$ we see that $\langle \hat{J}_x\rangle = \sigma E_0 \frac{\omega}{\omega + iDq_x^2}$. Similarly the charge fluctuation $\langle \rho_q\rangle = \sigma\phi_q \frac{-iq_x^2}{\omega + iDq_x^2}$, where $\phi_q = -iE_0/q_x$. Luttinger's argument is that in the fast or transport limit $\omega \to 0, q_x \to 0$ so that the diffusion term can be dropped. However, in the slow limit, the relations derived above show that $\frac{\sigma}{D} = -\langle \rho_q\rangle/\langle \phi_q\rangle$. The right hand side of this is easily computed from *thermodynamics*, whereby the Einstein relation $\frac{\sigma}{D} = e^2/(\partial\mu/\partial n)_T$ follows.

with $p_n = \frac{1}{Z}\exp(-\beta\varepsilon_n)$ the probability of the state n. In the limit $\vec{q} \to 0$, $K(\vec{q})$ tends to the Hamiltonian, and hence cannot mix states of different energy, hence we write $\lim_{\varepsilon_m \to \varepsilon_n} \frac{p_n - p_m}{\varepsilon_m - \varepsilon_n} \to \beta p_n$, whereby

$$\lim_{\vec{q}\to 0} \frac{\delta\langle K(\vec{q})\rangle}{\psi(\vec{q})} \to -\beta\left[\langle K^2\rangle - \langle K\rangle^2\right] \tag{21}$$

$$= -TC(T)\,. \tag{22}$$

The subtraction in Eq. (21) eliminates the disconnected pieces, which arises by taking the limit of $\vec{q} \to 0$. Comparing the final Eq. (22), with the standard thermodynamic definition of $C(T)$, we see that

$$\lim_{q\to 0} \frac{\delta\langle K(\vec{q})\rangle}{\psi(\vec{q})} = -T\frac{d}{dT}\langle K\rangle\,, \tag{23}$$

whereby

$$\lim_{\vec{q}\to 0} \psi(\vec{q},0) = -\lim_{\vec{q}\to 0} \frac{\delta T_q}{T}\,. \tag{24}$$

Comparing Eq. (24) and Eq. (19), we see that

$$\lim_{q\to 0} \left[L_{12}(q,0) - \hat{L}_{12}(q,0)\right] = 0\,. \tag{25}$$

From this relation, Luttinger concludes that L_{12} in the DC limit can be computed from $\hat{L}_{12}$. Thus the problem of computing thermal response is reduced to computing mechanical response to the field $\psi(\vec{x},t)$, and essentially treating[g] the $\lim_{\vec{q}\to 0} \psi(\vec{q},0) = \lim_{\vec{q}\to 0} \frac{\delta T_q}{T}$.

This is undoubtedly huge progress. However, as far as I can make out, this fine proof of Luttinger makes another implicit assumption, namely that

$$\lim_{\omega\to 0} \left[L_{12}(0,\omega) - \hat{L}_{12}(0,\omega)\right] = 0\,, \tag{26}$$

somehow follows from Eq. (25). This is assumed so despite the fundamental difference in the two limits, namely the slow (thermodynamic) and fast (transport) limits. The belief thus seems to be that the two functions L_{ij} and $\hat{L}_{ij}$ must be identical in the fast limit, if they are so in the slow limit.

In this work we need to define finite q, ω thermal response functions. Towards this end, we will in fact *extend the above to all* q, ω, and simply assume that

$$L_{ij}(q,\omega) = \hat{L}_{ij}(q,\omega)\,. \tag{27}$$

[g]The alert reader would have noted that this has an opposite sign from Eq. (24). The explanation of this slight "booby trap" is hopefully clear from the sequence of steps leading to Eq. (25)!

The RHS is computable within perturbation theory, and the LHS, although defined rigorously only in the regime of small q, ω by hydrodynamic type reasoning, is extended to all q, ω by this relation. This idea of extending the thermal functions seems reasonable, since the resulting functions agree with hydro-thermodynamics for small q, ω, and are guaranteed to satisfy general properties such as causality and Onsager reciprocity. With this, we can define all thermal response functions at all q, ω, and in the sequel we will work within this generalized Luttinger viewpoint.

3.2. *Finite ω thermal response functions*

With this preparation, we can return to exploring the thermal response Eq. (32) at finite frequencies. The timing of our quest seems fortuitous, since there is growing experimental interest in the transport of energy and heat pulses, requiring a knowledge of these variables, and of the approach to equilibrium.

We first need to define the heat current $\hat{J}^Q_x$. Towards this end, we take the time derivative of the first law of thermodynamics for fixed volume $T\frac{dQ}{dt} = \frac{dE}{dt} - \mu\frac{dn}{dt}$. Imagining a small volume with the flow of energy and heat as well as density, and applying this law locally, it is reasonable to identify the heat current as the energy current minus the particle current (times μ). Thus the heat current can be decomposed as the difference of two terms:

$$\hat{J}^Q_x = \hat{J}^E_x - \frac{\mu}{q_e}\hat{J}_x \,, \tag{28}$$

where $\hat{J}^E_x$ is the *energy current* and $\hat{J}_x$ the charge current. In a quantum mechanical system, the heat current operator is easiest computed from the commutator of the energy density operator with total energy as follows (setting $\hbar = 1$):

$$\hat{J}^Q_x = \lim_{q_x \to 0} \frac{1}{q_x} \left[K, K(q_x)\right] \,. \tag{29}$$

By inspection, a local heat current operator can also be written down provided the interactions are local, so that we can take Fourier components in a periodic box and write

$$\hat{J}^Q_x(\vec{q}) = \sum_x \hat{J}^Q_x(\vec{x}) \exp(i\vec{q}.\vec{x}), \quad \text{so that } \hat{J}^Q_x = \lim_{q \to 0} \hat{J}^Q_x(\vec{q}) \,. \tag{30}$$

For different models, the heat current is easy to compute using the above prescription, and many standard models are treated in [18].

Let us impose fields that vary as $\psi(\vec{x},t) = \psi_q \exp\{-i(q_x x + \omega t + i0^+ t)\}$, and similarly for the electric field with the electric potential $\phi(\vec{x},t) = \phi_q \exp\{-i(q_x x + \omega t + i0^+ t)\}$. Using the notation $\frac{1}{\Omega}\langle \hat{J}_x(q_x)\rangle = \delta\hat{J}_x$ and $\frac{1}{\Omega}\langle \hat{J}^Q_x(q_x)\rangle = \delta\hat{J}^Q_x$, we find from Eq. (15,16)

$$\delta\hat{J}_x = L_{11}(q_x,\omega)(iq_x\phi_q) + L_{12}(q_x,\omega)(iq_x\psi_q)\,, \tag{31}$$

$$\delta\hat{J}^Q_x = L_{21}(q_x,\omega)(iq_x\phi_q) + L_{22}(q_x,\omega)(iq_x\psi_q)\,. \tag{32}$$

These responses are to be computed for a Hamiltonian perturbed by a single fourier component as

$$K_{tot} = K + [\rho(-q_x)\phi_q + K(-q_x)\psi_q]\exp(-i\omega t + 0^+ t)\,, \tag{33}$$

where $\rho(q)$ is the charge density fluctuation operator.

We can reduce the calculations of all L_{ij} to essentially a single one, with the help of some notation. Keeping q_x small but non zero, we define currents, densities and forces in a matrix notation as follows:

	i=1	i=2
	Charge	Energy
$\mathcal{I}_i$	$\hat{J}_x(q_x)$	$\hat{J}^Q_x(q_x)$
$\mathcal{U}_i$	$\rho(-q_x)$	$K(-q_x)$
$\mathcal{X}_i$	$E^x_q = iq_x\phi_q$	$iq_x\psi_q\,.$

(34)

The perturbed Hamiltonian Eq. (33) can then be written as

$$K_{tot} = K + \sum_j Q_j e^{-i\omega_c t}, \quad \text{where} \quad Q_j = \frac{1}{iq_x}\mathcal{U}_j\mathcal{X}_j\,. \tag{35}$$

We denote $\omega_c = \omega + i0^+$ above and elsewhere. From standard linear response theory[12] applied to Eq. (35), we readily extract the induced current response

$$\langle \mathcal{I}_i\rangle = -\sum_j \chi_{\mathcal{I}_i,Q_j}(\omega_c)\,, \tag{36}$$

where the susceptibility for any two operators $\chi_{A,B}(\omega_c)$ can be expressed as (with $A_{nm} \equiv \langle n|A|m\rangle$)

$$\begin{aligned}
\chi_{A,B}(\omega_c) &= -i\int_0^\infty dt\; e^{i\omega t - 0^+ t}\langle [A(t),B(0)]\rangle \\
&= \sum_{n,m}\frac{p_m - p_n}{\varepsilon_n - \varepsilon_m + \omega_c}A_{nm}B_{mn} \\
&= -\frac{1}{\omega_c}\left[\langle [A,B]\rangle + \sum_{n,m}\frac{p_m - p_n}{\varepsilon_n - \varepsilon_m + \omega_c}A_{nm}([B,K])_{mn}\right]\,.
\end{aligned} \tag{37}$$

The last line of Eq. (37) follows from an integration by parts of the first line, and the averages $\langle\rangle$ are carried out over the ensemble where the external fields are dropped.

From Eq. (36), using the notation in Eqs. (34,37), the generalized Onsager coefficients

$$L_{ij}(q_x,\omega) = \frac{1}{\Omega} \lim_{\mathcal{X}_j \to 0} \langle \mathcal{I}_i \rangle / \mathcal{X}_j \tag{38}$$

are written down immediately

$$L_{ij}(q_x,\omega) = \frac{1}{i\Omega\omega_c} \left[\langle [\mathcal{I}_i, \mathcal{U}_j] \rangle \frac{1}{q_x} + \frac{1}{q_x} \sum_{n,m} \frac{p_m - p_n}{\varepsilon_n - \varepsilon_m + \omega_c} (\mathcal{I}_i)_{nm} ([\mathcal{U}_j, K])_{mn} \right]. \tag{39}$$

We now use the continuity equation to write $[\mathcal{U}_j, K] \to q_x \mathcal{I}_j^\dagger$, whereby

$$L_{ij}(q_x,\omega) = \frac{i}{\Omega\omega_c} \left[-\langle [\mathcal{I}_i, \mathcal{U}_j] \rangle \frac{1}{q_x} - \sum_{n,m} \frac{p_m - p_n}{\varepsilon_n - \varepsilon_m + \omega_c} (\mathcal{I}_i)_{nm} (\mathcal{I}_j^\dagger)_{mn} \right]. \tag{40}$$

We next proceed to taking the limit of small q_x. Here the inconvenient-looking first term in Eq. (40) tends to a finite limit in *all cases*, owing to a simple but important point: $K(-q_x)$ tends to the Hamiltonian K in the limit $q_x \to 0$. Using this and the cyclic invariance of the trace, it follows Ref. [18] that for a generic operator P, that $\langle [P, K(-q_x)] \rangle \propto q_x$ with a well defined coefficient. In case of L_{11}, the operator $\mathcal{U}_1$ is the charge density operator, which in the uniform limit *commutes with* $\mathcal{I}_i$, and hence the limit of the ratio is well defined more trivially. This is the content of the "Identity I" in Ref. [18]. Thus we find in the uniform limit, we can set $\mathcal{I}_j^\dagger = \mathcal{I}_j$ and for arbitrary frequencies the Onsager functions read as

$$L_{ij}(\omega) = \frac{i}{\Omega\omega_c} \left[\langle \mathcal{T}_{ij} \rangle - \sum_{n,m} \frac{p_m - p_n}{\varepsilon_n - \varepsilon_m + \omega_c} (\mathcal{I}_i)_{nm} (\mathcal{I}_j)_{mn} \right], \tag{41}$$

$$\langle \mathcal{T}_{ij} \rangle = - \lim_{q_x \to 0} \langle [\mathcal{I}_i, \mathcal{U}_j] \rangle \frac{1}{q_x}. \tag{42}$$

The operators $\mathcal{T}_{ij}$ are not unique, since one can add to them a "gauge operator" $\mathcal{T}_{ij}^{gauge} = [P, K]$ with arbitrary P, without affecting the thermal average, due to the Identity-I discussed above. These fundamental operators play a crucial role in the subsequent analysis, since they determine the high frequency behaviour of the response functions. These important operators are written in a more familiar representation[18] as follows:

$$\begin{array}{|c|}\hline \text{Stress tensor} \\ \mathcal{T}_{11} \\ \mathcal{T}^{xx} \\ -\frac{d}{dq_x}\left[\hat{J}_x(q_x),\rho(-q_x)\right]_{q_x\to 0} \\ \hline \text{Thermal operator} \\ \mathcal{T}_{22} \\ \Theta^{xx} \\ -\frac{d}{dq_x}\left[\hat{J}^Q_x(q_x),K(-q_x)\right]_{q_x\to 0} \\ \hline \text{Thermoelectric operator} \\ \mathcal{T}_{12}=\mathcal{T}_{21} \\ \Phi^{xx} \\ -\frac{d}{dq_x}\left[\hat{J}_x(q_x),K(-q_x)\right]_{q_x\to 0} \\ \hline \end{array} \tag{43}$$

The thermoelectric operator can also be written as

$$\Phi^{xx} = \mathcal{T}_{21} = -\frac{d}{dq_x}\left[\hat{J}^Q_x(q_x),\rho(-q_x)\right]_{q_x\to 0}, \tag{44}$$

and its equivalence to the form given in Eq. (43) amounts to showing $\mathcal{T}_{12} = \mathcal{T}_{21}$, modulo the addition of a "gauge operator" discussed above. This task is more nontrivial than one might naively anticipate, and requires the use of Jacobi's identity as discussed later.

Several aspects of Eqs. (41,43) are worth mentioning at this point.

3.3. *Onsager reciprocity at finite frequencies*

We first note that the celebrated reciprocity relations of Onsager are extended to finite ω here. These require in the present case (with no magnetic fields)

$$L_{ij}(\omega) = L_{ji}(\omega)\,. \tag{45}$$

One part of the above dealing with the second term of Eq. (41) goes back to Onsager's famous argument: in the absence of a magnetic field we can choose a real phase convention for the quantum wave functions such that the product $(\mathcal{I}_i)_{nm}(\mathcal{I}_j)_{mn}$ are real. Invariance under complex conjugation then implies invariance under the exchange $i \leftrightarrow j$.

The full (frequency dependent) function shows reciprocity only if we can show that $\mathcal{T}_{ij} = \mathcal{T}_{ji}$, since this is the first part of Eq. (41). This identity requires the use of the Jacobi identity and can be proved as follows. Consider $\mathcal{T}_{12}$ which requires the first order term in q of the expectation of $[\hat{J}_x(q), K(-q)]$. Now we use $\hat{J}_x(q) = 1/q[K, \rho(q)]$ to lowest order in q, so that

$$\langle \mathcal{T}_{12} \rangle = -\left(\frac{d}{dq} \frac{1}{q} \left[\langle [K, \rho(q)], K(-q)] \rangle \right] \right)_{q \to 0} \tag{46}$$

$$= \left(\frac{d}{dq} \frac{1}{q} \langle [[\rho(q), K(-q)], K] + [K(-q), K], \rho(q)]] \rangle \right)_{q \to 0} \tag{47}$$

$$= \left(\frac{d}{dq} \langle \left[[\hat{J}_x^Q(-q), \rho(q)] \right] \rangle \right)_{q \to 0} \tag{48}$$

$$= \langle \mathcal{T}_{21} \rangle. \tag{49}$$

We used Jacobi's identity to go to Eq. (47) from Eq. (46), and dropped the first term in Eq. (47) using the Identity-I.[18] Eq. (48) follows on using the definition of the heat current Eq. (29). Thus we have *reciprocity for all* ω. A generalization to include magnetic fields can be readily made, but we skip it here.

3.4. *General formulas for $L_{ij}(\omega)$*

By using a simple algebraic identity with partial fractions,[18] we can convert Eq. (41) to the following Kubo[13] type expression:

$$L_{ij}(\omega) = \frac{i}{\omega_c} \mathcal{D}_{ij} + \frac{1}{\Omega} \int_0^\infty dt\, e^{i\omega_c t} \int_0^\beta d\tau\, \langle \mathcal{I}_i(t - i\tau) \mathcal{I}_j(0) \rangle, \tag{50}$$

$$\mathcal{D}_{ij} = \frac{1}{\Omega} \left[\langle \mathcal{T}_{ij} \rangle - \sum_{nm} \frac{p_n - p_m}{\varepsilon_m - \varepsilon_n} (\mathcal{I}_i)_{nm} (\mathcal{I}_j)_{mn} \right]. \tag{51}$$

The stiffnesses $\mathcal{D}_{ij}$ are discussed in detail in Ref. [18], and are in general non zero for all non dissipative systems such as superfluids and superconductors. For a superconductor $\mathcal{D}_{11}$ is the Meissner stiffness, so that the superfluid density can be defined in terms of it.[18] In a superfluid or a highly pure crystal supporting second sound, the stiffness $\mathcal{D}_{22}$ is related to the second sound velocity. A detailed treatment of this relationship, and a generalized hydrodynamical treatment will be published later. For dissipative systems, these stiffnesses vanish, and on dropping them from Eq. (50) we get back the familiar Kubo type formulas.[12,13]

3.5. *High frequency behaviour*

The high frequency behaviour of these functions is easily found from Eq. (41) as

$$\lim_{\omega \gg 0} L_{ij}(\omega) = \frac{i}{\omega \Omega} \langle \mathcal{T}_{ij} \rangle + O(1/\omega^2) \,. \tag{52}$$

Thus these fundamental operators determine the high frequency response, and we will pursue the consequences later.

3.6. *Sum rules for electrical and thermal conductivity*

It is worth noting that these relations imply sum rules as well, for the thermal response functions. To see this, note that the causal nature of the Onsager coefficients and an asymptotic fall off as inverse frequency provides a dispersion relation

$$\Re e L_{ij}(\omega) = \frac{1}{\pi} \mathcal{P} \int_{-\infty}^{\infty} \frac{d\nu}{\nu - \omega} \Im m L_{ij}(\nu) \,, \tag{53}$$

$$\Im m L_{ij}(\omega) = \frac{1}{\pi} \mathcal{P} \int_{-\infty}^{\infty} \frac{d\nu}{\omega - \nu} \Re e L_{ij}(\nu) \,. \tag{54}$$

We see at high frequencies from Eq. (52) and Eq. (54) and assuming the reality of the averages $\langle \mathcal{T}_{ij} \rangle$:

$$\lim_{\omega \gg 0} \omega \; \Im m L_{ij}(\omega) = \frac{\langle \mathcal{T}_{ij} \rangle}{\Omega} = \int_{-\infty}^{\infty} \frac{d\nu}{\pi} \Re e L_{ij}(\nu) \,. \tag{55}$$

This relation gives all the interesting sum rules in this problem. More explicitly we find:

$$\int_{-\infty}^{\infty} \frac{d\nu}{2} \Re e \sigma(\nu) = \frac{\pi \langle \tau^{xx} \rangle}{2\Omega} \,, \tag{56}$$

$$\int_{-\infty}^{\infty} \frac{d\nu}{2} \Re e \kappa(\nu) = \frac{\pi \langle \Theta^{xx} \rangle}{2T\Omega} \,. \tag{57}$$

These are known as follows: (a) Eq. (56) is the well known lattice plasma or f-sum rule[19] with the RHS equalling $\omega_p^2/8$ with ω_p as the effective plasma frequency. (b) Eq. (57) is the thermal sumrule[18] equalling $\pi C_\mu(T) v_{eff}^2/(2d\Omega)$ found recently where $C_\mu(T), v_{eff}$ are the specific heat and an effective velocity respectively. From our earlier discussion, we see that the thermal conductivity has a correction for mobile carriers Eq. (12), so that we can define a finite frequency object

$$\kappa_{zc}(\omega) = \frac{1}{T} \left[L_{22}(\omega) - \frac{L_{12}(\omega)^2}{L_{11}(\omega)} \right] , \tag{58}$$

which also satisfies causality, and falls off at high frequencies as inverse ω, and therefore satisfies dispersion relations of the type Eq. (54). Thus by the same argument, and using the high frequency limits of all the coefficients Eq. (52), we infer a sum rule for this case as

$$\int_{-\infty}^{\infty} \frac{d\nu}{\pi} \Re e \kappa_{zc}(\nu) = \frac{1}{T\Omega}\left[\langle\Theta^{xx}\rangle - \frac{\langle\Phi^{xx}\rangle^2}{\langle\tau^{xx}\rangle}\right]. \tag{59}$$

The second term in Eq. (59) is usually small for Fermi systems at low temperatures and can be neglected. We may write the RHS as $\pi C_N(T) v_{eff}^2/(2d\Omega)$, in terms of the more conventional specific heat. It is interesting to note[h] that the explicit dependence on the chemical potential in the RHS of Eq. (57) arising from the definition of $\hat{J}_x^Q$ in Eq. (29), is exactly cancelled in the RHS of Eq. (59). Thus the zero current sum rule can be computed without knowing the chemical potential exactly. For immobile carriers this problem is irrelevant, Eq. (57) can be used without worrying about the distinction between heat current and energy current.

3.7. *Dispersion relations for thermopower, Lorentz number and figure of merit*

Let us now turn to the main objects of study here namely

$$\begin{aligned} \text{Thermopower } \ S(\omega) &= \frac{L_{12}(\omega)}{TL_{11}(\omega)} \\ \text{Lorentz Number } \ \mathbf{L}(\omega) &= \frac{\kappa_{zc}(\omega)}{T\sigma(\omega)} \\ \text{Figure of Merit } \mathbf{Z}(\omega)T &= \frac{S^2(\omega)}{\mathbf{L}(\omega)}\,. \end{aligned} \tag{60}$$

The first two objects are very well known in transport theories Ref. [14–16], while the figure of merit ZT is a dimensionless measure of the efficacy of a thermoelectric device, with large values $ZT \sim 1$ at low T being regarded as highly desirable in many applications. Let us analyze these definitions and extract their dispersion relations. It is readily seen that these variables differ qualitatively from the conductivity or the thermal conductivity in their high frequency behaviour. Asymptotically each of these approaches a

[h] I thank Dr S. Mukerjee and Dr M. Peterson for interesting discussions of this point.

constant, which can be written down by inspection.

$$\text{High Freq Thermopower } S^* = \frac{\langle \Phi^{xx} \rangle}{T \langle \tau^{xx} \rangle}$$
$$\text{High Freq Lorentz Number } \mathbf{L}^* = \frac{\langle \Theta^{xx} \rangle}{T^2 \langle \tau^{xx} \rangle} - (S^*)^2$$
$$\text{High Freq Figure of Merit } \mathbf{Z}^* T = \frac{\langle \Phi^{xx} \rangle^2}{\langle \Theta^{xx} \rangle \langle \tau^{xx} \rangle - \langle \Phi^{xx} \rangle^2} . \tag{61}$$

As a result, we can write their dispersion relations readily, they are

$$\Re e S(\omega) = S^* + \frac{\mathcal{P}}{\pi} \int_{-\infty}^{\infty} \frac{d\,\nu}{\nu - \omega} \, \Im m S(\nu) \tag{62}$$
$$\Re e \mathbf{L}(\omega) = \mathbf{L}^* + \frac{\mathcal{P}}{\pi} \int_{-\infty}^{\infty} \frac{d\,\nu}{\nu - \omega} \, \Im m \mathbf{L}(\nu) \tag{63}$$
$$\Re e \mathbf{Z}(\omega) = \mathbf{Z}^* + \frac{\mathcal{P}}{\pi} \int_{-\infty}^{\infty} \frac{d\,\nu}{\nu - \omega} \, \Im m \mathbf{Z}(\nu). \tag{64}$$

These transport quantities are generally real at only two values of frequency, namely zero or infinity, and are very similar in mathematical structure to the Hall resistivity discussed in Eq. (8). The imaginary part is expected to go linearly at small ω, falling off over some finite interval in ω corresponding to the energy range of the contributing physical processes. Thus the difference between the DC transport and high frequency values can be expressed in all these cases as an integral over the imaginary part of these three variables divided by the frequency, and may be amenable to direct measurements, as in the case of the Hall effect.

4. Thermoelectric phenomena in correlated matter

4.1. *Limiting case of free electrons, S^* the Heikes Mott and Mott results*

We propose the use of the high frequency variables Eq. (61) in correlated matter, for reasons that are essentially the same as those for proposing the high frequency Hall constant, explained earlier. These variables are singled out by the fact that they have a finite limit at high ω, as compared to say $\kappa(\omega)$ or $L_{12}(\omega)$, which vanish in that limit. In particular we expect that these high frequency limits of the three variables listed in Eq. (61), are good indicators of the DC transport measurements in correlated matter, where we can use the projected $t-J$ model, whereas for the Hubbard model, these should be good only for intermediate to weak coupling. In the following, we will see the consequences of this proposal, and estimate its accuracy in some well controlled examples.

Let us begin by listing the three basic operators for the case of the free electron model with particle energy ε_k and velocity $v_p^x = \partial\varepsilon_k/\partial k_x$, where a small calculation of Eq. (43) shows

$$\begin{aligned} \tau^{xx} &= q_e^2 \sum_{p,\sigma} v_p^x c_{p,\sigma}^\dagger c_{p,\sigma} \,, \\ \Theta^{xx} &= \sum_{p,\sigma} \frac{\partial}{\partial p_x} \left\{ v_p^x (\varepsilon_p - \mu)^2 \right\} c_{p,\sigma}^\dagger c_{p,\sigma} \,, \\ \Phi^{xx} &= q_e \sum_{p,\sigma} \frac{\partial}{\partial p_x} \left\{ v_p^x (\varepsilon_p - \mu) \right\} c_{p,\sigma}^\dagger c_{p,\sigma} \,. \end{aligned} \tag{65}$$

The corresponding operators for various standard models are available in [18], and given their length it seems hardly worthwhile to reproduce them here. We merely mention that the operators involve the interaction parameters, just as the energy currents do, and have to be worked out for each model individually. The one exception is the τ^{xx} operator, which usually has the same form as in Eq. (65), due to the fact that interactions are velocity independent. We will see the explicit form of the Φ^{xx} operator for the $U = \infty$ Hubbard model below Eq. (85).

To get back to the free electron theory, we can form the thermal averages,

$$\begin{aligned} \langle \tau^{xx} \rangle &= 2q_e^2 \sum_p \; n_p \; \frac{d}{dp_x} \left[v_p^x \right] \,, \\ \langle \Theta^{xx} \rangle &= 2 \sum_p \; n_p \; \frac{d}{dp_x} \left[v_p^x (\varepsilon_p - \mu)^2 \right] \,, \\ \langle \Phi^{xx} \rangle &= 2q_e \sum_p \; n_p \; \frac{d}{dp_x} \left[v_p^x (\varepsilon_p - \mu) \right] \,. \end{aligned} \tag{66}$$

Here n_p is the Fermi function. At low temperatures, we use the Sommerfeld formula after integrating by parts, and obtain the leading low T behaviour:

$$\begin{aligned} \langle \tau^{xx} \rangle &= \Omega \; 2 \; q_e^2 \rho_0(\mu) \; \langle (v_p^x)^2 \rangle_\mu \,, \\ \langle \Theta^{xx} \rangle &= \Omega \; T^2 \; \frac{2\pi^2 k_B^2}{3} \; \rho_0(\mu) \; \langle (v_p^x)^2 \rangle_\mu \,, \end{aligned} \tag{67}$$

$$\langle \Phi^{xx} \rangle = \Omega \; T^2 \; \frac{2q_e \pi^2 k_B^2}{3} \; \left[\rho_0'(\mu) \langle (v_p^x)^2 \rangle_\mu + \rho_0(\mu) \frac{d}{d\mu} \langle (v_p^x)^2 \rangle_\mu \right] \,, \tag{68}$$

where $\rho_0(\mu)$ is the density of states per spin per site at the Fermi level μ and the primes denote derivatives w.r.t. μ, the average is over the Fermi surface as usual. These formulas are indeed very close to what we expect from Boltzmann theory. At low T we may form the high frequency ratios and get the leading formulas[i]

$$S^* = T\frac{\pi^2 k_B^2}{3q_e}\frac{d}{d\mu}\ln\left[\rho_0(\mu)\langle(v_p^x)^2\rangle_\mu\right], \tag{69}$$

$$L^* = \frac{\pi^2 k_B^2}{3q_e^2}. \tag{70}$$

It is therefore clear that the high frequency result gives *the same* Lorentz number. The thermopower formula is slightly different from the usual Mott transport formula[14,15] for energy momentum dependent relaxation rate:

$$S_{Mott} = T\frac{\pi^2 k_B^2}{3q_e}\frac{d}{d\mu}\ln\left[\rho_0(\mu)\langle(v_p^x)^2\tau(p,\mu)\rangle_\mu\right]. \tag{71}$$

A comparison between the two formulas for the thermopower reveals the nature of the high frequency limit: it ignores the energy dependence of the relaxation time, but captures the density of states. Thus this formalism is expected to be accurate whenever the scattering is less important than say the density of states and correlations.

Let us also note the general formula for the thermopower from Eqs. (60,50). On dropping the second term Eq. (51), we get the standard formulas appropriate for dissipative systems, where we can write the "exact" Kubo formula:[13]

$$S_{Kubo} = \left[\frac{\int_0^\infty dt\,\int_0^\beta d\tau\,\langle\hat{J}_x^E(t-i\tau)\hat{J}_x(0)\rangle}{\int_0^\infty dt\,\int_0^\beta d\tau\,\langle\hat{J}_x(t-i\tau)\hat{J}_x(0)\rangle} - \frac{\mu(0)}{q_e}\right] + \frac{\mu(0)-\mu(T)}{q_e}. \tag{72}$$

We have used Eq. (28) and further added and subtracted the $\frac{\mu(0)}{q_e}$ term for convenience, to arrive at Eq. (72). The Mott result Eq. (71) follows from this general formula in the limit of weak scattering, as textbooks indicate.[16] For narrow band systems, Heikes introduced another approximation popularized by Mott,[20,21] namely the Heikes Mott formula

$$S_{HM} = \frac{\mu(0)-\mu(T)}{q_e}. \tag{73}$$

This formula emphasizes the thermodynamic interpretation of the thermopower, this term can be loosely regarded as the entropy per particle[j].

---

[i]The reader is requested to ignore the irksome issue of the dimensionality of the argument of the logarithm. The logarithm is just a notational device to collect the coefficients in this formula and in Eqs. (71,81).

[j]Strictly speaking μ is a derivative of the entropy w.r.t. the number of particles, i.e. $\mu(T) = -T(\partial S(N,T)/\partial N)_{E,T}$.

This motivates us[18] to decompose the thermopower as

$$S_{Kubo} = S_{Tr} + S_{HM}, \tag{74}$$

thereby defining the "transport part" of thermopower as the first part of Eq. (72), as opposed to the thermodynamic part S_{HM}. Using the high frequency approximation Eq. (61), we approximate (only) the transport part in Eq. (74) and write

$$\begin{aligned} S^* &= S^*_{Tr} + S_{HM}\,, \\ S^*_{Tr} &= \frac{\langle \Phi'^{xx} \rangle}{T\langle \tau^{xx} \rangle}\,, \end{aligned} \tag{75}$$

with the Φ'^{xx} being computed with the $T = 0$ value of μ. The low T limit for the free particle case of this relation is given in Eq. (69). For a correlated many body system, it is much easier to work with this variable. The computational advantage in Eq. (75) over Eq. (72) is that the transport part is approximated by an equal time correlator as opposed to a dynamical correlator. This allows us to apply one of several possible techniques to the problem, such as exact diagonalization and also high T expansions.

4.2. *Kelvin's thermodynamical formula for thermopower*

It is interesting to discuss Kelvin's thermodynamic derivation of the thermopower.[22] In his seminal work, Onsager[11] discussed Kelvin's derivation of reciprocity given several decades earlier. He argued that the phenomenon of transport, including reciprocity, cannot be understood within equilibrium thermodynamics or statistical mechanics. Interestingly as late as 1966, Wannier wrote in his textbook:[23] "*Opinions are divided as to whether Kelvin's derivation is fundamentally flawed or not*". A detailed account of this debate and its resolution seem to be missing in literature.

Our discussion of the thermopower takes us to the brink of this old debate, and so we make a small excursion to obtain a thermodynamic approximation of the correct answer. This derivation captures the spirit of the Kelvin argument, and provides an approximate expression for the thermopower S. For deriving this, let us rewind to Eq. (38) of the finite q, ω

dependent Onsager coefficients $L_{ij}(q,\omega)$. Using Eqs. (34,36,37) we see that

$$L_{11}(q,\omega) = \frac{i}{\Omega q_x}\chi_{\hat{J}_x(q_x),\rho(-q_x)}(\omega)\,,$$

$$L_{12}(q,\omega) = \frac{i}{\Omega q_x}\chi_{\hat{J}_x(q_x),K(-q_x)}(\omega), \quad \text{hence}$$

$$S(q_x,\omega) = \frac{\chi_{\hat{J}_x(q_x),K(-q_x)}(\omega)}{T\ \chi_{\hat{J}_x(q_x),\rho(-q_x)}(\omega)}\,. \tag{76}$$

Onsager's prescription at this point is to take the transport limit, i.e. first let $q_x \to 0$ followed by the static limit, to get the exact formula.[12,13] We saw in the previous section that this ratio has another finite and interesting limit, leading to S^*, when we let $q_x \to 0$ followed by $\omega \gg 0$. It is interesting and amusing that in the opposite slow limit, i.e. $\omega \to 0$ followed by $q_x \to 0$, also has a finite and well defined result. This limit is what we identify with the Kelvin calculation and formula, since the objects that arise are purely equilibrium quantities. Thus

$$S_{Kelvin} = \lim_{q_x\to 0,\omega\to 0} S(q_x,\omega)\,,$$

$$S(q_x,\omega) = \frac{\chi_{[K,\rho(q_x)],K(-q_x)}(\omega)}{T\ \chi_{[K,\rho(q_x)],\rho(-q_x)}(\omega)} \tag{77}$$

$$= \frac{\chi_{\rho(q_x),K(-q_x)}(\omega)}{T\ \chi_{\rho(q_x),\rho(-q_x)}(\omega)}\,. \tag{78}$$

We have used the continuity relation $\hat{J}_x(q_x) = \frac{1}{q_x}[K,\rho(q_x]$ to go from Eq. (76) to Eq. (77). The next stage involves writing a Ward type identity,

$$\chi_{[K,\rho(q_x)],K(-q_x)}(\omega) = -\omega\chi_{\rho(q_x),K(-q_x)}(\omega) + \langle[K(-q_x),\rho(q_x)]\rangle\,, \tag{79}$$

and a similar one for the denominator, followed by realizing that the second term of the r.h.s. of Eq. (79) vanishes on using parity for any finite q in a system with inversion symmetry[k]. We can now take the static limit and get the equilibrium Kelvin result

$$S_{Kelvin} = \lim_{q_x\to 0}\frac{\chi_{\rho(q_x),K(-q_x)}(0)}{T\ \chi_{\rho(q_x),\rho(-q_x)}(0)}\,, \tag{80}$$

[k]The argument is trivial for the denominator since density fluctuations commute at different wave vectors. In the numerator, consider the real expectation $\gamma(q_x) = \langle[K(-q_x),\rho(q_x)]\rangle$. Clearly $\gamma^*(q_x) = \langle[\rho(-q_x),K(q_x)]\rangle = -\gamma(-q_x)$. But from parity $\gamma(-q_x) = \gamma(q_x)$ and hence the result $\gamma(q_x) = 0$.

where in the limit, the denominator is the thermodynamic compressibility, and the numerator is an equilibrium cross correlation function function between energy and charge density. It is straightforward to see that reciprocity holds in this sequence of limits as well.

In the case of free particles, it is easy to evaluate Eq. (80) at low T and we find

$$S_{Kelvin} = T\frac{\pi^2 k_B^2}{3q_e}\frac{d}{d\mu}\ln\left[\rho_0(\mu)\right] . \tag{81}$$

It is amusing to compare Eq. (69,71) and Eq. (81). Compared to the "exact" Mott formula that follows from the Onsager limiting procedure, S^* captures the answer except for the energy dependent relaxation rate. The Kelvin formula further approximates S^* by neglecting the energy dependence of the velocity average. Thus we conclude that the Kelvin approximation is inferior to the high frequency approximation, but does capture the density of states effects.

The above, rather formal manipulation with the limits, can be nicely visualized by working instead with open boundary conditions[1]. We apply a space-time varying gravitational field $\psi(\vec{x},t) = \delta\psi_0 \frac{x}{L}\exp\{-i\omega t\}$ together with a similar electrostatic potential $\phi(\vec{x},t)$, and compute the induced oscillating dipole moment $P = \sum_x x\rho(\vec{x})$ using perturbation theory. The gravitational field is again a proxy for temperature variation. By forming the ratio of the gravitational field amplitude $\delta\psi_0$ to the electrostatic amplitude $\delta\phi_0$ needed to produce a given dipole moment, we can extract the thermopower. The rigorously correct transport limit, as applied to this situation, requires *the thermodynamic limit to be taken before* $\omega \to 0$. If we compute the opposite limit instead, i.e. a finite system and a DC field, then the result maps to the above Eq. (80). Such a limiting process is tempting from the physical picture of the so called "absolute thermopower". In this case, one studies a single system with applied thermal gradients, which develops a voltage across its ends. This type of a picture was presumably behind the Kelvin derivation.

4.3. *Applications to sodium cobaltates in the Curie Weiss metallic phase*

At this point it is worthwhile to compare the results of various approximations in the important and current problem of sodium cobaltates Na_xCoO_2,

[1]B.S. Shastry unpublished (2007).

with $x \sim .68$. Recent interest in this system started with the observation of high $S \sim 80\mu V/K$ in this system by Terasaki,[2] which is unusual in that it is also a good metal. Wang, Rogado, Cava and Ong, in another important paper [3] found that this thermopower is strongly magnetic field dependent. They further found that the metallic conduction is coexistent with a Curie Weiss susceptibility characteristic of insulators. This has given rise to the nomenclature of a Curie Weiss metallic phase. The basic modeling

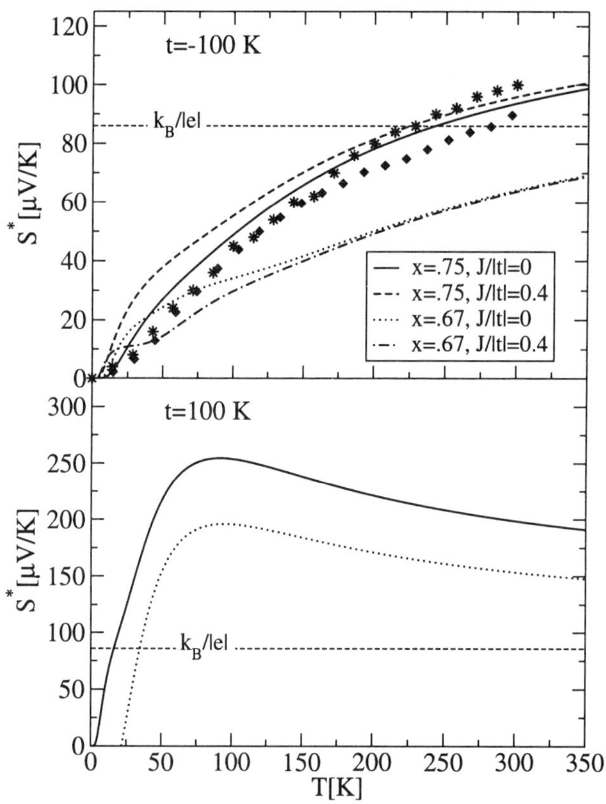

Fig. 3. **Upper Panel:** The computed thermopower versus T, compared to the experimental data of Ref. [2] (stars) and Ref. [3] (diamonds). The absolute scale is set by a single parameter $t = -100^0$K. The different curves correspond to various values of doping x and $J/|t|$. **Lower Panel:** This shows the effect of reversing the sign of hopping in this system. This is a prediction of this theory for a fiduciary hole doped sodium cobaltate type system. The peak value of $250\mu V/K$ can be further manipulated by changing material parameters J, x.

of this system, as suggested by Wang *et. al.*, is in terms of a strongly correlated Fermi system, with no double occupancy of holes. The holes move on a triangular lattice provided by the Co atoms, and the system may be regarded, to a first approximation, as a bunch of uncoupled 2D triangular lattice planes with a $t-J$ model description of correlated holes. After performing a particle hole transformation we can write the basic Hamiltonian as

$$H = -\sum_{i,j} t_{ij} \tilde{c}^{\dagger}_{\vec{r}_i,\sigma} \tilde{c}_{\vec{r}_j,\sigma} + J \sum \vec{S}_i . \vec{S}_j \, . \tag{82}$$

This model corresponds to the limit of $U \to \infty$. In this limit the Fermionic commutation relations need to be modified into the Gutzwiller-Hubbard projected operators[24]

$$\begin{aligned} \tilde{c}_{\vec{r},\sigma} &= P_G \; c_{\vec{r},\sigma} \; P_G \, , \\ \left\{ \tilde{c}_{\vec{r},\sigma}, \tilde{c}^{\dagger}_{\vec{r}',\sigma'} \right\} &= \delta_{\vec{r},\vec{r}'} \left\{ \delta_{\sigma,\sigma'} (1 - n_{\vec{r},\bar{\sigma}}) + (1 - \delta_{\bar{\sigma},\sigma'}) \tilde{c}^{\dagger}_{\vec{r},\sigma} \tilde{c}_{\vec{r},\bar{\sigma}} \right\} \\ &\equiv Y_{\sigma,\sigma'} \; \delta_{\vec{r},\vec{r}'} \, . \end{aligned} \tag{83}$$

The presence of the Y factor is due to strong correlations, and makes the computation nontrivial. The number operator $n_{\vec{r},\sigma}$ is unaffected by the projection. Let us consider the kinetic energy only, i.e. the t part, since this is expected to dominate in transport properties, at least far enough from half filling and for $t \gg J$. The addition of the J part can be done without too much difficulty, in fact the numerics discussed below include the full Hamiltonian.

Let us note down the expressions for the charge current and the energy current at finite wave vectors by direct computation:

$$\hat{K}(k) = -\sum_{\vec{r},\vec{\eta},\sigma} (t(\vec{\eta}) + \mu \delta_{\vec{\eta},0}) \; e^{i\vec{k}.(\vec{r}+\frac{1}{2}\vec{\eta})} \; \tilde{c}^{\dagger}_{\vec{r}+\vec{\eta},\sigma} \tilde{c}_{\vec{r},\sigma} \, , \tag{84}$$

$$\hat{J}_x(k) = i q_e \sum_{\vec{r},\vec{\eta},\sigma} \eta_x t(\vec{\eta}) \; e^{i\vec{k}.(\vec{r}+\frac{1}{2}\vec{\eta})} \; \tilde{c}^{\dagger}_{\vec{r}+\vec{\eta},\sigma} \tilde{c}_{\vec{r},\sigma} \, ,$$

$$\begin{aligned} \hat{J}^Q_x(k) = -\frac{i}{2} \sum_{\vec{r},\vec{\eta},\vec{\eta}',\sigma} & (\eta_x + \eta'_x) t(\vec{\eta}) t(\vec{\eta}') \; e^{i\vec{k}.(\vec{r}+\frac{1}{2}(\vec{\eta}+\vec{\eta}'))} \\ & \cdot Y_{\sigma',\sigma}(\vec{r}+\vec{\eta}') \; \tilde{c}^{\dagger}_{\vec{r}+\vec{\eta}+\vec{\eta}',\sigma'} \tilde{c}_{\vec{r},\sigma} - \frac{\mu}{q_e} \hat{J}_x(k) \, . \end{aligned}$$

Here $\vec{\eta}$ is the nearest neighbor vector on the triangular lattice.

We evaluate the thermopower operator as:

$$\Phi^{xx} = -\frac{q_e}{2} \sum_{\vec{\eta},\eta',\vec{\sigma},\sigma',\vec{r}} (\eta_x + \eta'_x)^2 \, t(\vec{\eta}) \, t(\vec{\eta'}) \, Y_{\sigma',\sigma}(\vec{r}+\vec{\eta}) \, \tilde{c}^{\dagger}_{\vec{r}+\vec{\eta}+\vec{\eta'},\sigma'} \tilde{c}_{\vec{r},\sigma}$$
$$-q_e\mu \sum_{\vec{\eta},\sigma} \eta_x^2 \, t(\vec{\eta}) \, \tilde{c}^{\dagger}_{\vec{r}+\vec{\eta},\sigma} \tilde{c}_{\vec{r},\sigma}. \quad (85)$$

This expression gives an idea of the complexity of the operators that arise in the theory. Let us first present some numerical results obtained by exact diagonalization[25,26] of small clusters of the triangular lattice. We can compute all eigenstates and matrix elements for up to 14 or 15 site clusters of the triangular lattice. We can therefore assemble the full dynamical conductivities from Eq. (60). The involved calculations are fully described in the papers, and we will content ourselves with displaying the main results. Firstly, consider the absolute scale of the thermopower S^* as a function of temperature, shown in Fig. (3). The upper panel in Fig. (3) shows that this comparison with experiment is quite successful on a quantitative scale. One can next ask, how good is the approximation of infinite frequency, purely in theoretical terms. To answer this we compute the frequency dependence of $S(\omega)$, as shown in Fig. (4). It is clear from this figure that the approximation of high frequency is excellent, and thus we are computing essentially the DC transport object, at least for clusters of these sizes.

4.4. *High temperature expansion for thermopower*

The lower panel of Fig. (3) shows the remarkable enhancement of the computed thermopower at low and intermediate T's, achieved by flipping the sign of hopping from the upper panel. Let us next discuss this remarkable effect of the sign of hopping on the transport part of S. Let us perform a simple computation at high T that throws light on this phenomenon. We focus on the kinetic energy which is expected to dominate the transport contributions. Let us compute the thermopower S^* from Eqs. (85,61)

$$S^* = -\frac{\mu}{q_e T} + \frac{q_e \Delta}{T\langle \tau^{xx} \rangle}, \quad (86)$$

where

$$\Delta = -\frac{1}{2} \sum_{\vec{\eta},\vec{\eta'},\vec{r}} (\eta_x + \eta'_x)^2 \, t(\vec{\eta}) \, t(\vec{\eta'}) \, \langle Y_{\sigma',\sigma}(\vec{r}+\vec{\eta}) \, \tilde{c}^{\dagger}_{\vec{r}+\vec{\eta}+\vec{\eta'},\sigma'} \tilde{c}_{\vec{r},\sigma} \rangle . \quad (87)$$

The computation of the different parts proceeds as follows: we show readily that (for the hole doped case) using translation invariance and with n as

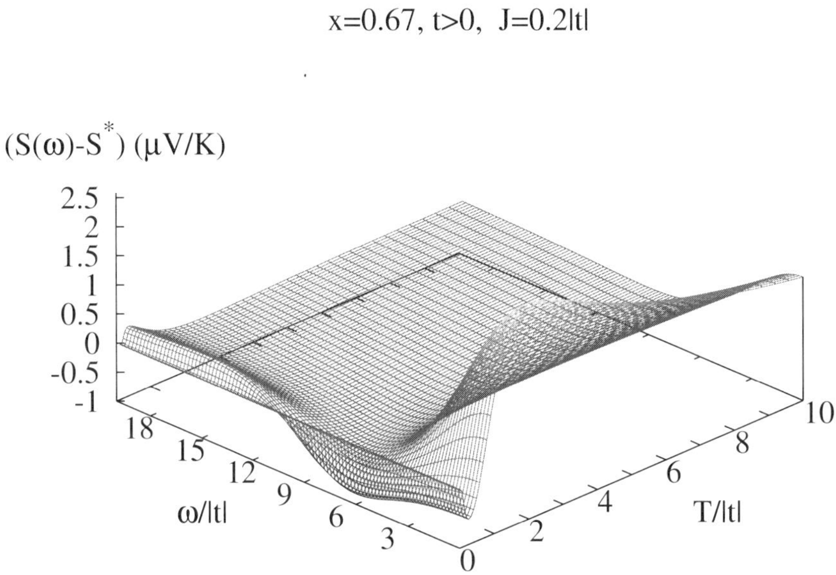

Fig. 4. The frequency dependence of the thermopower computed numerically for a typical set of parameters close to those of experimental interest in Na_xCoO_2 (the sign of hopping is flipped relative to that in Fig. (3) in this and all other figures by using a p-h mapping). Recall that the scale of $S \sim 100\mu V/K$, therefore the frequency dependence is indeed very small $\sim 3\%$ at most.

the number of particles per site at high T,

$$\langle \tau^{xx} \rangle = 6\Omega q_e^2 t \langle \tilde{c}_1^\dagger \tilde{c}_0 \rangle \sim 3\Omega q_e^2 \beta t^2 n(1-n). \tag{88}$$

The structure of the term Eq. (87) is most instructive. At high temperatures, for a square lattice we need to go to second order in βt to get a contribution with $\eta_x + \eta'_x \neq 0$, to the expectation of the hopping $\langle \tilde{c}^\dagger_{\vec{r}+\vec{\eta}+\vec{\eta}',\sigma'} \tilde{c}_{\vec{r},\sigma} \rangle$. For the triangular lattice, on the other hand, we already have a contribution at first order. For the triangular lattice, corresponding to each nearest neighbor, there are precisely two neighbors where the third hop is a nearest neighbor hop. A short calculation gives

$$\Delta \sim -3\Omega t^2 \sum_{\sigma,\sigma'} \langle Y_{\sigma',\sigma}(\vec{\eta}) \tilde{c}^\dagger_{\vec{\eta}+\vec{\eta}',\sigma'} \tilde{c}_{\vec{0},\sigma} \rangle. \tag{89}$$

The spins must be the same to the leading order in βt where we generate a hopping term $\tilde{c}^\dagger_{\vec{0},\sigma} \tilde{c}_{\vec{\eta}+\vec{\eta}',\sigma}$ from an expansion of $\exp(-\beta K)$, and hence a simple estimation yields

$$\Delta = -\frac{3}{2}\Omega t^3 \beta n(1-n)(2-n) + O(\beta^3)\,. \tag{90}$$

This together with $\mu/k_B T = \log(n/2(1-n)) + O(\beta^2 t^2)$ gives us the result for $0 \leq n \leq 1$

$$S^* = \frac{k_B}{q_e}\left\{\log[2(1-n)/n] - \beta t \frac{2-n}{2} + O(\beta^2 t^2)\right\}, \tag{91}$$

and

$$S^* = -\frac{k_B}{q_e}\left\{\log[2(n-1)/(2-n)] + \beta t \frac{n}{2} + O(\beta^2 t^2)\right\} \tag{92}$$

for $1 \leq n \leq 2$ using particle hole symmetry.[18]

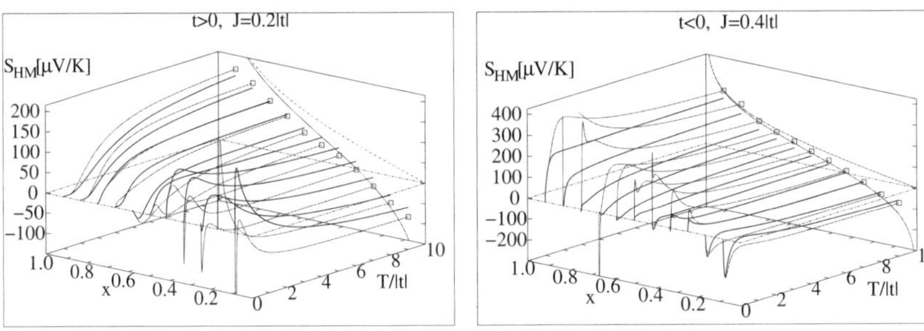

Fig. 5. The thermopower versus filling $x = 1 - n$ and temperature T in the $t - J$ model from numerical studies[25,26] on clusters. In both cases the lower curves correspond to the Heikes Mott formula Eq. (73) and upper to the high frequency result of Eq. (61). Left: The case of the sodium cobaltates, i.e. electron doping, where the two estimates are very close. Right: The fiduciary hole doped cobaltate. The two curves in the high T limit corresponds to the first term in Eq. (91) and from the uncorrelated chemical potential. For the case on right, the Heikes Mott formula misses the enhancement that the high frequency formula predicts. Such enhanced values of the thermopower are very exciting in the current quest for better materials.

We observe that the thermodynamic contribution, i.e. the Heikes Mott part from the $\mu(T)$ dominates at very high T. The approach to this value is governed by the correction term arising from transport part. This transport term is $O(\beta t)$ for the triangular lattice, whereas it is only $O(\beta t)^2$ for the square lattice due to the existence of closed loops of length three in the former. The high T expansion made here clearly identifies the role of the lattice topology here. The other important consequence is the dependence upon the sign of the hopping in the transport term. To be specific, for electron doping the thermopower in Eq. (92) shows that S approaches its high T limit from below as long as $t < 0$, as we find for sodium cobaltates.[2,3] On the other hand, if we could flip the sign of the hopping, as in a fiduciary hole doped Cobalt Oxide layer, *the high T value would be reached from above.* Since the S must vanish at low T, this observation implies that we must find a maximum in $S(T)$ at some intermediate T. This then motivates the calculation for a fiduciary system with the flipped sign of hopping. As seen in Fig. (3,5), numerical results are very encouraging, leading to a thermopower that is $\sim 250\mu V/K$, and should act as an incentive to the materials community who could seek this type of doping. Crystal structures containing triangular loops are clearly favourable, and this includes several 3D structures as well, such as the FCC and HCP lattices.

4.5. *Lorentz number and figure of merit for the triangular lattice $t - J$ model*

We briefly indicate the dependence of the Lorentz number $\mathbf{L}^*$ and the figure of merit $\mathbf{Z}^*T$ as computed by us[25,26] in the case of the triangular lattice, with parameters appropriate for sodium cobaltates at $x \sim .68$. Fig. (6) indicates the dependence of these important parameters on x, T for the $t-J$ model clusters of size up to 14. The frequency dependence was estimated to be small and of the same scale as that of $S(\omega)$, therefore the results are good indicators of the DC values. We must keep in mind however, that the

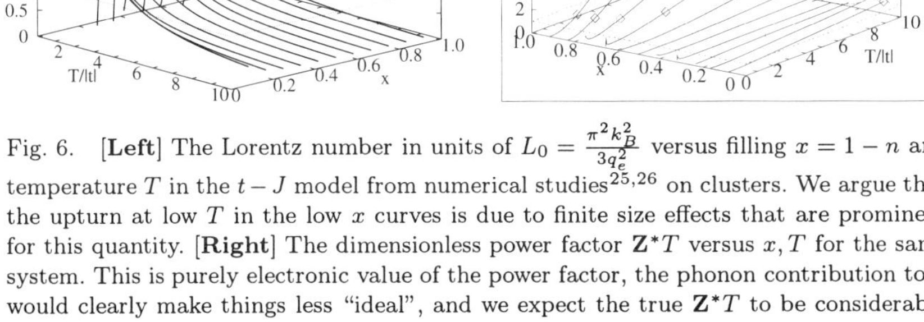

Fig. 6. [**Left**] The Lorentz number in units of $L_0 = \frac{\pi^2 k_B^2}{3q_e^2}$ versus filling $x = 1 - n$ and temperature T in the $t - J$ model from numerical studies[25,26] on clusters. We argue that the upturn at low T in the low x curves is due to finite size effects that are prominent for this quantity. [**Right**] The dimensionless power factor $\mathbf{Z}^*T$ versus x, T for the same system. This is purely electronic value of the power factor, the phonon contribution to κ would clearly make things less "ideal", and we expect the true $\mathbf{Z}^*T$ to be considerably decreased.

finite size effects are quite substantial for these small clusters, and hence the behaviour at low T is particularly subject to corrections. However, it is clear that several interesting trends emerge from this study, perhaps a striking one is that the proximity of the Mott Hubbard insulating state $x \sim 0$ is not very favourable for good thermoelectric behaviour with a large Z^*T. This is despite the enhancement of S itself, due to Mott Hubbard correlations that lead to a logarithmic divergence of S near half filling.[10] Rather, the proximity of almost filled bands or almost empty bands seems to be more favourable, and indeed the experience with doping in Na_xCoO_2 seems to bear out this finding rather well.

Acknowledgement

BSS was supported by grant NSF-DMR 0706128, and by grant DOE-BES DE-FG02-06ER46319. It is a pleasure to thank Dr. M. Peterson and Dr. S. Mukerjee for helpful discussions. I thank the organizers of the Winter School for a stimulating meeting. I thank Prof. A. Shakouri and Prof. N.P. Ong for providing me with experimental inputs that have been invaluable. It is a pleasure to acknowledge stimulating comments from Prof. P.W. Anderson regarding the Kelvin-Onsager debate.

References

1. P.W. Anderson, Nature **235**, 1196 (1987).
2. I. Terasaki, Y. Sasago and K. Uchinokura, Phys. Rev. **B56**, R12685 (1997).
3. Y. Wang, N.S. Rogado, R.J. Cava and N.P. Ong, Nature **423**, 425 (2003).
4. B.S. Shastry, B.I. Shraiman and R.R.P. Singh, Phys. Rev. Lett. **70**, 2004 (1993).
5. R.E. Peierls, Ann. Physik **3**, 1055 (1929).
6. A.T. Zheleznyak, V.M. Yakovenko and H.D. Drew, Phys. Rev. **B57** 3089 (1998).
7. Jan O. Haerter and B.S. Shastry, arXiv:07081651.
8. B. Kumar and B.S. Shastry, Phys. Rev. **B69**, 059901 (2004), (Erratum) Phys. Rev. **B68**, 104508 (2003).
9. Y. Wang, N.S. Rogado, R.J. Cava and N.P. Ong, arXiv:cond-mat/0305455.
10. J.O. Haerter, M.R. Peterson and B.S. Shastry, Phys. Rev. Lett. **97**, 226402 (2006), Phys. Rev. **B74**, 245118 (2006).
11. L. Onsager, Phys. Rev. **37**, 405 (1931), Phys. Rev. **38**, 2265 (1931).
12. J.M. Luttinger, Phys. Rev. **A135**, 1505 (1964), Phys.Rev. **A136** 1481 (1964).
13. R. Kubo, J. Phys. Soc. Japan **12**, 5 (1957).
14. N. Ashcroft and N.D. Mermin, *Solid State Physics*, (Harcourt Brace Jovanovich College Publishers, Fort Worth, 1976).
15. J.M. Ziman, *Principles of the Theory of Solids*, (Cambridge University Press, Cambridge, 1969).

16. G.D. Mahan, *Many Particle Systems* (Plenum, New York, 1990).
17. A.L. Fetter and J.D. Walecka, *Theory of Many Particle Systems* (Dover, New York 2003).
18. B.S. Shastry, Phys. Rev. **B73** 085117 (2006); (Erratum) Phys. Rev. **B74** 039901 (2006). An updated version with several errors removed is available at `http://physics.ucsc.edu/~sriram/papers_all/ksumrules_errors_etc/evolving.pdf`.
19. R. Bari, D. Adler and R.V. Lang, Phys. Rev. **B2**, 2898 (1970); E. Sadakata and E. Hanamura, J Phys. Soc. Japan **34**, 882 (1973); P.F. Maldague, Phys. Rev. **B16**, 2437 (1977).
20. P. Chaikin and G. Beni, Phys. Rev. **B13**, 647 (1976).
21. R.R. Heikes, *Thermoelectricity* (Wiley-Interscience, New York, 1961).
22. W. Thomson (Lord Kelvin), Proc. Roy. Soc. Edinburgh: p. 123 (1854), Collected Papers I, pp. 237-41.
23. G.H. Wannier, Statistical Physics, pp.506 (Dover Publications, NY 1966).
24. The Hubbard operators are defined in J. Hubbard, Proc. Roy. Soc. **A276**, 238 (1963); **A277**, 237 (1964); **A281**, 401 (1964); **A285**, 542 (1964).
25. M. Peterson, S. Mukerjee, B.S. Shastry and J.O. Haerter, Phys. Rev. **B76** 125110 (2007).
26. M. Peterson, B.S. Shastry and J.O. Haerter, arXiv:0705.3867; Phys. Rev. **B76** 125110 (2007)).

CARRIER CONCENTRATION INDUCED FERROMAGNETISM IN SEMICONDUCTORS

T. STORY

Institute of Physics, Polish Academy of Sciences
Al. Lotników 32/46, 02-668 Warsaw, Poland

In semiconductor spintronics ferromagnetic semiconductor materials with new magnetically controlled electrical or optical functionalities are needed to facilitate an integration of standard electronic functions of semiconductors with magnetic memory function. All classical semiconductors, like GaAs, are nonmagnetic materials, in which ferromagnetic properties can be induced by substitutional incorporation of magnetic ions (usually Mn^{2+} ions) and heavy doping with conducting carriers up to 10^{20}-10^{21} cm^{-3}. The experimental observations of carrier concentration induced ferromagnetism are discussed for three groups of semiconductor materials: p-$Sn_{1-x}Mn_xTe$, p-$Pb_{1-x-y}Sn_yMn_xTe$, and p-$Ge_{1-x}Mn_xTe$ (diluted magnetic semiconductors of IV-VI group, in which paramagnet - ferromagnet and ferromagnet - spin glass transitions are found for very high hole concentration), p-$Ga_{1-x}Mn_xAs$, p-$In_{1-x}Mn_xAs$, and p-$In_{1-x}Mn_xSb$ (diluted magnetic semiconductors of III-V group, in which ferromagnetism appears due to Mn ions providing both local magnetic moments and acting as acceptor centers), and n-$Eu_{1-x}Gd_xTe$ (magnetic mixed crystals, in which the substitution of Gd^{3+} ions for Eu^{2+} ions creates very high electron concentration and transforms antiferromagnetic, insulating EuTe compound into ferromagnetic n-type semiconductor alloy). For each of these materials systems the key physical features are discussed concerning: local magnetic moments formation, magnetic phase diagram as a function of magnetic ions and carrier concentration as well as Curie temperature engineering. Various theoretical models proposed to explain the effect of carrier concentration induced ferromagnetism in semiconductors are presented involving mean field approaches based on Zener and Ruderman-Kittel-Kasuya-Yosida (RKKY) mechanisms.

Outline

(1) Introduction: ferromagnetic semiconductors – materials review
(2) Diluted magnetic (semimagnetic) semiconductors (DMS) – crystal structure, magnetic doping, and basic physical concepts
(3) Carrier concentration induced ferromagnetism in IV-VI DMS: p-(Sn,Mn)Te, p-(Pb,Sn,Mn)Te, and p-(Ge,Mn)Te

(4) Magnetic phase diagram of DMS materials governed by RKKY interaction
(5) Ferromagnetism in III-V DMS induced by conducting holes: p-(Ga,Mn)As, p-(In,Mn)As, and p-(In,Mn)Sb
(6) Antiferromagnet-ferromagnet transition in n-(Eu,Gd)Te induced by conducting electrons
(7) Summary – search for new ferromagnetic semiconductors for spintronics

1. Introduction: ferromagnetic semiconductors – materials review

Most of ferromagnetic materials are metals (with canonical examples of Fe, Co, and Ni from 3*d* elements and Gd from 4*f* elements), metallic alloys (e.g., permalloy NiFe) or intermetallic compounds. Ferromagnetism is also observed in a number of materials with semiconductor-like electron band structure. The list of ferromagnetic semiconductors is given, e.g. in classical monographs on magnetic semiconductors.[1,2] It contains textbook examples of nonmetallic ferromagnetic materials such as europium chalcogenides (EuO and EuS), various spinels (e.g., $CdCr_2Se_4$), and magnetic oxides.

In semiconductor spintronics the key materials issue concerns ferromagnetic semiconductors that would exhibit new magnetically controlled electrical or optical functionalities and permit an integration (in a single multilayer heterostructure) of standard electronic functions of semiconductors with magnetic memory function. Although all classical semiconductor materials, such as Si or GaAs, are nonmagnetic, upon substitutional incorporation of a few atomic percents of magnetic ions (usually Mn^{2+} ions) and heavy doping with conducting carriers up to 10^{20}-10^{21} cm^{-3}, a ferromagnetic transition can be induced in such diluted magnetic semiconductors (also known as semimagnetic semiconductors). In this lecture the experimental observations of carrier concentration induced ferromagnetism will be discussed for three model groups of semiconductor crystals:

p-$Sn_{1-x}Mn_xTe$, p-$Pb_{1-x-y}Sn_yMn_xTe$, and p-$Ge_{1-x}Mn_xTe$ – classical diluted magnetic semiconductors of IV-VI group, in which paramagnet – ferromagnet and ferromagnet – spin glass transitions are found for very high hole concentration;

p-$Ga_{1-x}Mn_xAs$, p-$In_{1-x}Mn_xAs$ and p-$In_{1-x}Mn_xSb$ – currently the most actively studied diluted magnetic semiconductors of III-V group, in which ferromagnetism appears due to Mn ions providing both local magnetic moments and acting as acceptor centers;

n-$Eu_{1-x}Gd_xTe$ – magnetic (concentrated) mixed crystals, in which the substitution of Gd^{3+} ions for Eu^{2+} ions creates very high electron concentration and transforms antiferromagnetic EuTe (insulating compound) into ferromagnetic n-type semiconductor alloy.

For each of these materials systems the key physical features will be discussed concerning: local magnetic moments formation, magnetic phase diagram as a function of magnetic ions and carrier concentration as well as Curie temperature engineering. Various theoretical models proposed to explain the effect of carrier concentration induced ferromagnetism in semiconductors will be briefly discussed involving mean field approaches based on Zener and Ruderman-Kittel-Kasuya-Yosida (RKKY) mechanisms and new proposals aiming at proper theoretical description of such features of diluted magnetic semiconductors as magnetic (site) disorder, electronic disorder due to doping, and competition between ferromagnetic and antiferromagnetic exchange interactions.

2. Diluted magnetic (semimagnetic) semiconductors (DMS) – crystal structure, magnetic doping, and basic physical concepts

Diluted magnetic semiconductors (DMS) are substitutional solid solutions of the well-known II-VI (e.g., CdTe, ZnTe, ZnSe), IV-VI (e.g., PbTe, SnTe, GeTe), and III-V (e.g. GaAs, InAs or InSb) semiconductors and magnetic semiconductors like, e.g., MnTe, MnSe, MnS as well as EuTe and EuSe.[3–7] In DMS materials magnetic ions are randomly allocated over the sites of cation (metal) sublattice of the host semiconducting compound. The probability that a given cation site is occupied by magnetic ion is x, where x is the content of magnetic ions. The usual way of specifying the composition of DMS materials is analogous to other substitutional alloys: e.g. $Ga_{1-x}Mn_xAs$ or (Ga,Mn)As or simply GaMnAs. The latter may indicate that not all the Mn ions occupy the lattice sites (substitutional ions) but there may also be Mn ions located in the interstitial position in the crystal or even in nano-sized regions with locally different atomic coordination. This site (composition) randomness is an inherent property of DMS materials. The simplest model, successfully applied in the analysis of optical and electrical properties of II-VI and IV-VI DMS, is the virtual crystal approximation (VCA), in which the strict lattice periodicity is restored in DMS materials by assigning the same x-dependent potential to each cation site. This approach may not be valid for materials in which magnetic ion produces very strong local distortion of crystal lattice or potential.

Apart from this technological definition, the concept of diluted magnetic semiconductor involves the formation of the well-localized magnetic moments by $3d$ or $4f$ magnetic ions, and the existence of the *sp-d* or *sp-f* exchange interaction between magnetic ions and conducting carriers. The $3d$ or $4f$ electron orbitals of magnetic ions preserve in DMS materials their local atomic character creating localized magnetic moments. In DMS materials one identifies two distinguished spin subsystems: the system of local magnetic ions and the system of delocalized conducting electrons or conducting holes. These two systems are coupled by *sp-d* or *sp-f* exchange interaction which contributes to the total energy of the system:

$$\hat{H}_{\text{exch}} = -\Sigma J_{sd}\,\hat{\mathbf{S}} \cdot \hat{\sigma} \;.$$

Here $\hat{\mathbf{S}}_i$ and $\hat{\sigma}$ represent the spin of local magnetic moments and quasi-free carriers, respectively. J_{sd} is the *sp-d* exchange integral (exchange constant) – a quantitative measure of the density of exchange energy. Two microscopic mechanisms are known to determine the strength of the *sp-d* and *sp-f* exchange interaction. The, so called, potential exchange results from the direct overlap of the wave functions of conducting carriers and $3d$ or $4f$ electrons of magnetic ions. This mechanism gives positive (ferromagnetic) sign of the exchange integral and is rather weak. The other mechanism (kinetic exchange) involves the mixing (hybridization) of the electronic states of conducting carriers and $3d$ electrons. This mechanism is known to produce quite strong exchange interaction of antiferromagnetic sign. In II-VI and III-V DMS materials with Mn the typical values of the exchange integrals are $J_{sd} = 0.2$ eV for electrons and $J_{pd} = -1$ eV for holes. In IV-VI DMS the exchange integrals are smaller, typically of the order of $10 - 100$ meV.[3–9]

The characteristic property of DMS materials is that the incorporation of even relatively high concentration of magnetic ions does not lead to qualitative changes in the band structure or in the crystal structure of these materials. Being qualitatively similar to the case of non-magnetic host material, the band structure and the crystal structure of DMS materials is characterized by the composition dependent parameters such as the energy gap $E_g(x)$ or the lattice parameter $a_0(x)$. The important element of the electron structure of DMS is the location of the density of states (DOS) derived from the $3d$ or $4f$ magnetic orbitals of magnetic ions. The scheme of the situation encountered in IV-VI semiconductors with Mn and with Eu is presented in Fig. 1. The DOS labeled $3d$ and $4f$ corresponds to the half (spin up) of the total DOS due to these orbitals. The other half (spin down) of the total DOS is shifted up in the energy scale by few eV and

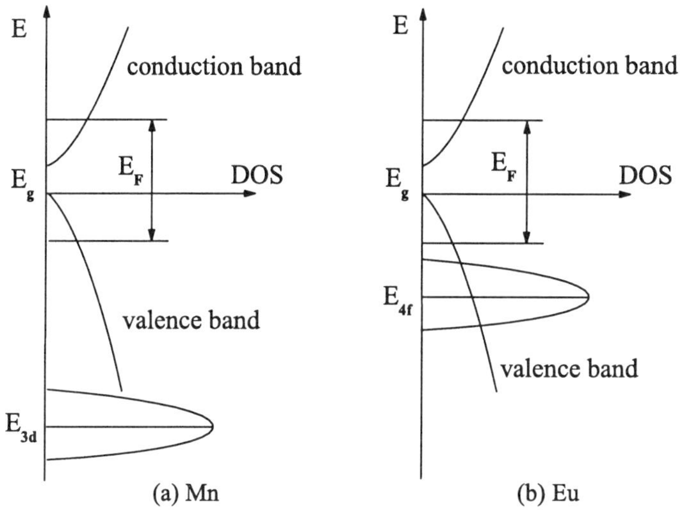

Fig. 1. A scheme of the electron band structure of DMS materials presenting the location of density of states (DOS) contribution due to magnetic $3d$ (for Mn^{2+} ions) or $4f$ (for Eu^{2+} ions) orbitals. The DOS for group IV-VI DMS material: $Pb_{1-x}Mn_xTe$ is presented in Fig. 1a and for $Pb_{1-x}Eu_xTe$ in Fig. 1b. The range of Fermi level positions encountered in various DMS crystals (depending on doping or annealing conditions) is indicated with double arrow lines labeled E_F.

is expected to be located far above the bottom of the conduction band.[4] The large energy difference between both contributions is due to the strong on-site electron-electron interactions and correlations. In both cases the contributions due to magnetic orbitals are located far below the top of the valence band. It means that magnetic orbitals do not contribute to the DOS in the energy range covered by the Fermi level positions encountered in canonical DMS materials, even for heavy doped crystals. Therefore, Mn^{2+} and Eu^{2+} are expected to be electrically neutral. Mn^{2+} and Eu^{2+} ions in DMS materials possess spin-only magnetic moments expected for free ions.

The presence of the *sp-d* exchange coupling between the system of local magnetic ions and valence band or conduction band electrons results in a number of new optical and electrical effects.[3–8] One of the spectacular consequences of the presence of the *sp-d* exchange interaction is anomalously large spin splitting of the conduction and valence band states in DMS materials (usually expressed in terms of the effective g-factor of carriers g^*):

$$g^* = g_0 + \frac{J_{sd} M(H,T)}{g_M \mu_B H} \,.$$

Here g_M is the g-factor of magnetic ions (usually g_M=2.0), and g_0 is the

g-factor of carriers in non-magnetic host semiconductor. Since the magnetization $M(H,T)$ of DMS materials is temperature and magnetic field dependent, the same dependence is observed in magneto-optical properties of these materials. This effect becomes very large at helium temperatures and corresponds to the effective g^*-factors of the order of 100.

All II-VI, IV-VI, III-V and group IV canonical semiconductors (not doped with magnetic ions) are standard diamagnetic materials with very low magnetic susceptibility of $\chi = -3 \cdot 10^{-7}$ emu/g. The paramagnetic, ferromagnetic and spin glass magnetic properties observed in DMS materials are due to the system of local magnetic moments formed by magnetic ions such as, e.g.:

Mn^{2+} (electron configuration $3d^5$, $S = 5/2$, $L = 0$);
Eu^{2+} and Gd^{3+} (electron configuration $4f^7$, $S = 7/2$, $L = 0$);
Cr^{3+} (electron configuration $3d^3$, $S = 3/2$, $L = 3$);
Cr^{2+} (electron configuration $3d^4$, $S = 2$, $L = 2$, $J = 0$) – van Vleck ion.

The presence of magnetic ions in a given charge and spin state can be detected in electron paramagnetic resonance (EPR) measurements by the observation of their characteristic spectra.[4] For example, Mn^{2+} ions show a characteristic 6-line isotropic structure which is due to the hyperfine interaction between the electronic and the nuclear magnetic moments of Mn^{2+} ions.[4,10] The identification of the charge state of magnetic ion in DMS matrix establishes the magnetic moment carried out by each ion as well as helps to predict the electrical activity of the ion. For example, Mn^{2+} ions which substitute Cd^{2+} or Sn^{2+} in CdTe or SnTe are expected to be electrically neutral, whereas the Mn ion substituting Ga^{3+} in GaAs becomes an acceptor center. The situation is qualitatively different in DMS materials with Gd or Cr ions which frequently (not always) are in Gd^{3+} and Cr^{3+} charge state showing donor character.[4,11]

3. Carrier concentration induced ferromagnetism in IV-VI DMS: p-(Sn,Mn)Te, p-(Pb,Sn,Mn)Te, and p-(Ge,Mn)Te

Magnetic properties of IV-VI DMS with Mn depend strongly on the concentration of conducting holes. The crystals of PbMnTe, PbMnSe, and PbMnS with relatively low concentration of carriers (typically $10^{17} - 10^{18}$ cm^{-3}) are Curie-Weiss paramagnets exhibiting rather weak antiferromagnetic *d*-*d* nearest-neighbor exchange interaction with the *d*-*d* exchange integral of the order of -0.5 K.[4,5,12–15] The superexchange mechanism proceeding via the electronic states of p-orbitals of neighboring anions is believed to be responsible for this interaction. This mechanism leads to short range interaction and is expected to dominate in semiconducting crystals with low carrier concentration. The important physical parameters determining the strength of the superexchange mechanism are the energetic location of magnetic $3d$ or $4f$ energy states with respect to band edges of a semiconductor, and the hybridization parameters characterizing the efficiency of the overlap and mixing of magnetic and non-magnetic orbitals.

IV-VI DMS materials with Mn possesing very high concentration of conducting holes $p \geq (2-3) \cdot 10^{20}$ cm^{-3} are ferromagnets. These are ternary and quaternary DMS alloys based on SnTe, GeTe and their alloys with PbTe such as: $Sn_{1-x}Mn_xTe$,[3–5,8,16–18] $Pb_{1-x-y}Sn_yMn_xTe$,[3–5,16,18–20] $Ge_{1-x}Mn_xTe$,[21,22] and $Pb_{1-x-y}Ge_yMn_xTe$.[23] The ferromagnetic properties of SnTe and GeTe based DMS materials with Mn were studied experimentally by ac magnetic susceptibility, magnetization, magnetic specific heat, neutron scattering, ferromagnetic resonance, thermoelectric power, and anomalous Hall effect measurements. The carrier concentration induced ferromagnetic transition is observed in p-(Pb,Sn,Mn)Te bulk crystals in the entire available composition range. Fig. 2 illustrates this effect in $Pb_{1-x-y}Sn_yMn_xTe$ ($y = 0.72$) crystals with Mn content $x = 0.12$ and varying carrier concentration.[16,18–20,24,25] At high temperatures the Curie-Weiss, $\chi(T) = C/(T - \Theta)$, dependence of magnetic susceptibility is observed with the ferromagnetic Curie-Weiss temperature Θ. It indicates the dominance of the ferromagnetic *d-d* interactions. The strength of the ferromagnetic exchange interactions in IV-VI DMS materials depends on the concentration of carriers what results also in the carrier concentration dependence of the ferromagnetic transition (see the shift of the parameter Θ in Fig. 2). By varying the carrier concentration one can change the Curie temperature of PbSnMnTe crystals with a constant Mn content (which can be determined experimentally from the Curie constant C – see Fig. 2, bottom panel). The $\Theta(p)$ dependence (normalized per 1 at.% of Mn ions) for the crystals of $Pb_{1-x-y}Sn_yMn_xTe$ with Mn content up to 8 at. % is summarized in Fig. 3. The characteristic feature of this dependence is the existence of a certain threshold carrier concentration $p_c \cong 3 \cdot 10^{20}$ cm^{-3}, above which the IV-VI DMS materials with Mn show ferromagnetic properties. For carrier concentrations lower than the threshold value $p < p_c$, the crystals are paramagnetic similarly to the low carrier concentration materials like (Pb,Mn)Te. This very characteristic feature is related to the electron band structure of PbSnMnTe presented in Fig. 3a. For conducting

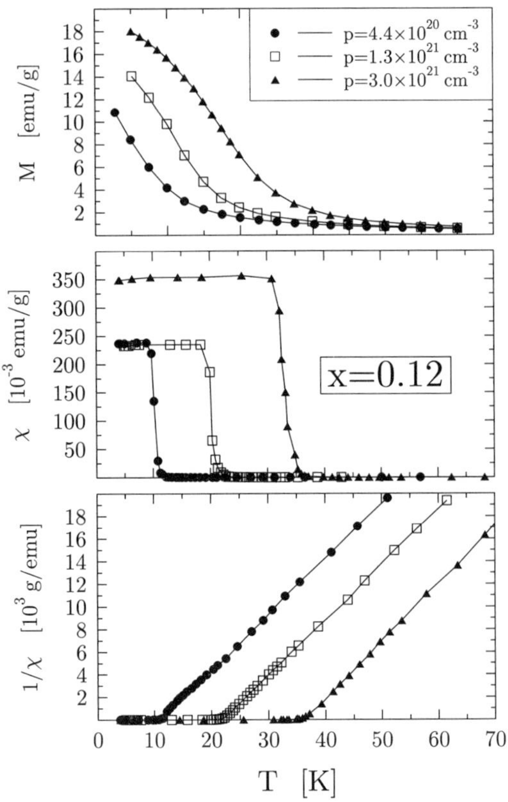

Fig. 2. The temperature dependence of the magnetization, ac magnetic susceptibility, and the inverse of the magnetic susceptibility of $Pb_{1-x-y}Sn_yMn_xTe$ bulk crystals with $x = 0.12$ and $y = 0.72$ for various concentrations of conducting holes p.[20]

hole concentration $p > p_c$ the heavy holes from the 12-valley Σ-band start to contribute to charge transport in IV-VI DMS and strongly enhance the carrier concentration mediated exchange interactions.[24]

The ferromagnetic *d-d* interspin exchange interactions in IV-VI DMS are due to the Ruderman-Kittel-Kasuya-Yosida (RKKY) mechanism which proceeds via quasi-free carriers from the valence band. The RKKY interaction has long-range character and is known to result in *d-d* or *f-f* exchange interactions of different signs depending on the value of the characteristic parameter $2k_F R_{ij}$, where k_F is the Fermi wave vector and R_{ij} is the interspin distance (see Fig. 4). The strength of the RKKY interaction depends on the concentration of carriers (via $k_F \sim p^{1/3}$), the *sp-d* exchange integral

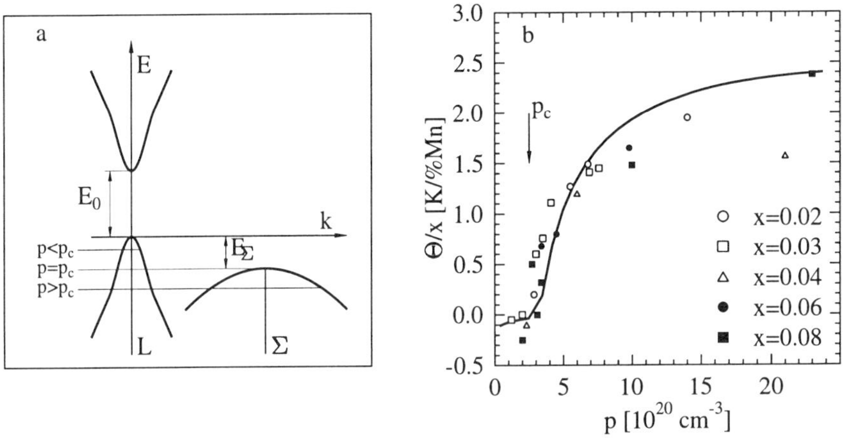

Fig. 3. The model of the band structure of $Pb_{1-x-y}Sn_yMn_xTe$ crystals explaining the origin of the threshold carrier concentration p_c: L indicates the band of light holes whereas Σ shows the band of heavy holes. In Fig. 3b the carrier concentration dependence of the reduced (per 1 at. % of Mn) paramagnetic Curie temperature Θ is presented for $Pb_{1-x-y}Sn_yMn_xTe$ bulk crystals with $y = 0.72$ and various Mn content. The solid line shows the theoretical calculations based on the RKKY interaction via heavy and light holes.

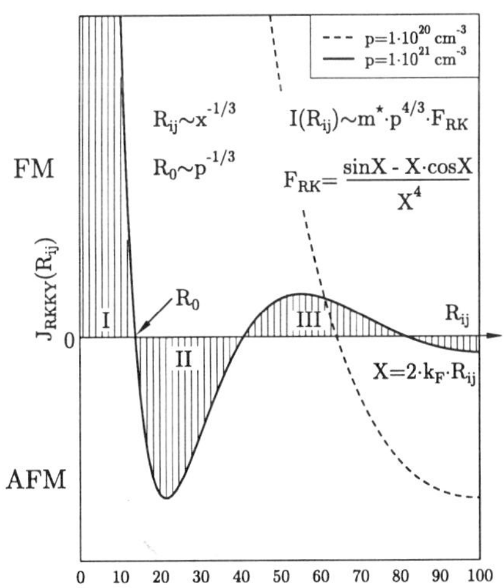

Fig. 4. The plot of the RKKY *d*-*d* exchange integral versus interspin distance R_{ij} (expressed in terms of the number of coordination sphere) for two carrier concentrations indicated in the figure.

J_{pd}, and the effective mass of carriers m^*:

$$I_{ij} = \frac{J_{sd}^2 (k_F a_0)^6}{64 E_F \pi^3} F_{RK}(X) e^{-R_{ij}/\lambda}$$

$$F_{RK}(X) = \frac{\sin(X) - X\cos(X)}{X^4}$$

$$X = 2k_F R_{ij} \text{ and } E_F = \hbar^2 k_F^2 / 2m^* ,$$

where λ is the carrier mean free path and a_0 is the lattice parameter.

In more detailed analysis of the RKKY mechanism in IV-VI DMS materials one has to take into account the anisotropic and multi-valley character of the valence band structure of PbSnMnTe. It results, in particular, in anisotropic (cubic) *d-d* interspin interactions.[26] Another important aspect of the realistic description of the RKKY mechanism in IV-VI DMS is the incorporation of the effects brought about by a disorder present in both electronic and magnetic subsystems. One of the consequences of the electronic disorder is that the exchange integrals are now statistical quantities characterized by their mean value and variance. The existence of the distribution of the values of the exchange integrals reflects the statistical distribution of the sources of electronic disorder (e.g., non-magnetic defects). The detailed analysis of this effect is presented in.[27]

Carrier concentration induced ferromagnetism was also demonstrated in about 1 micron thick monocrystalline layers of $Sn_{1-x}Mn_xTe$ ($x \leq 0.04$) grown by molecular beam epitaxy on BaF_2 (111) substrates. The control of the concentration of carriers in the range $5 \cdot 10^{19}$ cm^{-3} $\leq p \leq 2 \cdot 10^{21}$ cm $^{-3}$ was achieved by adjusting the molecular flux from Te effusion cell in addition to the standardly employed SnTe and Mn cells. Depending on the concentration of conducting holes both ferromagnetic and paramagnetic SnMnTe layers were grown.[28]

The strong influence of carrier concentration on magnetic properties was also observed in $Ge_{1-x}Mn_xTe$ ($x \leq 0.98$) layers grown by ionized-cluster beam, rf sputtering, and molecular beam epitaxy techniques on BaF_2(111) and glass substrates.[29–32] Upon controlling GeTe, MnTe, and Te fluxes the concentration of conducting holes was reduced from $2.6 \cdot 10^{21}$ cm^{-3} down to $5.4 \cdot 10^{19}$ cm^{-3} (Fig. 5).

In contrast to p-(Pb,Sn,Mn)Te and p-(Sn,Mn)Te IV-VI DMS materials, in p-(Ge,Mn)Te layers the ferromagnetic transition is observed in the entire range of carrier concentration studied. The concentration of carriers

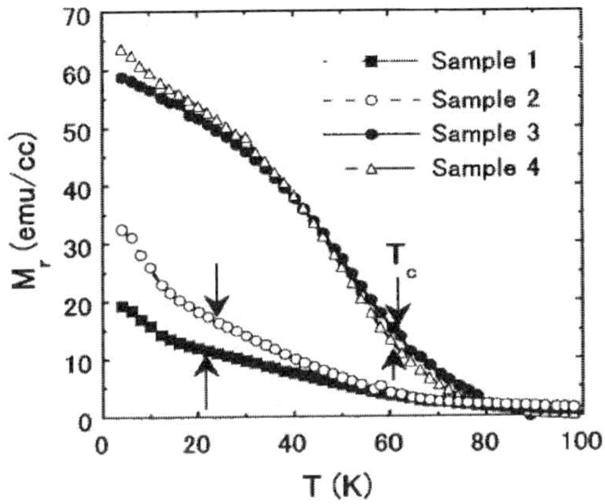

Fig. 5. The temperature dependence of the remanent magnetization of the crystalline layers of $Ge_{1-x}Mn_xTe$ ($x = 0.4$) grown on BaF_2 substrate under varying technological conditions resulting in different conducting hole concentrations: $p = 5.4 \cdot 10^{19}$ cm^{-3} (sample 1), $p = 1.7 \cdot 10^{20}$ cm^{-3} (sample 2), $p = 2.3 \cdot 10^{21}$ cm^{-3} (sample 3), and, $p = 2.6 \cdot 10^{21}$ cm^{-3} (sample 4).

strongly influences the anomalous Hall constant, negative magnetoresistance and the character of the temperature dependence of the magnetization of the layers.[29,30]

4. Magnetic phase diagram of DMS materials governed by the RKKY interaction

The RKKY interaction is known to be responsible both for the ferromagnetic properties of magnetic materials (e.g. rare earth metals like Gd) and for spin glass properties (e.g. diluted metallic alloys like Cu:Mn). The analysis of the RKKY interaction in IV-VI DMS shows that within the range of parameters available in these materials the effect of the transformation of RKKY ferromagnet into RKKY spin glass may be observed. This idea was realized experimentally in SnMnTe and PbSnMnTe bulk crystals with relatively low Mn content. In Fig. 6a the basic magnetic properties are compared for two samples of $Pb_{1-x-y}Sn_yMn_xTe$ ($y = 0.72$) with the same Mn content ($x = 0.02$) and with different carrier concentrations: the sample with $p = 6.8 \cdot 10^{20}$ cm^{-3} is ferromagnetic, whereas in the sample with $p = 1.4 \cdot 10^{21}$ cm^{-3} the ferromagnetic order breaks down and spin glass properties are observed.[27,33] Another experimental evidence for the carrier induced transition from the ferromagnetic to spin glass state is presented in Fig. 6b for the $Sn_{0.96}Mn_{0.04}Te$ samples with three different concentrations of holes. For the ferromagnetic sample ($p = 7.5 \cdot 10^{20}$ cm^{-3}) there is practically no frequency dependence of ac magnetic susceptibility in the range of 10 Hz – 20 kHz whereas in the sample with very high carrier concentration ($p = 2.4 \cdot 10^{21}$ cm^{-3}) a strong reduction of the magnetic susceptibility and shift of the temperature of characteristic cusp is clearly observed.[34] The very strong experimental evidence for ferromagnet-spin glass transition is also provided by neutron diffraction experiments showing the standard mean-field-like temperature dependence of spontaneous magnetization in ferromagnetic sample and almost no spontaneous magnetic moment for spin glass sample.[35]

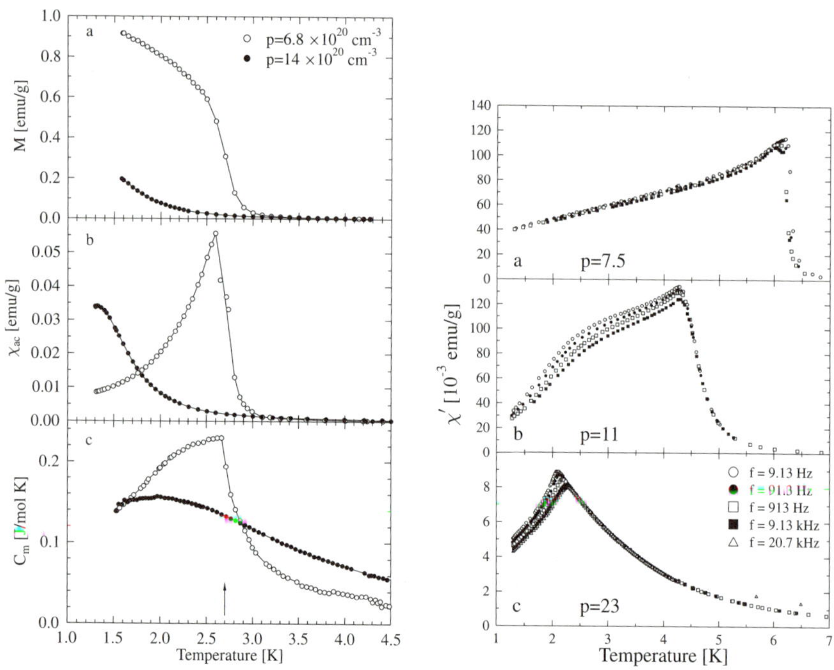

(a) (b)

Fig. 6. The temperature dependence of the magnetization (M), the ac magnetic susceptibility (χ_{ac}), and the magnetic contribution to the specific heat (C_{m}) in $\mathrm{Pb}_{1-x-y}\mathrm{Sn}_y\mathrm{Mn}_x\mathrm{Te}$ bulk crystals ($y = 0.72$, $x = 0.02$) for two carrier concentrations. In Fig. 6b the temperature dependence of the real part of the ac magnetic susceptibility of $\mathrm{Sn}_{1-x}\mathrm{Mn}_x\mathrm{Te}$ ($x = 0.04$) is presented for varying ac magnetic field frequency f.

The simple model of this effect is based on the properties of the RKKY interaction. For the ferromagnetism to be observed one requires that the parameter $2k_F R_{ij}$ for the average interspin distance parameter $R_{ij} = R$ corresponds to ferromagnetic region of the Ruderman-Kittel function (see Fig. 4). That leads to the condition $R \ll R_0$, where the characteristic distance $R_0 \sim 1/k_F \sim 1/p^{1/3}$ corresponds to the first switch of the RKKY interaction from ferromagnetic to antiferromagnetic interaction. In the other limit, $R \gg R_0$, the oscillatory character of the RKKY interaction is expected to lead to the spin glass order.[27,33]

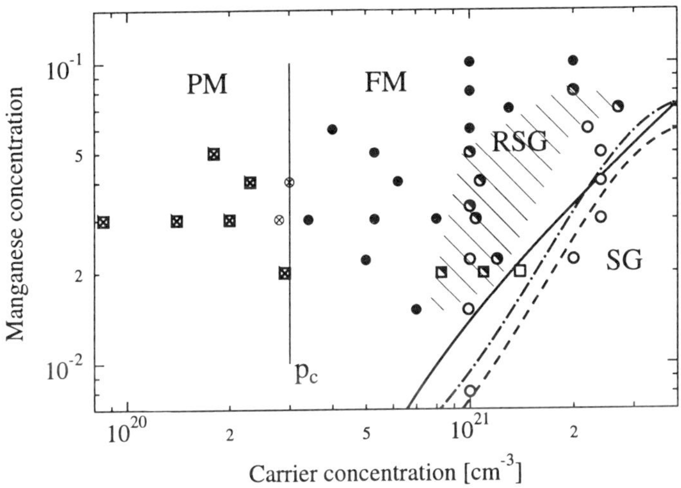

Fig. 7. The Mn concentration - carrier concentration (x-p) magnetic phase diagram of $Sn_{1-x}Mn_xTe$ and $Pb_{1-x-y}Sn_yMn_xTe$ crystals: FM – ferromagnets (full circles), PM – paramagnets (crossed squares and crossed circles), SG – spin glasses (open circles and open squares), RSG – re-entrant spin glasses (half-full circles and squares). The lines present the theoretical calculations in various models explained in the text.

The variety of magnetic properties of IV-VI DMS is summarized in the form of x-p magnetic phase diagram presented in Fig. 7. Depending on the concentration of carriers (p) and the content of Mn ions (x) the ferromagnetic (FM), paramagnetic (PM), spin glass (SG) and the mixed (so-called, re-entrant spin glass, RSG) phase may be observed. This magnetic phase diagram of the RKKY governed diluted magnetic systems can be derived in another way based on the analysis of the statistical distribution of the algebraic sum of interspin exchange couplings in a random substitutional magnetic alloy. The physical criterion for the ferromagnetic phase to be established is analogous to the well known Sherrington-Kirkpatrick condition. The mean value of the distribution of the sum of exchange integrals must be positive (ferromagnetic) and larger than the variance of the distribution.

Physically, it corresponds to the dominance of ferromagnetic couplings in the spectrum of relevant exchange interactions. In the opposite case, ferromagnetic and antiferromagnetic interactions compete one with the other and spin glass state is expected. The theoretical analysis of the x-p magnetic phase diagram of PbSnMnTe and SnMnTe crystals performed within the Sherrington-Kirkpatrick-like model is given in [27].

5. Ferromagnetism in III-V DMS induced by conducting holes: p-(Ga,Mn)As, p-(In,Mn)As, and p-(In,Mn)Sb

The solubility of Mn ions in bulk crystals of GaAs and other III-V semiconductors grown under thermodynamic equilibrium conditions is limited to just a doping level ($x < 0.001$). The world wide research activity on (Ga,Mn)As and (In,Mn)As started only after the important technological discovery that a high crystal quality layers of III-V DMS with Mn content up to 9 at. % can be grown by low temperature molecular beam epitaxy technique (LT-MBE).[36,37] In III-V DMS materials Mn ions substitute Ga^{3+} or In^{3+} ions in the zinc blende crystal lattice. In contrast to canonical II-VI DMS, like $Zn_{1-x}Mn_xTe$, in which Mn^{2+} ions are isovalent with nonmagnetic Zn^{2+} cations, Mn in (Ga,Mn)As usually does not appear as Mn^{3+} ion but forms a complex Mn^{2+}-like center. Based on the EPR studies of GaAs:Mn a model of Mn center in GaAs was established that involves an electron strongly bound to $3d^4$ core of Mn ion with a hole (h) weakly localized around such a $(3d^4$+e)+h acceptor center.[38,39] The ionization energy of this hole is about 100 meV. In (Ga,Mn)As layers with Mn content $x = 0.01 - 0.09$ a metallic conductivity is observed.[36,40] The Mn acceptor states are expected to merge in a band overlapping with the valence band of GaAs. Therefore, Mn ions act as single acceptor centers responsible for the p-type conductivity of (Ga,Mn)As and (In,Mn)As. The incorporation of 1 at. % of Mn generates about 10^{20} cm^{-3} of conducting holes. In the theoretical analysis of magnetic and electronic properties of (Ga,Mn)As layers it is usually assumed that (Ga,Mn)As is a semiconductor with a degenerate hole gas. The holes occupy the electron states below the top of the valence band.[41–43] The nonequilibrium growth regime inherent to LT-MBE technology has important consequences for the incorporation of Mn ions in III-V semiconductor matrices. Apart from the standard substitutional lattice position (i.e. the acceptor centers described above), Mn also appears at the interstitial positions, in which it acts as a double donor center compensating p-type conductivity.[44–46] Additionally, the high concentration of native defects (Ga-As anti-site centers) is expected in LT-MBE grown GaAs. These defects are also electrically active (double donor) centers. Recently, a very efficient way of reducing the number of interstitial Mn ions and increasing hole concentration in (Ga,Mn)As was discovered. It is a simple annealing procedure during which Mn ions occupying interstitial positions are moved out from the bulk of the layer and become passivated at the surface. This process is very important in the optimization of magnetic properties of (Ga,Mn)As layers.[43–47]

The ferromagnetic transition in (Ga,Mn)As, (In,Mn)As, and (In,Mn)Sb layers is very well established experimentally by ac magnetic susceptibility, magnetization, anomalous Hall effect, electrical resistivity, ferromagnetic resonance, and neutron diffraction experiments.[36,40–42,47,48] The maximal Curie temperature is about $T_C = 170$ K for (Ga,Mn)As,[43,47] $T_C = 60$ K for (In,Mn)As,[37] and $T_C = 15$ K for (In,Mn)Sb.[49,50] The Mn content dependence of the Curie temperature in (Ga,Mn)As layers is presented in Fig. 8a for both as grown and annealed layers.[47] The importance of the magnetic and electrical optimization of the layers by annealing procedure is directly illustrated in Fig. 8b where the shift of the ferromagnetic transition temperature is clearly observed on the temperature dependence of magnetization of 50 nm thick (Ga,Mn)As layer subjected to various annealing conditions.[45] It is important to note that the annealed (Ga,Mn)As layers posses higher ferromagnetic Curie temperature, magnetic moment per 1 Mn ion close to expected for Mn^{2+} ion, and the conducting hole concentration approximately equal to the concentration of Mn acceptors.

As ferromagnetism in III-V DMS materials is observed in heavy doped p-type materials, it was from the very beginning associated with carrier concentration induced effect via the RKKY interaction. In canonical metallic spin glass systems as well as in IV-VI DMS with Mn the exchange energy J_{pd} is smaller than the Fermi energy E_F, justifying the perturbation theory analysis with small parameter J_{pd}/E_F, i.e. the RKKY model calculations. In (Ga,Mn)As the other physical limit appears to hold: the exchange coupling $J_{pd} = -1.2$ eV is bigger then the Fermi energy $E_F = 0.2 - 0.3$ eV. Therefore, the first mean field theoretical calculations of the Curie temperature of (Ga,Mn)As and other III-V, II-VI and group IV DMS materials with Mn were based on the Zener mean field model of ferromagnetism extended to take into account the important details of the valence band structure of these semiconductors.[41,42] In this model the free energy of the systems of local magnetic moments and quasi free holes coupled by the *p-d* exchange interaction is analyzed. Below the critical temperature T_C, the minimum energy of such a system corresponds to the nonzero magnetization (at zero

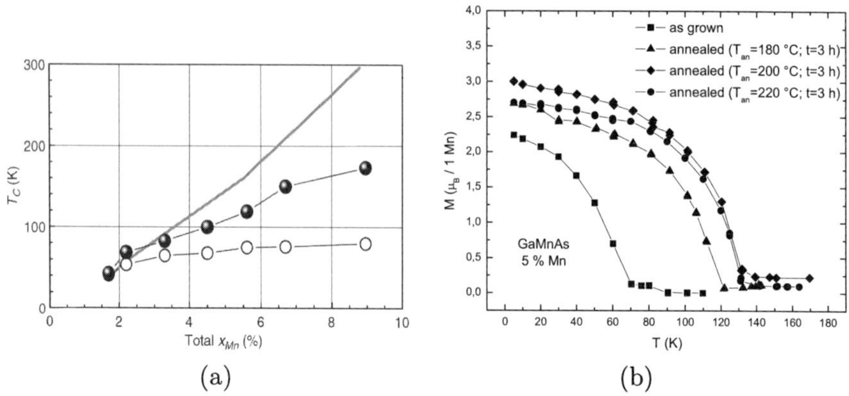

Fig. 8. The Mn content dependence of ferromagnetic Curie temperature of (Ga,Mn)As: open circles correspond to as grown layers whereas full circles present the data for annealed (optimized) layers.[47] The solid line shows the predictions of the mean-field Zener model.[42] In Fig. 8b the temperature dependence of the magnetization (per 1 Mn ion) is presented for the as grown and annealed (Ga,Mn)As layer with Mn content $x = 0.05$.

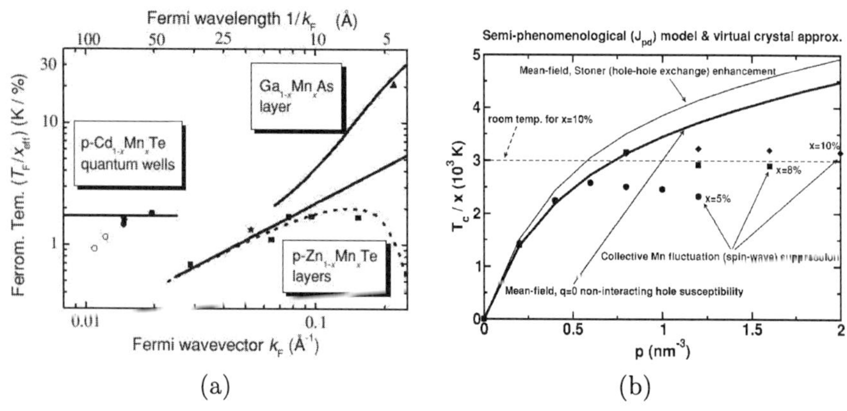

Fig. 9. The ferromagnetic Curie temperature (per 1 at. % of Mn) calculated in the Zener mean field model as a function of Fermi wave vector for various III-V and II-VI DMS systems: thick layers of p-$Ga_{1-x}Mn_xAs$ and p-$Zn_{1-x}Mn_xTe$ and quantum wells of p-$Cd_{1-x}Mn_xTe$.[51] In Fig. 9b the predictions of various theoretical models for the ferromagnetic Curie temperature of (Ga,Mn)As (per 1 at. %) are compared for different hole concentrations p.[43]

external magnetic field) of both the subsystem of local magnetic moments of Mn ions (ferromagnetic state with spontaneous magnetization) and the subsystem of degenerate gas of conducting holes (exhibiting spontaneous spin splitting of the valence band states). In the Zener model any details of the interspin distance dependence of the *d-d* exchange interactions are neglected and the Curie temperature $T_C \sim xJ_{pd}^2 m^* k_F$ is proportional to magnetic ions content and to carrier concentration $p^{1/3}$ (via k_F).[42,43]

In both the Zener and the RKKY models the Curie temperature increases linearly with increasing content of magnetic ions x. The key materials parameters are the *p-d* exchange integral, the carrier concentration (Fermi wave vector) and the density of states effective mass at the Fermi level. This model successfully explains experimentally observed basic chemical trends and provides the good estimation of the Curie temperature (see Fig. 9a). Because of the above mentioned physical limitation of this model in neglecting magnetic fluctuations, chemical disorder and the competition of ferromagnetic and antiferromagnetic interactions, a variety of other theoretical models have recently been developed as summarized in [43] and in Fig. 9b.[43,52–56]

In Fig. 10 the effect of light induced ferromagnetism in (In,Mn)As layers is presented. This effects was discovered in $In_{0.94}Mn_{0.06}As$/GaSb heterostructure grown by LT-MBE on GaAs substrate.[57,58] The active part of the structure is the (In,Mn)As/GaSb interface with band offset alignment and band bending effects resulting in the energy band scheme presented in Fig. 10a. The light generates electron hole pairs e-h in the entire heterostructure. The direction of the electric field in the heterojunction is such that it drives electrons away into the GaSb layer and GaAs substrate but it forces holes to accumulate in the DMS layer of (In,Mn)As. The light induced local increase of hole concentration leads to the appearance of magnetic hysteresis loops (see Fig. 10b).

In Fig. 11 the spectacular observation of carrier induced ferromagnetism achieved by electrical (gate voltage) control of carrier concentration in (In,Mn)As layer is demonstrated. In this experiment the increase of carrier concentration necessary for the generation of the ferromagnetic state was realized in the field effect transistor with the conducting channel of p-(In,Mn)As layer, in which the concentration of holes could be increased or decreased applying negative or positive gate voltage, respectively (see Fig. 11). The magnetic properties of the heterostructre were experimentally studied by anomalous Hall effect measurements providing an efficient magnetization measurement technique for such a micro device geometry. The changes of the carrier concentration achieved experimentally were clearly large enough to generate, depending on the gate voltage, the ferromagnetic magnetization loops or just paramagnetic response.[59]

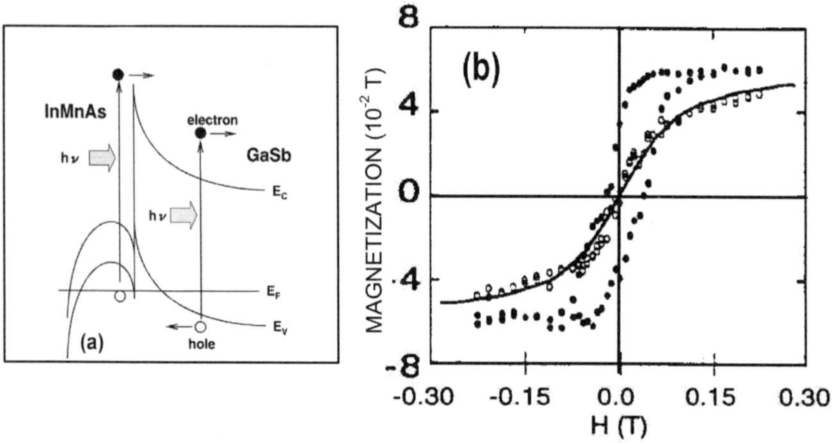

Fig. 10. The light induced ferromagnetism in (In,Mn)As/GaSb heterostructure. Due to the specific band alignment, presented in Fig. 10a, the holes generated after illumination drift in the electric field of the heterojunction towards the InMnAs layer locally increasing the hole concentration. In Fig. 10b the magnetization of the (In,Mn)As/GaSb heterostructures is presented before illumination (solid line, paramagnetic state) and after illumination (full dots, ferromagnetic state).[57,58]

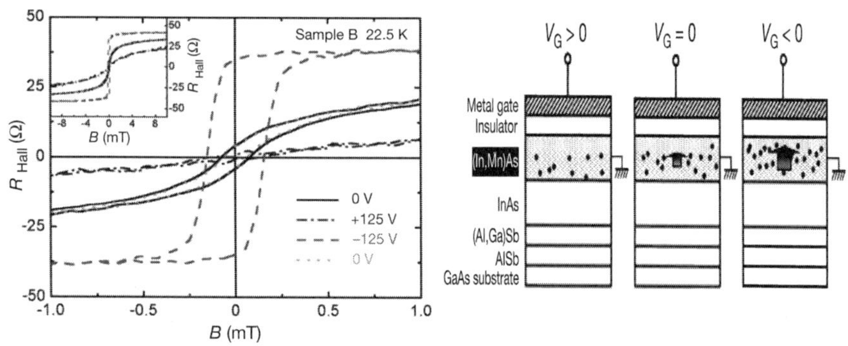

(a) (b)

Fig. 11. The electric-field control of carrier concentration induced ferromagnetism in (In,Mn)As field effect transistor. In Fig. 11a the magnetic hysteresis loops observed in anomalous Hall effect measurements are presented for various electrical bias conditions. In Fig. 11b the model of this effect is illustrated. The ferromagnetic state is achieved for a negative gate voltage conditions $V_G < 0$ with an increased conducting hole concentration in (In,Mn)As DMS channel.[59]

6. Antiferromagnet-ferromagnet transition in n-(Eu,Gd)Te induced by conducting electrons

Europium monochalcogenides EuX (X=O, S, Se, Te) are the well known magnetic semiconductor compounds which crystallize in the rock salt structure. In these compounds Eu^{2+} ions have electronic configuration [Xe] $4f^7$ resulting in the spin-only magnetic moments of 7 Bohr magnetons, as expected from the Hund's rule. The strongly localized character of the $4f$ orbitals makes these materials model Heisenberg magnetic systems. In stoichiometric, undoped EuX crystals (insulators) the exchange interaction between 12 nearest Eu-Eu neighbors is ferromagnetic whereas the 6 next nearest Eu-Eu neighbors are usually coupled antiferromagnetically by the superexchange interaction via anions. In this group of materials, EuTe is the only antiferromagnet (MnO-like antiferromagnetic type II structure) with the Néel temperature $T_N = 9.6$ K, while both EuO and EuS are ferromagnets with the Curie temperature $T_C = 69$ K and 16.5 K, respectively. EuSe exhibits more complicated state with various magnetic phases at low temperatures.[1,2,60–63]

Substitution of divalent Eu^{2+} ions by trivalent Gd^{3+} ions, which have the same $4f^7$ electronic configuration, practically does not affect the system of local magnetic moments in the crystal matrix but generates quasi-free electrons in the conduction band. Consequently, in (Eu,Gd)Te crystals the concentration of quasi-free carriers of the order of $n = 10^{20} - 10^{22}$ cm^{-3} can be obtained. The additional, carrier-induced magnetic interactions brought about by the carriers offer a unique possibility to transform EuTe from an antiferromagnetic insulator to a ferromagnetic (Eu,Gd)Te material with n-type metallic conductivity. The mechanism playing the key role in this phenomenon is believed to be the RKKY indirect exchange interaction via conducting electrons. As for low carrier concentrations this mechanism is mostly of ferromagnetic character, the ferromagnetic state is expected in Eu chalcogenides doped with Gd. The maximal ferromagnetic Curie temperature observed experimentally in optimally Gd doped Eu chalcogenides vary from $T_C = 150$ K for oxides to about $T_C = 10$ K for tellurides.[1,60,61,64,65] However, if Gd content in EuTe exceeds about 60 at. %, the character of the magnetic ordering changes from ferromagnetic to antiferromagnetic again. This behavior was observed for bulk materials with all terminal GdX compounds being antiferromagnetic materials with metallic type of conductivity. This evolution of magnetic properties of Eu chalcogenides is summarized in Fig. 12. Upon increasing the rock salt lattice parameter from 0.5141 nm in EuO to 0.6598 nm in EuTe, the ferromagnetic order is replaced with an antiferromagnetic state in insulating compounds whereas Gd-doped materials are all ferromagnets.

In ferromagnetic state the n-type (Eu,Gd)Te crystals are expected to exhibit close to 100 % electron spin polarization related to large splitting of

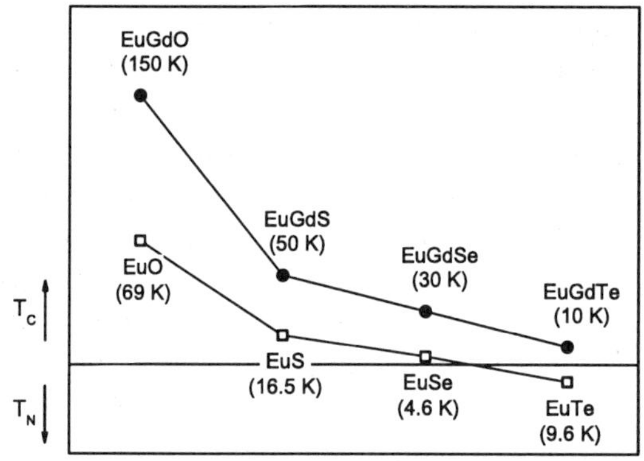

Fig. 12. A scheme of the evolution of magnetic properties in the family of Eu chalcogenides: the open symbols correspond to the undoped (insulating) compounds while the full symbols describe the metallic (n-type) mixed crystals with optimal content of Gd (about 10 at. %). The ferromagnetic (T_C) and antiferromagnetic (T_N) transition temperatures are indicated.

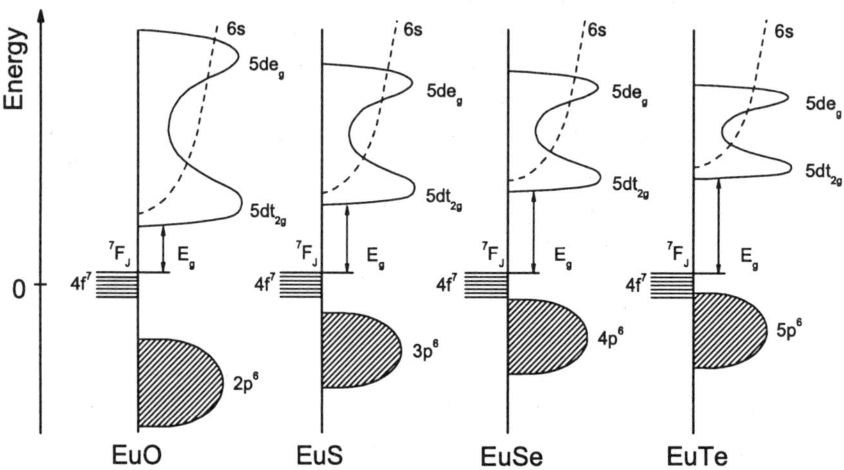

Fig. 13. A model of the electron band structure of the family of Eu chalcogenides. The density of electronic states (DOS) derived from the magnetic $4f^7$ spin up orbitals of Eu is located in the band gap region whereas the $5d$ orbitals of Eu contribute to the conduction band DOS.

the $5d$-$6s$ conduction band states due to $5d$-$4f$ exchange interaction. This is illustrated in Fig. 13 where the scheme of the electron density of states is presented for the family of Eu chalcogenides.[60,61] In these materials the band gap E_g varies in the range $1 - 2$ eV. The Eu $4f^7$ (spin up) states are located in the band gap region forming a narrow band of strongly correlated electrons that is completely filled for $4f^7$ configuration of Eu^{2+} ions. The spin down $4f^7$ states are located at much higher energies (well above the bottom of the conduction band). This feature together with the expected epitaxial compatibility of (Eu,Gd)Te to the well known nonmagnetic semiconductor materials such as PbTe or CdTe makes (Eu,Gd)Te a key-element of new all-semiconductor spin injection spintronic heterostructures.[66]

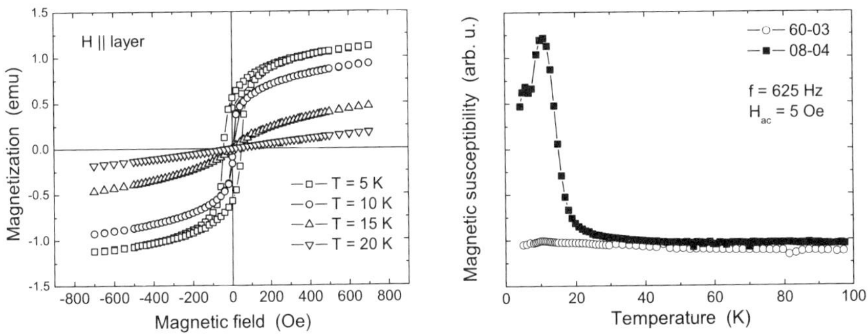

(a) (b)

Fig. 14. The temperature dependence of the hysteresis loops of ferromagnetic n-(Eu,Gd)Te epitaxial layers. Fig. 14b The temperature dependence of the ac magnetic susceptibility in n-type conducting (full squares) and insulating (open circles) (Eu,Gd)Te layers.

In (Eu,Gd)Te the ferromagnetic state is induced by conducting electrons. In all previously discussed DMS systems conducting holes (exhibiting much stronger *p-d* exchange coupling) were responsible for the ferromagnetic state. The spectacular magnetic transformation from ferromagnetic to antiferromagnetic state was recently observed in monocrystalline (Eu,Gd)Te layers grown by molecular epitaxy on of BaF_2 (111) substrate with Gd content about $1-5$ at. %.[66,67] Depending on Gd content and crystal stoichiometry, (Eu,Gd)Te layers were obtained with either metallic or insulating electrical properties. For n-type (Eu,Gd)Te layers a clear ferromagnetic transition was observed at $T_C = 11-15$ K (see magnetic hysteresis loops in Fig. 14a), while insulating (Eu,Gd)Te layers exhibit antiferromagnetic transition at $T_N \simeq 10$ K. In Fig. 14b the temperature dependence of the ac magnetic susceptibility χ is compared for conducting and insulating layers. Dramatically different behavior of the $\chi(T)$ curves depending on the type of conductivity, provides a strong experimental evidence for the quasi-free electron induced ferromagnetic ordering in (Eu,Gd)Te alloys. The possibility to obtain insulating (Eu,Gd)Te layers (despite of n-type doping of EuTe with Gd) is related to the stoichiometry control achieved during the molecular beam epitaxy growth under excess Te conditions resulting in insulating state of (Eu,Gd)Te layers similar to undoped EuTe.

Ferromagnetic transition is also reflected in electrical properties of (Eu,Gd)Te layers. The temperature and magnetic field dependence of resistivity shows at the Curie temperature a factor of 10 increase of resistance and about 90 % negative magnetoresistance effect at magnetic field of 0.6 T.[67] These effects are due to the spin-dependent critical scattering of electrons on magnetic ions spin subsystem.

7. Summary – search for new ferromagnetic semiconductors for spintronics

Although the well defined ferromagnetic state was found in many magnetic semiconductors and, in particular, it can be induced by conducting carriers in diluted magnetic (semimagnetic) semiconductors, the ferromagnetic semiconductor material for truly industrial applications is still not available. The basic requirements concern: (1) the compatibility with the existing industrial Si- or GaAs-based microelectronic technologies, (2) the well defined, robust ferromagnetic properties at room temperature, and (3) the strong coupling of magnetic state to electrical and optical properties of a semiconductor.

Nowadays, $Ga_{1-x}Mn_xAs$ remains the best experimentally studied and theoretically examined ferromagnetic semiconductor material. Despite the fact that this material can only be grown under very specific (low temperature molecular beam epitaxy) technological conditions, it reveals very good compatibility with technological standards of GaAs microelectronics. It shows well defined magnetic properties (which are also preserved in GaMnAs microstructures) as well as exhibits variety of large tunneling and anisotropic magnetoresistance effects. The applicational potential of GaMnAs is, however, strongly limited by its too low ferromagnetic Curie temperature, which, despite world-wide technological efforts, remains below 200 K. For spintronic applications many other materials were also examined revealing ferromagnetic features, e.g. in SiMn, GeMn, (Ga,Fe)N, (Ga,Mn)N, (Ga,Gd)N, (Zn,Mn)O, and (Zn,Co)O – to name just technologically the most important materials. In most (all?) of these new materials, careful examination of their magnetic and structural properties showed the principal role of nano inclusion of various ferromagnetic compounds, instead of the intrinsic ferromagnetic state of the material. This conclusion concerns in particular Si – industrially the most important semiconductor material, in which transition metals were introduced either by various thin layer deposition techniques or by ion implantation resulting, so far, in multiphase materials with magnetic inclusions.

Acknowledgments

This work was supported by the research project 0992/T02/2007/32 of the Ministry of Science and Higher Education (Poland) granted for the period 2007-2010.

References

1. S. Methfessel and D.C. Mattis, *Magnetic Semiconductors*, Springer, Berlin, 1968.
2. E.I. Nagaev, *Physics of Magnetic Semiconductors*, Mir, Moscow 1983.
3. W.J.M. de Jonge and H.J.M. Swagten, J. Magn. Magn. Mat. **100**, 322 (1991).
4. *Lead chalcogenides: Physics and Applications*, D.R. Khokhlov (ed.), Taylor and Francis, New York 2003.
5. R.R. Gałązka, J. Magn. Magn. Mat. **140-144**, 13 (1995).
6. T. Dietl, in *Handbook on Semiconductors*, Ed. T.S. Moss, vol. 3b Ed. by S. Mahajan, p. 1251 (Elsevier Science B. V., Amsterdam 1994).
7. *Semiconductors and Semimetals* vol. 25 *Diluted Magnetic Semiconductors* Ed. by J. K. Furdyna and J. Kossut, Academic Press, Boston 1988.
8. W.D. Dobrowolski, J. Kossut and T. Story in *Handbook of Magnetic Materials*, Ed. K.H.J. Buschow, vol. 16, p. 189, Elsevier, Amsterdam 2003.
9. G. Bauer, H. Pascher and W. Zawadzki, Semicond. Sci. Technol. **7**, 703 (1992).
10. T. Story, C.H. W. Swüste, P.J.T. Eggenkamp, H.J.M. Swagten and W.J. M. de Jonge, Phys. Rev. Lett. **77**, 2802 (1996).
11. T. Story, M. Górska, A. Łusakowski, M. Arciszewska, W. Dobrowolski, E. Grodzicka, Z. Gołacki and R.R. Gałązka, Phys. Rev. Lett. **77**, 3447 (1996).
12. M. Escorne, A. Mauger, J.L. Tholence and R. Triboulet, Phys. Rev. **B29**, 6306 (1984).
13. M. Górska, J.R. Anderson, J.L. Peng and Z. Gołacki, J. Phys. Chem. Sol. **56**, 1253 (1995).
14. M. Górska, J.R. Anderson, G. Kido, S.M. Green and Z. Gołacki, Phys. Rev. **B45**, 11702 (1992).
15. M. Górska and J.R. Anderson, Phys. Rev. **B38**, 9120 (1988).
16. T. Story, Acta Phys. Pol. **A91**, 173 (1997).
17. M. Escorne and A. Mauger, Solid State Commun. **31**, 893 (1979).
18. P. Łazarczyk, T. Story, A. Jędrzejczak, R.R. Gałązka, W. Mac, M. Herbich and A. Stachow-Wójcik, J. Magn. Magn. Mat. **176**, 233 (1997).
19. T. Story, R.R. Gałązka, R.B. Frankel and P.A. Wolff, Phys. Rev. Lett. **56**, 777 (1986).

20. P. Łazarczyk, T. Story, M. Arciszewska and R.R. Gałązka, J. Magn. Magn. Mat. **169**, 151 (1997).
21. R.W. Cochrane, M. Plischke and J.O. Ström-Olsen, Phys. Rev. **B9**, 3013 (1974).
22. R.W. Cochrane, F.T. Hedgcock and J.O. Ström-Olsen, Phys. Rev. **B8**, 4262 (1973).
23. T. Hamasaki, Solid State Commun. **32**, 1069 (1979).
24. T. Story, G. Karczewski, L. Świerkowski and R.R. Gałązka, Phys. Rev. **B42**, 1115 (1990).
25. H.J.M. Swagten, W.J.M. de Jonge, R.R. Gałązka, P. Warmenbol and J. T. Devreese, Phys. Rev. **B37**, 9907 (1988).
26. T. Story, P.J.T. Eggenkamp, C.H.W. Swüste, W.J.M. de Jonge and L.F. Lemmens, Phys. Rev. **B45**, 1660 (1992).
27. P.J.T. Eggenkamp, H.J.M. Swagten, T. Story, V.I. Litvinov, C.H.W. Swüste and W.J.M. de Jonge, Phys. Rev. **B51**, 15250 (1995).
28. A.J. Nadolny, J. Sadowski, B. Taliashvili, M. Arciszewska, W. Dobrowolski, V. Domukhovski, E. Łusakowska, A. Mycielski, V. Osinniy, T. Story, K. Świątek, R.R. Gałązka and R. Diduszko, J. Magn. Magn. Mat. **248**, 13 (2002).
29. Y. Fukuma, H. Asada, N. Nishimura and T. Koyanagi, J. Appl. Phys. **93**, 4034 (2003).
30. Y. Fukuma, H. Asada, M. Arifuku and T. Koyonagi, Appl. Phys. Lett. **80**, 1013 (2002).
31. W.Q. Chen, K.L. Teo, S.T. Lim, M.B.A. Jalil and T.C. Chong, Appl. Phys. Lett. **90**, 142514 (2007).
32. W.Q. Chen, K.L. Teo, M.B.A. Jalil and T. Liew, J. Appl. Phys. **99**, 08D515 (2006).
33. W.J.M. de Jonge, T. Story, H.J.M. Swagten and P.J.T. Eggenkamp, Europhys. Lett. **17**, 631 (1992).
34. P.J.T. Eggenkamp, T. Story, H.J.M. Swagten, C.H.W.M. Vennix, C.H.W. Swuste and W.J.M. de Jonge, Semicond. Sci. Technol. **8**, S152 (1993).
35. C.W.H.M. Vennix, E. Frikkee, P.J.T. Eggenkamp, H.J.M. Swagten, K. Kopinga and W.J.M. de Jonge, Phys. Rev. **B48**, 3770 (1993).
36. H. Ohno, Science **281**, 951 (1998).
37. H. Munekata, H. Ohno, S. von Molnar, A. Segmuller, L.L. Chang and L. Esaki, Phys. Rev. Lett. **63**, 1849 (1989).
38. A.K. Bhattacharjee and C. Benoit a la Guillaume, Sol. State Commun. **113**, 17 (2000).
39. J. Schneider, U. Kaufman, W. Wilkening and M. Baeumler, Phys. Rev. Lett. **59**, 240 (1987).
40. F. Matsukura, M. Sawicki, T. Dietl, D. Chiba and H. Ohno, Physica **E21**, 1032 (2004).
41. T. Dietl, H. Ohno, F. Matsukura, J. Cibert and D. Ferrand, Science **287**, 1019 (2000).
42. T. Dietl, H. Ohno and F. Matsukura, Phys. Rev. **B63**, 195205 (2001).
43. T. Jungwirth, K.Y. Wang, J. Masek, K.W. Edmonds, J. Koning, J. Sinova,

N.A. Goncharuk, A.H. MacDonald, M. Sawicki, A.W. Rushforth, R.P. Campion, L.X. Zhao, C.T. Foxon and B.L. Gallagher, Phys. Rev. **B72**, 165204 (2005).
44. K.W. Edmonds, P. Bogusławski, K.Y. Wang, R.P. Campion, S.N. Novikov, N.R.S. Farley, B.L. Gallagher, C.T. Foxon, M. Sawicki, T. Dietl, M. Buongiorno Nardelli and J. Bernholc, Phys. Rev. Lett. **92**, 037201 (2004).
45. M. Adel, L. Ilver, J. Kanski, V. Stanciu, P. Svendlidh, J. Sadowski, J.Z. Domagała, F. Terki, C. Hernandez and S. Charar, Appl. Phys. Lett. **86**, 112501 (2005).
46. S.J. Potashnik, K.C. Ku, S.H. Chun, J.J. Berry, N. Samarth and P. Schiffrer, Phys. Rev. **B66**, 012408 (2002).
47. M. Sawicki, J. Magn. Magn. Mat. **300**, 1 (2006).
48. H. Kępa, J. Kutner-Pielaszek, A. Twardowski, C.F. Majkrzak, J. Sadowski, T. Story and T. Giebultowicz, Phys. Rev. **B64**, 121302R (2001).
49. M. Csontos, G. Mihaly, B. Janko, T. Wojtowicz, X. Liu and J. Furdyna, Nature Materials **4**, 447 (2005).
50. T. Wojtowicz, G. Cywinski, W.L. Lim, X. Liu, M. Dobrowolska, J.K. Furdyna, K.M. Yu, W. Walukiewicz, G.B. Kim, M. Cheon, X. Chen, S.M. Wang and H. Luo, Appl. Phys. Lett. **82**, 4310 (2003).
51. T. Dietl, Semicond. Sci. Technol. **17**, 377 (2002).
52. A. Kaminski and S. Das Sarma, Phys. Rev. Lett. **88**, 247202 (2002).
53. M. Berciu and R.N. Bhatt, Phys. Rev. **B69**, 045202 (2004).
54. S. Hilbert and W. Nolting, Phys. Rev. **B71**, 113204 (2005).
55. E.Z. Meilikhov, Phys. Rev. **B75**, 045204 (2007).
56. J. Schliemann and A.H. MacDonald, Phys. Rev. Lett. **88**, 137201 (2002).
57. S. Koshihara, A. Oiwa, M. Hirasawa, S. Katsumoto, Y. Iye, C. Urano, H. Takagi and H. Munekata, Phys. Rev. Lett. **78**, 518 (2001).
58. H. Munekata, A. Oiwa and T. Slupinski, Physica **E13**, 516 (2002).
59. H. Ohno, D. Chiba, F. Matsukura, T. Omiya, E. Abe, T. Dietl, Y. Ohno and K. Ohtanoi, Nature **408**, 944 (2000).
60. P. Wachter, in K.A. Gschneider, L. Eyring (Eds.), *Handbook on Physics and Chemistry of Rare Earth*, North Holland, Amsterdam, 1979, vol. 2, p. 507.
61. A. Mauger and C. Godart, Phys. Rep. **141**, 51 (1986).
62. R. Kirchschlager, W. Heiss, R.T. Lechner, G. Bauer and G. Springolz, Appl. Phys. Lett. **85**, 69 (2004).
63. R.T. Lechner, G. Springolz, T.U. Schulli, J. Stangl, T. Schwarzl and G. Bauer, Phys. Rev. Lett. **94**, 157201 (2005).
64. S. von Molnar, J. Supercond. **16**, 1 (2003).
65. F. Holtzberg, T.R. McGuire and S. Methfessel, J. Appl. Phys. **37**, 976 (1966).
66. P. Dziawa, S. Wrotek, W. Domuchowski, K. Dybko, L. Kowalczyk, E. Łusakowska, A. Makosa, A. Morawski, V. Osinniy, B. Taliashvili, Z. Tkaczyk, T. Wosinski, J. Wróbel and T. Story, J. Alloys and Compounds **423**, 208 (2006).
67. P. Dziawa, V. Osinniy, B. Taliashvili, V. Domukhovski, E. Łusakowska, K. Dybko, W. Dobrowolski, L. Kowalczyk, T. Story and M. Goiran, *The Institute of Physics Conference Series*, vol. 187, 21 (2006).

INDEX

d-d exchange interaction, 332–334, 341
g-factor, 331
sp-d exchange, 331, 334
(ESR) electron spin resonance, 83
ab initio, 277

Airy function, 178
all-electron methods, 283, 288
alloy charge transfer insulator, 23
alloy Mott insulator, 22
Anderson Hamiltonian, 8
Anderson model, 8
Anderson-Falicov-Kimball Hamiltonian, 10
Anderson-Falicov-Kimball model, 10
Anderson-Hubbard Hamiltonian, 10
Anderson-Hubbard model, 9
annealed disorder, 10
area theorem, 130
arithmetic average, 11
arithmetic mean, 11
atomic structure, 185, 188, 193
average, 10

ballistic transport, 175
band insulator, 5
band structure, 168, 170, 175, 176, 181, 183, 187, 190, 191
barrier potential, 177, 178
basis incompleteness of LCAO, 280
basis set size, 286, 288
biexciton, 123, 126, 127, 132–135, 137, 154
binary alloy disorder, 9
Bloch sphere, 127
Bloch states, 183, 188, 189, 277
Bloch-waves, 183
Bogolyubov transformation, 64, 72, 77
Boltzmann approach, 293, 295
boundary conditions, 178, 189

carbon, 201
carbon nanotubes, 160
Central Limit Theorem, 11
charge current, 179, 305, 319
charging energy, 91, 113
cluster code, 288
coherence length, 176
coherent potential approximation (CPA), 274, 290
coherent transport, 175, 176, 186
conductance, 163, 164, 166, 168, 171, 174, 180, 182–184
conduction band, 163, 190
conductivity, 172, 301, 302, 311
conductivity, electrical, 228
continuity equation, 303, 307
continuous disorder, 9
controlled-NOT gate, 126
Cooper-pair, 90–92
coplanar waveguide resonator, 100, 104
core states, 278
correlated matter, 294, 296, 312
correlation, 2
correlation expansion, 151
critical current, 91, 94
crystal orientation, 169, 172
crystalline phase, 168
crystallographic growth direction, 167

Cs_2CoCl_4, 38
Cs_2CuCl_4, 81
Curie temperature, 23, 162, 169, 190, 193, 327, 333, 335, 341–343, 345, 348
current operator, 297, 300, 301, 305

density functional theory, 3
density functional theory (DFT), 272
density of states, 165
DFT, 3
diamond chain, 238–240, 245, 247–249, 252, 253, 255, 257, 261
dielectric constant, 184
diluted magnetic semiconductors, 328, 329, 348
dimer dynamic structure factor, 48
dimerized XY chain, 38
Dirac equation, 208, 214, 216
Dirac operator, 210, 216
Dirac-Kohn-Sham implementation, 288
disclination, 204, 205, 209–212, 215
disorder, 167, 175, 185, 186
dispersion relation, 298, 310–312
DMFT, 1
domain-wall resistance, 188
doped Mott insulator, 294
DOS (density of electronic states), 206, 207
Drude, 296
Drude theory, 296
dynamic limit, 304, 317
dynamic structure factor, 35, 37, 38, 46, 48, 49, 56, 60–66, 69–76, 78, 82, 83
dynamical conductivity, 296
dynamical mean-field function, 18
dynamical mean-field theory, 1
dynamical resistivity, 296
dynamical thermal conductivity, 293
Dyson equation, 16
Dzyaloshinskii-Moriya interaction, 35, 38, 40, 70–72, 76, 83

effective mass, 176, 178, 180, 187, 190
effusion cell, 336
electrical conductivity, 293
electrochemical potential, 179
energy current, 301, 305, 311, 313, 319
equal time correlation function, 293
equation of state, 226
Euler's theorem, 204, 215
europium chalcogenides, 328
exchange energy, 177
exchange Hamiltonian, 12
exciton, 122–127, 131–136, 138, 140–142, 144, 145, 148, 152, 153
external bias, 165, 167, 168, 171, 172, 177, 179, 185, 187, 190–193

Fermi energy, 165, 168, 177, 183, 184
Fermi surface, 173, 181, 182
Fermi surfaces, 290
Fermi wave vector, 334, 342, 343
Fermi-Dirac function, 179
ferromagnetic electrodes, 160, 162, 164, 166, 169, 175, 177, 178
ferromagnetic material, 162, 166, 169
ferromagnetic materials, 160
ferromagnetic semiconductor, 160, 162, 163, 169, 175, 186, 193
field-theory model, 201, 208
figure of merit, 293, 311, 323
flat-band ferromagnetism, 265, 266
fluctuations, 185
four-fermion excitation continuum, 35, 38, 52–56, 64
Frank index, 204, 209, 211
free electron gas, 175
frustrated bilayer, 240, 241, 246, 248, 249, 252, 253, 257–260
frustrated lattices, 237–239, 243, 248, 251, 262, 266
full-potential local-orbital (FPLO) code, 272
full-potential methods, 272
fullerene, 202, 215, 217, 219

gauge field, 209
generalized mean, 11
geometric mean, 11
giant magnetoresistance, GMR, 160, 161
grain boundaries, 185
graphite, 202
gravitational field, 301, 317
Green function, 16
Green's function, 181

Hölder mean, 12
half filling, 294, 295, 319, 324
half-metallic, 171, 174
half-metallic ferromagnets, 168
half-metalls, 168
half-voltage, 168
Hall, 296, 297, 299, 312
Hall constant, 293, 296, 297, 299, 312
hard-hexagon model, 258, 260
hard-monomer, 255, 259
Hartree-Fock mean-field, 3
heat current, 300, 301, 305, 311
heavy hole, 122, 187
Heikes, 314, 322
Heisenberg model, 238, 239, 247, 263
Hellman-Feynman approach, 283
high frequency, 293, 297, 307, 310–312, 314, 320, 322
high-temperature superconductors, 160
hole, 122
hole concentration, 190–192
Hooft-Polyakov monopole, 215
Hubbard Hamiltonian, 7
Hubbard model, 6, 237, 239, 262, 263, 265–267, 297, 298, 312, 313
Hund rule, 297
hybridization, 330, 333

impurity, 167, 175
incident plane wave, 178
incomplete LCAO basis, 280
interdiffusion, 185
interface, 161, 168, 171–176, 182, 184, 185, 189, 193
interface roughness, 175, 182, 185
interface states, 183, 184, 186
interface transparency, 172
interlayer, 8, 173, 174
ionized cluster beam epitaxy, 336

Jacobi identity, 309
Jordan-Wigner transformation, 35, 38, 39, 41, 42, 44, 45, 56, 76, 82
Josephson energy, 91
Josephson equations, 90, 91
Josephson junction, 90, 94
Julliere's formula, 166, 168, 171, 187
Jupiter, 230

kagomé lattice, 238, 240, 241, 246–253, 255, 257, 258, 260, 265, 266
Kelvin, 315–317
Kitaev model, 44
Kohn-Sham equations, 276
Korringa-Kohn-Rostoker (KKR) method, 183
Kubo formulae, 296, 301, 314
Kubo-Greenwood formula, 228
Kubo-Landauer formula, 181

Landauer-Büttiker formula, 182, 188
large dimensional limit, 14
lattice-gas model, 253, 254, 257
LDA, 3
LDEA, 18
light hole, 122, 187
linear combination of atomic orbitals (LCAO), 273, 277
linear combination of local orbitals (LCLO), 278, 284
linear independence, 248, 252
linear response, 296, 300–302, 306
local density approximation, 3, 272
local density approximation (LDA), 277
local Dyson equation approximation, 18
local spin density, 182

localized electron states, 237, 238, 263–265, 267
localized magnon states, 237, 238, 241, 243, 245–255, 258, 261, 263
log-normal PDF, 12
logic devices, 162
Lorentz number, 293, 314, 323
LSDA+U, 290

magnetic cooling, 261
magnetic impurities, 174
magnetic materials, 160
magnetic phase diagram, 327, 329, 337, 339
Magnetic random access memory, MRAM, 162, 167, 171, 172
magnetic sensors, 161
magnetic tunnel junctions, 159
magnetic tunnel junctions (MTJ), 161–163, 166–168, 171, 173, 175, 182, 183, 186, 187
magnetization, 160, 164, 166, 171, 177, 179, 184, 187, 191
magnetization curves, 240, 249, 250, 261
magneto-optical potential, 7
magnetocaloric effect, 262, 266
magnetoresistance, 159
magnon excitation, 168
majority-spin, 184
manganites, 193
matter, warm dense, 223
Mermin–Wagner theorem, 261
metal-insulator interface, 182
metric connection, 208, 213
minimum basis, 282, 286
minority-spin, 183
MIT, 4
molecular beam epitaxy, 336, 340, 348
molecular mean-field, 13
most probable value, 10, 11
Mott, 322
Mott insulating state, 294, 295
Mott insulator, 4, 294
Mott transition, 229
Mott-Hubbard metal insulator transition, 4

Néel temperature, 345
nanotube, 202–206, 210, 212, 219, 220
nearsightedness, 286
negative TMR, 172, 184, 185
non-orthogonal basis, 278

one-dimensional hard dimer, 253, 256, 259, 264, 266
one-sheet hyperboloid, 206, 212
Onsager reciprocity, 305
optimized linear combination of atomic orbitals (OLCAO), 281
orbital polarization correction, 290
organic semiconductors, 160
oscillations, 170, 173, 181, 191
over-binding, 272

parabolic bands, 177, 187
partition function, 13
Pauli blocking, 123
PDF, 9
Peierls phase factor, 297
Pfaffian, 59
phase difference, 90, 93
phase qubit, 90–92
phonon, 138
photodiode, 124
pinholes, 174
plane wave, 178
planetary interior, 230
polarization states, 285, 286
potential energy, 177
probability distribution function, 9
Pulay terms, 283

quantum dot, 119
quantum point contact, 8
quantum spin, 238, 250, 262
quantum-well, 173, 174
qubit, 126, 148
qubit subspace, 92, 100, 101, 109, 116

quenched disorder, 10

Rabi oscillations, 124, 126, 129
  two-photon, 132, 134
radial basis functions, 286
radial wave functions, 279
Ramsey fringes, 125
re-entrant spin glass, 339
reflection coefficients, 178, 179, 183
relative conductance, 187
residual entropy, 241, 252–254, 257, 261
resistivity, 296, 312
RKKY interaction, 328, 334, 335, 337–339, 341, 345
rotating wave approximation, 129
roughness interface, 167
Ruderman-Kittel function, 339
RWA, 129

sawtooth chain, 237, 239, 240, 243, 247–253, 256, 257, 262, 263, 265, 267
scalar relativistic method, 288
scattering states, 181, 189
Schrödinger equation, 189
Seebeck coefficient, 301
selection rules, 122
self-averaging, 11
self-energy, 16
semi-core states, 279
semiconductor flash memories, 161
semimagnetic semiconductors, 328, 329, 348
Sherrington-Kirkpatrick condition, 339
simulations, quantum molecular dynamics, 225, 227, 234
slave Boson, 295
slave Fermion , 295
sodium cobaltate, 293, 294, 317, 323
Sommerfeld model, 293
spectral density, 147, 182
spheroidal fullerene, 217–219
spin channel, 182
spin connection, 208, 210, 213
spin field effect transistor (SFET), 160
spin glass, 327, 328, 332, 337, 339, 341
spin injection, 160, 163, 183
spin transport, 188
spin wave, 80, 81
spin-dependent reflection, 174
spin-dependent tunneling (SDT), 164, 182, 183
spin-flip tunneling, 174
spin-orbit coupling, 122, 176, 187, 188
spin-Peierls transition, 241, 251
spin-polarized, 172, 174, 176
spinodal decomposition, 162, 193
spintronics, 160, 161, 163, 193
square-lattice XY model, 38, 76, 78
SQUID, 93, 94, 97–101
stacking faults, 185
static mean-field approximation, 13
strongly correlated electron systems, 4
SU(2) symmetry, 263
sum rule, 293, 310, 311
susceptibility, 306, 318

thermal conductivity, 301, 310, 311
thermal response, 299, 302, 304, 305, 310
thermopower, 295, 314, 315, 317, 318, 320, 323
thin film, 8
tight-binding Hamiltonian, 182, 188–190
tight-binding method, 162, 181, 190
time-domain interference, 126
transport coefficients, 293, 295
transverse dynamic structure factor, 45, 47, 61, 63, 64, 66, 73, 76–78
triangular lattice, 293, 299, 319–323
trimer dynamic structure factor, 51, 55, 64
trion, 123, 154, 155
tunnel barrier, 162, 167, 169, 170

tunneling current, 164, 165, 168, 173, 175, 176, 182, 184, 187, 193
tunneling magnetoresistance, TMR, 161, 163, 191
two-band model, 180
two-dimensional Jordan-Wigner transformation, 42, 77
two-fermion excitation continuum, 64, 72
two-sheet hyperboloid, 210, 212
typical behavior, 11

valence band, 163, 187, 190, 191
valence states, 278
van Hove singularities, 46, 53, 54, 66, 74, 75
van Hove singularity, 215
virtual crystal approximation, 329

wave functions, 170, 177, 178, 180
wave vector, 176–178, 188
Weiss mean-field, 3, 13
Wick-Bloch-de Dominicis theorem, 45, 51, 58, 78

Zeeman splitting, 217, 218
Zener model, 329, 342, 343
Zener-Esaki diode, 163, 188